WHAT WE LEAVE BEHIND

ALSO BY DERRICK JENSEN

*Railroads and Clearcuts: Legacy of Congress's 1864
Northern Pacific Land Grant*

*Listening to the Land: Conversations About Nature,
Culture, and Eros*

A Language Older Than Words

Standup Tragedy (live CD)

The Culture of Make Believe

The Other Side of Darkness (live CD)

Strangely Like War: The Global Assault on Forests

Walking on Water: Reading, Writing, and Revolution

*Welcome to the Machine: Science, Surveillance, and the
Culture of Control*

Endgame, Volume 1: The Problem of Civilization

Endgame, Volume 2: Resistance

*Thought to Exist in the Wild: Awakening From the
Nightmare of Zoos*

*As the World Burns: 50 Simple Things You Can Do to
Stay in Denial*

*How Shall I Live My Life: On Liberating the Earth
From Civilization*

Now This War Has Two Sides (live CD)

Songs of the Dead

ALSO BY ARIC MCBAY

Peak Oil Survival: Preparation for Life After Gridcrash

WHAT WE LEAVE BEHIND

DERRICK JENSEN AND ARIC MCBAY

SEVEN STORIES PRESS
NEW YORK LONDON TORONTO MELBOURNE

SPECIAL THANKS TO THE WALLACE GLOBAL FUND FOR THEIR ONGOING SUPPORT.

Seven Stories Press
140 Watts Street
New York, NY 10013
www.sevenstories.com

In Canada: Publishers Group Canada, 559 College Street, Suite 402, Toronto, ON M6G 1A9

In the UK: Turnaround Publisher Services Ltd., Unit 3, Olympia Trading Estate, Coburg Road, Wood Green, London N22 6TZ

In Australia: Palgrave Macmillan, 15–19 Claremont Street, South Yarra, VIC 3141

College professors may order examination copies of Seven Stories Press titles for a free six-month trial period. To order, visit http://www.sevenstories.com/textbook or send a fax on school letterhead to (212) 226-1411.

Book design by Jon Gilbert

Library of Congress Cataloging-in-Publication Data

Jensen, Derrick, 1960-
  What we leave behind / Derrick Jensen and Aric McBay. -- 1st ed.
      p. cm.
  Includes bibliographical references.
  ISBN 978-1-58322-867-8 (pbk.)
  1. Refuse and refuse disposal--Popular works. 2. Pollution--Popular works. I. McBay, Aric.
II. Title.
TD175.J445 2009
304.2'8--dc22
                              2008047287
Printed in the USA

9 8 7 6 5 4 3 2 1

CONTENTS

Preface vii

PART I

Decay 3
Waste 17
Garbage 25
Sustainability 55
Sustainability™ 61
Compartmentalization and its Opposite 81
Plastic 99
Mining 119
Medicine 129
Toxic Gifts 139
Bodies 147

PART II

Morality 153
Taking it Personally 179
Morality Revisited 185
Legacy 191
The Real World 199
Taking it Personally, Volume II 211

Magical Thinking 219

Complexity 237

Despair 249

Powerlessness 257

Growing Up 267

PART III

The Future: Business as Usual 283

Technotopia 299

Technotopia: Producing Waste 327

Technotopia: Industry 339

Collapse 359

Fighting Back 379

The Living 399

Notes 403

Bibliography 425

Index 437

About the Authors 454

PREFACE

Industrial civilization is incompatible with life. It is systematically destroying life on this planet, undercutting its very basis. This culture is, to put it bluntly, murdering the earth. Unless it's stopped—whether we intentionally stop it or the natural world does, through ecological collapse or other means—it will kill every living being.

We need to stop it.

/ / /

There are many reasons for this culture's ubiquitous destructiveness (and of course, especially at this point, understanding or articulating the reasons for this destructiveness are only important insofar as that understanding or articulation helps us to stop the horrors). In my book *A Language Older Than Words*, I used the lens of domestic violence to explore the more personal aspects of this culture's destructiveness. *The Culture of Make Believe* explored the culture's system of rewards, and how the dominant economic system inescapably gives rise to (and requires) atrocity. In *Endgame* I showed how the culture can never be sustainable or anything other than destructive so long as it requires the importation of resources, that is, so long as it requires more than the local landbase provides (and now that I put it this way, it makes me wonder why it took more than a thousand pages to say this elsewhere: duh). In *Welcome to the Machine*, the emphasis was on this culture's relentless drive for control, which leads inevitably to standardization—*standardization* is a nice way of saying the destruction of all diversity—and ultimately the murder of all that is living (dead creatures are much easier to control than live ones).

This book takes a different approach to this culture's destructiveness.

Or rather, it takes two different approaches: two approaches because this book is not, in fact, one book. It began as one book, and partway through, it speciated.[1] It became two books: one about shit, decay, regeneration, life, death, and the suffocation of the planet under the weight of this culture's waste products; and the other about sustainability, denial, reformism, magical thinking, resistance, death, life, and the processes that many of us undergo in pushing past our denial to confront the reality we face, and from there to acting to defend life on this planet.

Aric and I thought, *Fine, so we have two books. Just pull them apart, put different titles and covers on them, and let's move on.* But when we tried to pull them apart, both died. Further, we found ourselves unable to discern their precise boundaries. The books were intertwined, interdependent.

All of this bothered us greatly, and time and again we tried to tear apart the books: this one here, that one there. But we slowly began to realize that the problem was not with the book but rather with our way of thinking about the book. The book was merely manifesting the miscible edges and intermingling that we were writing about. The book wasn't about shit, decay, regeneration, sustainability, resistance, or any of the other topics we thought it was about. It was about the spaces in between those topics. It was about their interplay, the tangling and untangling of topics, where one moved into another and another moved into the one.

If the language sounds at least a little sexual, that's because it's supposed to (and not just because at some point we write for a few pages about orgasms): when you make love, where do you stop and where does the other start? You're still two distinct beings, and yet there's something else happening, too, right?

That's really the point. We finally gave up on attempting to impose our will on this book and allowed it to teach us to think more realistically. For the real world never has boundaries sharp as books, sharp as scientific equations, sharp as a bottom line. Where do salmon stop and streams begin? Where do spotted owls stop and ancient forests begin? What happens if you try to separate the two? What happens if you try to treat them as utterly distinct? Or, if you don't care about salmon or spotted owls, consider this: there are a hundred times more bacteria in your body than there are your "own" cells. Many of these bacteria are absolutely crucial to "your" continued existence. Without them "you" die. Without "you," they die.

Where do you stop and the bacteria start? You're distinct beings, and yet there's something else happening, too, right?

As in sex, as in the rest of the real world, the real action happens in between.

/ /

In part because this book is as much about identity as it is about decay, Aric and I struggled all through the writing with how we would deal with the fact that this book has two authors, and each one of us tells stories in the first person. We knew right off we didn't want to promote the pseudo-objectivity and phony distance-masquerading-as-perspective so standard among so much formal (academic, philosophical, and journalistic) discourse by writing only in the third person. We also, for obvious reasons, eschewed the royal *We*. We considered and eventually rejected the use of italics to set off one of us while the other used Roman. Finally, we began to understand that because the book is not so much about identification as identity itself, the only appropriate action would be for us to leave each *I* as *I* and let readers use context or sleuthing (admittedly the sleuthing won't be too tough) to figure out who this or that *I* is, presuming it really matters.

And in many ways it doesn't. We must begin to remember not only how to distinguish, differentiate, separate, categorize—all of which this culture kind of teaches us how to do—but also how to recognize (and let ourselves fall into) those places where boundaries dissolve. I write this right now looking out a window—an artificial, more or less impermeable barrier— at the wind moving through and among and between redwood needles. The branches dance and dip and rise again. I shift, and see the ground, where fallen needles decay and feed grasses who live and die and feed the soil who feeds trees who live and die and dance in the wind. And where does each start and stop? Does it really matter?

I step outside, see a pond, feel the cool, moist wind on my cheek and hands, see wind rippling the surface of the water, see the moving dimples where water striders walk, hear frogs, smell soil and redwoods, smell cedars and spring.

When I smell cedars, redwoods, soil, I take tiny parts of them into my body, hold them, sense them, analyze them, join with them, remember and recognize them. Who's to say, at this point, where cedar, redwood, soil ends, and I begin? Who's to say where I end and they begin?

I walk to the grave of a dog I buried last week. He was seventeen. I'd known and loved him—and for that matter still know and love him—for far more than half of my adult life. I buried him next to another dog, whom I also knew and also loved—and still know and love—for fourteen years, and whom I buried a year ago. They both, of course, are still a part of me, as is the beloved cat buried not ten feet away.

As well as being part of me, they're also each a part of the forest here, more fully and intimately now than I am, since I only live here. And each day they grow more fully and intimately a part of the forest than they were the day before. Once again, where does one stop, and the other begin? Does it matter?

The real action, as always, is in between.

How we perceive the world affects how we think about the world. How we think about the world affects how we perceive the world. How we perceive and think about the world affects how we behave in and toward the world. This culture's behavior is killing the planet. We need to change our behavior. This means we need to change, among other things, how we perceive and think about the world.

And at base, that's what this book is about. It is about remembering how to think realistically—that is, to think like life itself, to think with boundaries as permeable as deep, soft, rich soil. To begin to think again like life itself will be one step toward protecting that life, on this beautiful and extraordinary planet, our one and only home.

PART I

DECAY

I have always looked upon decay as being just as wonderful and rich an expression of life as growth. —HENRY MILLER

AS LONG AS I CAN REMEMBER I've been fascinated by decay, by how things fall apart, or how they don't. As a young child I used my tiny fingers to explore rotting wood and the creatures who lived inside. I loved watching ants quickly find dropped food, and soon began leaving crumbs on purpose so I could watch the ants struggle to carry off bread and meat and fruit and vegetables (and potato chips and Cheetos and other junk too, for which I should now apologize). When I was a little older I would go to the same spot week after week to watch what happened to a fallen leaf or broken twig, or to horse, cow, or dog manure. I remember going year after year to see changes in piles of wood, junked swing sets, thrown-out appliances. And I remember trips to ghost towns in the mountains of Colorado, where I explored mines and houses and stores that a hundred years before were bustling with humans, but now were weathered, falling in, bustling instead with insects, lizards who sunned themselves on planks and skittered away when I approached, and leaves skipping away in the wind. I remember mine tailings, some scarred over and blanketed with plants trying to eke out an existence, some shaped into scabrous piles of rocks still incapable of supporting any life whatsoever, and some with open wounds still bleeding tainted, discolored, poisonous water into their surroundings. I remember wondering how long these wounds would last.

In my early teens I read an article—I think it was in *Smithsonian* magazine but I could easily be wrong—about what would happen to some of the iconic creations of civilization if humans disappeared tomorrow. The article focused on the Great Pyramid of Cheops, which would last a long time before it eroded; Hoover Dam, which thankfully wouldn't last too long before it collapsed; and the Sears Tower in Chicago, which without supervision would, if I recall correctly, fall in on itself in a matter of decades.

The point is that I don't remember many articles I read as an early teen, but I remember that one.

I also remember an assignment given by my eighth-grade science teacher. We were supposed to observe two processes—they could be whatever we wanted—differentiated by one substantial variable. A friend taped a bean to the back of a television for a month, then grew it and another bean side by side to see if the radiation harmed the TV bean (it didn't seem to). Someone else—and this is unsurprising coming from a junior high student—soaked one pot of beans overnight and didn't soak another, cooked both, ate them at different times, and tested for differences in his flatulence. For my own experiment I went to a market and bought a big fish, cut it in two, put half in a bucket of water and the other half on the floor, then watched the two decay. I didn't tell my mom about the experiment, and I didn't set it up far from where we lived, but rather, stupidly, in an outbuilding attached to the house. A couple of days into the experiment, my mom began to wonder if a mouse had died in the walls, and the day after that she started asking, "What is that *smell*?" I told her. She—and this speaks volumes about my childhood and why, for better or worse, I turned out the way I did—didn't make me abandon the experiment, but rather move it to where she couldn't smell it.

Although decay fascinated me, I have to admit moving the fish wasn't fun. The fish in the bucket was easy: just carry the bucket, the contents of which had turned the color and consistency of two-thirds melted chocolate ice cream (but smelled altogether different), to another spot. The fish on the floor was more problematic. It had already become a pulsing, flowing mound of liquefied fish and maggots (who later became the fattest, slowest, tamest flies I've ever seen). I carried this mass on a coal scoop, stopping several times to dampen my gag reflex. That night I dreamt again and again of maggots.

Notwithstanding my fascination with decay, it was not a pleasant night.

/ / /

The health of the land begins with shit, with dead bodies, with body parts that fall to the ground. It begins with death, decomposition, decay. It begins with eating, metabolizing, excreting. That's how it has always been, since the beginning of life. You feed me, I feed the soil, the soil feeds everyone, the soil feeds me, I feed you, you feed the soil, and so on.

/ / /

Here's another way to look at it: I eat you, the soil eats me, everyone eats the soil, I eat the soil, you eat me, the soil eats you.

It's all the same.

/ / /

Our relationship—both personal and collective—with shit, and more broadly with our waste products, reveals much about our relationship with the land—with our habitat—and much about why and how this culture is killing the planet. In the case of shit, this culture has turned what was a gift from us to our habitat—a gift of fertile soil, given in response to the nourishment our habitat gives us—into something toxic, something harmful. Something shameful. And that is a terrible shame.

/ / /

A few years ago I started shitting outside. I did this in part because I'm in love with the frogs where I live.

Each December and January I count the nights until the frogs start singing. If they're a little late I start to worry that this might be the year the worldwide amphibian die-off comes home. And then a few nights later I begin to relax, as first one frog starts singing, then another, then over the next few nights more and more until they're loud enough to make normal (human) conversation impossible.

There are two songs I hear: the soprano of tiny green Pacific tree frogs, and sometimes the bass güiro-like chuckle of northern red-legged frogs.

Frogs eat slugs, among many other creatures. I noticed that slugs love to eat dog shit, and I presumed (correctly, it ends up) that they would like mine just as much. I figured that feeding slugs would in turn feed frogs, so instead of flushing all these nutrients down the toilet I decided to let them enter the forest's food stream.

/ / /

I probably would have started shitting in the forest much sooner had my eighth-grade basketball coach not described (incorrectly, it ends up) the proper defensive position as that of "taking a dump in the woods." Considering how wrong I later found he was on everything from athletics to philosophy to social skills, it didn't surprise me much when I discovered he'd been wrong about this, too.

But for many years, whenever I thought about "taking a dump in the woods" (admittedly, not often), I pictured myself on the balls of my toes, knees only slightly bent, thighs and buttocks taut and quivering, ready for any quick movement (except, of course, the relevant one).

The first time I actually tried shitting in the woods, the real position was obvious, comfortable, and natural—far more so than not only the defensive position in basketball, but also sitting on a toilet.

Try it yourself sometime.

/ / /

At this remove, I'm not sure which image strikes me as funnier: that of trying to shit from a standard basketball defensive posture (and hoping the shit doesn't run down my leg) or that of basketball players trying to play defense while squatting, knees apart, thighs resting comfortably against their calves.

At least they would still be on the balls of their feet.

/ / /

What is shit actually made of? Human shit is usually around three-quarters water. That portion increases, obviously, with diarrhea. Of the portion that isn't water about a third is indigestible materials like cellulose—carbohydrates too complex to be broken down by humans: what we normally call "fiber." Another third consists of dead bacteria, who until they die largely help our digestion and provide us with some nutrients we can't make for ourselves. The remaining third is a mixture of fats, mucous, salts, protein, live bacteria, and dead cells from our own bodies.

Poop is usually brown because of a substance called *bilirubin*. When red blood cells in your body get worn out, your spleen breaks them down and extracts the iron. Some of that iron is reused in your body, and some combines with bilirubin to become a brown pigment which is secreted into your intestines and then excreted into the larger world.

The exact composition of shit depends on what you eat. If, for example, you eat more meat (which contains sulfide-rich proteins) your poop and farts will be smellier as those sulfides are released—hydrogen sulfide is the substance that makes rotten eggs and farts smell bad—unless you're one of those men so high and mighty that your shit doesn't stink, or one of those women so ladylike your farts smell like gardenias.

The average person makes about 1300 pounds of poop a year. The global population of 6.5 billion humans produces more than 7 million tons of poop a day.[2]

/ / /

We're called to this book, and to this discussion, by many driving questions. What is waste, really? Is there good waste, or ways to waste well? When and how did shit become waste? More to the point, when and how did waste become waste? How is "waste" dealt with in the natural world? How are cultural relationships with shit and waste mirrored in attitudes towards wasted people, castes, and classes? How did the dominant culture come to cause such massive wastage of materials, people, and lands, and what will happen to that waste when this culture is gone?

Answering these questions is not a mere matter of academic or intellectual curiosity. These questions go to the root of our relationships with our biological selves, with the future, with the land, with sex, with death, with the sacred, and with all other living creatures. If we want to live in a world that is not being laid to waste—where living creatures are not viewed as garbage or where lives are callously wasted—we have to find good answers to these questions, and soon.

/ / /

The next question: what happens when the shit hits the land?

If it's here, chances are good the dogs eat it. They follow me outside, heads lolling, faces grinning, tails wagging slowly. When I squat they sidle round behind me, and I have to put my hands on their shoulders to keep them from nuzzling in too close.

After I finish they move in to clean it up, just as they do with the tootsie rolls the cats leave behind.

I know this is supposed to bother me, but it doesn't. I know that much of decomposition consists of consuming, digesting, and excreting, and I don't much care who does it.

/ / /

Of course I don't let the dogs lick my face. But I didn't let them do that before. I grew up in the country, and I *know* what dogs like to eat.

/ / /

The word *shit* comes from *skheid*, meaning "to separate." Its connotations are of separation from the body. The more polite word *excrement* comes from a similar meaning in the Latin *excrementum*, which is from *excernere*, meaning "to sift out," or "to separate out." Originally it referred to any secretion of the body and wasn't used strictly for shit until the mid-eighteenth century. The word *crap* didn't refer to shit until about 160 years ago. Prior to that, it and its ancestor words referred to chaff or siftings from grain or various kinds of leftovers and residue. The word *piss* has always had its current meaning, but was originally acceptable to use in polite conversation the way we might say *urinate*.[3] The word *feces* originally meant dregs or sediment. The origins of these words are descriptive rather than emotional, and don't reflect a particularly negative attitude toward shit.

Neither do the words for some of the places we shit. *Toilet* referred to a dressing room until a little more than a century ago. *Latrine* and *lavatory* are both from Latin *lavare*, which simply means "to wash." (However, the word *lavare* is closely related to the word *lotium*, which means "urine." This is because early Romans, who did not have soap, collected urine in which to wash their clothes.)

Words for garbage generally have similarly descriptive or neutral origins. *Trash* seems to find its root in Scandinavian words for fallen leaves and twigs or rags. Trash wasn't used to mean household garbage until about a hundred years ago. Similarly, *litter* is what falls to the forest floor. The origin of *garbage* is also neutral; it originally meant the leftover parts of a slaughtered animal, the entrails. The word *rubbish* seems to be related to the word *rubble* meaning broken bits of stone. *Junk* used to mean old but reusable bits of rope, cloth, glass, and other materials. The origins of the word *refuse* may be the most interesting of the bunch. Although its current meaning has held for more than half a millennium, it can be traced back to the Latin *refundere*, which means to pour back, to give back, or to restore.

A modern and polite phrase for shit, *human waste*, has a somewhat different origin. The word *waste* has many definitions, nearly all with negative connotations. Waste as both a noun and a verb emerged about eight centuries ago from the Latin words *vastus* ("empty, desolate") and *vastare* ("to lay waste"). That same root gave us the words *devastation* and *devastate*, meaning "to lay waste completely."

Waste is used to refer to the physical products of waste: *put that in the wastebasket*. It is the process of weakening or decay: *her body wasted away from cancer*. It is the destruction or obliteration of person or place: *The gangsters wasted the snitch in the alleyway*; *Man, he was so wasted last night*; *Dresden was laid to waste*. And *wastelands* refer to places which are not or cannot be used by agriculture or industry, or places that cannot support life *because* of agriculture or industry.

Various meanings of waste hold two common threads: utility and destruction. Fundamentally, something that is wasted is either not being used to its full (usually economic) potential, or has been used up and cannot be used any more. *A waste of time; A waste of money; A wasted life.*

At the center of utility, and of utilitarian viewpoints, is the user. The labeling of anything as waste derives from the perspective of that user.

／／／

Here's another way to say it: waste is a matter of perspective. Last week a cow was slaughtered on the farm where I live. The guts, from which, remember, comes the original word for garbage, were buried in a field (although I didn't know this at the time).

A few nights later I heard coyotes howling as I fell asleep. I often hear them at night, but this time they seemed much closer than usual.

The next morning I went to get a wheelbarrow full of compost from that field and smelled something unpleasant—like rotting meat. I looked closer, and saw that that the coyotes had dug up the entrails and scattered them as they ate. Since the guts were exposed to air, and since it was an unusually warm day, the entrails were already covered in flies eating and laying eggs.

Now here's what I mean by waste being a matter of perspective: to the humans who slaughtered the cow, those guts were waste, garbage by both the original and modern definitions. To the coyotes they were a meal, worth the trouble of smelling out and digging up. To me the smell of rotting guts was nauseous. To the flies it was an enticing aroma signaling a place to eat their fill, a place for their children to grow.

/ / /

It's a couple of weeks later. The guts are long gone. Or perhaps I should say I no longer see or smell them. But the nutrients from the poop of those who ate the guts have filtered into the soil, enriched it. The worms and the soil microbes have benefited, as will the cows who will someday eat these grasses, and the humans and other predators who will someday eat these cows.

That's life.

/ / /

What does it actually *mean* for shit to break down anyway? The dogs don't always eat it, and I've seen what happens when they don't. In the summer, flies arrive within a few minutes. They buzz, land, crawl all over. I presume they're eating, but they never seem to make a dent. I also never see maggots later, so I guess they don't find my feces suitable for a nursery.

Over the next several days, the shit crusts over, turns dark. Sometimes slugs find it and gnaw off chunks, revealing that the inside is still the original color. The pile shrinks as it loses moisture, but still it may last through the summer.

Winters here are wet, with pouring rain that can easily accumulate a half-dozen inches in a day. When that happens, shit *disappears*. It's broken down mechanically, carried in rivulets or dissolved and carried into the soil.

But me not seeing it doesn't mean it's not there. Two experiences taught me this.

When I first started shitting outside I already had a theoretical understanding that shit is good for soil. So I started casting about for some place to bless with these nutrients. Pretty quickly I thought of the slightly worn path below my clothesline. The line itself runs about ten yards from one redwood tree to another, wraps around branches, turns ninety degrees to the left, then stretches another ten yards to another redwood. Because I damage the plants through the summer by walking on them, I thought I might compensate through the winter by giving them nutrients. So I started pooping at the far end of the path, each time moving a few inches toward the near side. Because it was winter, the poop not eaten by dogs was within a day or two beaten into the ground by rain. It disappeared. But—and here's the point—it reappeared the next spring as especially lush and vibrant—happy—foliage. I could see a stark line between where I pooped and where I didn't.

The other experience started a couple of summers later. I got a pretty bad prostate infection, and the urologist put me on heavy doses of antibiotics—mainly cipro and levaquin. It took me a few months to realize that this time my waste products weren't helping, but instead harming the plants, which were dead or dying. The soil in the two main spots where I relieved myself became bare, something pretty rare in this temperate rainforest. The spots remained bare for the next two years.

/ / /

I know, I know. My sample size is only one, and I certainly can't point to reams of studies showing that waste materials containing metabolized, partially metabolized, and unmetabolized cipro and levaquin kills bacteria in soils, but I know what I saw. And what I saw was dead ground.

/ / /

How can killing bacteria kill plants? The answer goes straight to the question of how things break down, and especially how shit breaks down.

When shit falls to the ground or is buried in the soil, its nutrients are not all immediately available to plants. Plants drink up water, and many nutrients dissolved in water through their roots, but larger molecules and water-insoluble molecules can't be taken up so easily. Unlike animals, plants (with the exception of carnivorous plants like Venus flytraps) have no digestive systems, meaning other creatures in the soil must first break down those larger molecules to free up certain nutrients and make them available for plants to drink through their roots. This is a process called "mineralization," where complex molecules are broken into simple, mineral forms. The process of reducing the body of a plant or animal or an animal's droppings into mineral form can involve many creatures, from coyotes and vultures to earthworms and slugs to bacteria and fungi.

When an animal dies it immediately begins to break down its own tissues, in a process called autolysis, in which the digestive enzymes in each cell break it down from the inside.

For example, the cells in your brain will begin to digest themselves after as little as four minutes without oxygen. That's why people whose hearts stop or who drown sometimes get brain damage or die between the time 911 is called and paramedics arrive to administer oxygen and perform CPR. If the brain didn't deliberately start to digest itself but instead went into a state of hibernation (as it can in some cases of hypothermia), it could, in theory, last longer without actually dying. Instead, a cell's last act is to digest itself, and liberate its nutrients so they can be used by others.

The next step in a body's decomposition is putrefaction, in which bacteria start to break down and eat the body. The body's own bacteria, such as those who live in the digestive tract, get started very quickly and are a major factor in animal decomposition. (This means, by the way, that newborn babies who die before ever eating and who therefore do not have bacteria in their digestive tracts may, if in a relatively dry environment, mummify rather than rot.)

And then there are those on the outside who also help to break down the dead, scavengers happy to make a meal of either plant and animal bodies or shit. When an animal like a vulture eats part of a dead body it gets a meal, and at the same time it and the bacteria in its digestive tract help to mineralize the organic matter, which brings the nutrients a few steps closer to being available to plants again.

In decomposition, along each step of the way, every creature wins. Each scavenger, from the biggest to the most microscopic, gets its meal, and each scavenger produces a smaller meal for the one after. And when those nutrients become fully mineralized by bacteria, they get taken up by plants and the plants themselves are eventually eaten for that cycle to begin again.

*   *   *

A very close and mutually beneficial relationship exists between plants and the bacteria in the soil around them. Plants shelter the soil with their leaves, moderate the temperature, and increase moisture levels in the soil around their roots. Further, contrary to common belief, roots aren't simply one-way siphons sucking water from soil. Plants excrete many different mineral and organic substances (including sugars, vitamins, and amino acids) through their roots, and those substances attract and nourish bacterial populations of many different species. The plants themselves grow and shed roots during their lifetimes, much as we grow and shed hair, and those shed roots and root cells also act as nourishment for beneficial bacteria.

Different kinds of plants release different kinds of exudates through their roots. In essence, they are soliciting companionship from specific kinds of bacteria. For example, One corn plant in a sterile nutrient solution excretes 57 mg of sugar and 84 mg of acids in twenty days of growth.[4] We both found that profoundly sad. The corn was not in its community—in common language it was lonely—and by excreting all that sugar, it was calling out for bacterial companionship. Because it was in a sterilized environment, it was calling out in vain.

Conversely, different populations of bacteria produce different kinds of soil environments conducive to different kinds of plants. Both plants and bacteria ask for and encourage specific partners. They choose those they wish for companions.

The length of the roots on any given plant, laid end to end, would be hundreds of miles long. If you include the tiny "root hairs"—very thin projections from the roots that allow plants to absorb more water and nutrients from the soil—the total length may be measured in tens of thousands of kilometers. Even so, plants still need help from their microbial neighbors to be healthy and get all of their required nutrients.

In some cases bacteria and plants have such a cozy relationship that the bacteria will actually take up residence inside a plant's roots. In legumes (the family of plants including peas, beans, peanuts, alfalfa, and clover) tiny nodules form to house symbiotic bacteria, which "fix" nitrogen for the plants. This is because nitrogen (N_2) in the air cannot be used by plants. But bacteria can convert gaseous nitrogen into compounds like ammonium (NH_4+) that plants can use. The bacteria get food and housing, and the plant gets the nitrogen it needs to grow and thrive. And when that plant dies or sheds leaves, the nitrogen in its body is given to the soil for other plants (like those without nitrogen-fixing bacteria) to use.

Even those bacteria who don't actively fix nitrogen take up nitrogen in their bodies and keep it from turning back into a gas and leaving the soil. Similarly, many minerals are taken up into the bodies of soil microorganisms: this is called the "immobilization" of nutrients. If those minerals which are dissolved in water in the soil aren't taken up by plants or other creatures, they will be washed away into rivers or deep into the subsoil where fewer creatures live. Immobilization helps to keep nutrients available in the soil for many different creatures.

Bacteria help plants in plenty of other ways as well. Many bacteria are even capable of breaking down pollutants or pesticides in the soil. Beneficial bacteria on a plant's roots will also "out compete" other pathogenic (disease-causing) bacteria by taking up all of the available space. When beneficial bacteria are killed there is a niche for pathogenic bacteria to move in to, which is what happens to humans in cases like "thrush" (a yeast infection in the mouth) or a *C. difficile* infection (an intestinal infection that slips in once antibiotics have killed off the natural microorganisms).

These mutually beneficial relationships in the soil aren't limited to the partnership between plants and bacteria, of course. Fungi also form many essential relationships in forest soils. And insects and earthworms also help by breaking down organic matter and tunneling through the soil to produce channels for the movement of water and air. Plants need air to reach their roots—although they breathe in carbon dioxide and breathe out oxygen at the leaves, their roots must breathe in oxygen and breathe out carbon dioxide. Just like us, roots die without oxygen.

Now at last to the antibiotics. The antibiotics cipro and levaquin are both "broad-spectrum" antibiotics, meaning that instead of targeting any specific type or family of bacteria, they have an effect on a wide variety of different families and species. A significant amount of both cipro and levaquin passes through the body into urine and feces.[5] In the case of cipro, some 20 to 35 percent of any given dose will pass through and out of the digestive tract in full-strength unmetabolized form. This allows it to pass into the soil.

The essential point of all the discussion above is that a community of plants cannot grow and thrive without a community of bacteria in the soil.[6] The broad-spectrum antibiotics I was taking killed the bacteria in, on, and around the roots of the grass. That unraveled the many tight and mutually beneficial relationships the grasses depend on—and in this case it killed them.

WASTE

Water and air, the two essential fluids on which all life depends, have become global garbage cans. —JACQUES COUSTEAU

MUCH OF WHAT IS KNOWN ABOUT ancient civilizations comes from digging up their waste. Long after empires fall, after flags and flesh rot, garbage remains to tell stories about those civilizations. Although some texts have been retained, much of history is essentially constructed by historians through the act of reading trash.

Societies that produced little or no nondegradable waste are consequently less interesting to many archeologists and historians. Stories told by trash are "hard fact," but oral histories passed down by indigenous peoples are often relegated to the status of allegory or myth. In fact, there's a good chance that a society that "failed" to produce any lasting waste may not even exist in the orthodox historical record.

Within our current system, the life span of any particular artifact as waste is usually far longer than its life span as a useful tool. Let's say I go to a food court at a mall and eat a meal with a disposable plastic fork. Let's say I use the fork for five minutes before one of the tines breaks (as always seems to happen) and I throw it out. The fork goes in the garbage and is buried in the landfill. Let's say this particular type of plastic takes five thousand years to break down (we'll talk more later about what it means for plastic to break down). For every minute I used the fork it spends a thousand years as waste: a ratio of one to 526 million, a number so large it's hardly meaningful to human minds. On a scale that's easier to fathom, if we compressed the fork's five thousand year existence to one year, the fork would have spent only six one-hundredths of a second as an object useful to me.

We can also take into account the millions of years previous that the carbon the plastic fork was made of spent as oil deep underground. In that even longer time frame the useful life of the fork is an imperceptibly short instant sandwiched between a very long time spent in the ground as oil and a very long time spent in the ground as waste. This is true for almost every physical item civilization produces, from cars to computers to fast food containers—they spend many eons in the ground as iron ore or coal or sand, are used a staggeringly short time, and then left as waste for thousands of years or longer.

It's pretty easy to argue that, from a long-term perspective (and indeed from a short-term perspective), industrial civilization is essentially a complicated way of turning land into waste. It is, in all truth, "laying waste" to the earth.

/ / /

What actually happens when we throw away something disposable? Say, something as ubiquitous as a plastic grocery bag?

In many countries, so many plastic bags are released into the environment that they've become widely regarded as a major menace. Because they're so light, they can blow in the wind for great distances until they become snagged on something like a fence post or a sign. The sight of these bags fluttering in the wind has given rise to any number of disparaging nicknames. In China they're called "white pollution." In the US some know them as "urban tumbleweed." And the sight of them is so common in South Africa that they've been sardonically dubbed the "national flower."

But the problem runs deeper than mere aesthetics. Plastic bags mistaken for food can and often do kill animals who eat them. In the oceans, plastic now outweighs living creatures in many areas, a subject to which we'll return a bit later. And in several South Asian countries, clots of discarded plastic bags have blocked drainage culverts used to carry human sewage.[7]

The impacts are worsened by the mind-boggling scale of plastic bag manufacturing. Americans use about 100 billion polythene bags per year.[8] That seems big, and it is, but that number is dwarfed by global production of plastic bags, which exceeded 5 trillion in 2002.[9] Five trillion. If you stopped sleeping and did nothing but watch plastic bags be produced at a rate of one per second, it would take nearly two thousand lifetimes to observe only one year's worth of plastic bag production.[10]

Public outcries against the impacts of plastic bags on the environment have led to attempts at changing the chemical makeup of bags to try to make them less harmful. Some UV-degradable plastic grocery bags are treated with an additive called TDPA which accelerates their breakdown from sunlight. However, the plastic bags only break down into smaller, almost invisible, pieces of plastic. The polymer itself still takes at least a thousand years to break down. In the meanwhile, these small pieces of plastic are often confused for food by animals, especially marine animals. Eating the plastic can poison them, or obstruct or fill their digestive tracts, killing them.

Many countries have begun to take strict measures against discarding plastic bags. Because thinner bags wear out and are discarded sooner, South Africa has outlawed the sale of plastic bags less than 80 microns thick.[11] (Plastic grocery bags in North America are usually about 18 microns thick.) Taiwan and Bangladesh, two of the countries experiencing blockage of sewage drains, have banned free distribution of bags from stores.[12] India has taken especially strong action against plastic bags, with some states blaming them for the obstruction of drainage systems and the worsening of floods during monsoon rains. Two Indian states have banned the production, sale, and even use of plastic bags.[13] In the northern state of Himachal Pradesh, a person using a plastic bag could be imprisoned for as long as seven years or fined up to two thousand dollars.[14]

And if you think you can turn to disposable paper bags instead, consider that disposable paper bags use more energy and produce more waste than plastic bags.[15] The problem isn't just with the plastic. It's with the entire approach of disposability.

In the case of shopping bags, a superficial solution is easily available. Cloth bags are common, inexpensive, and indefinitely reusable. But this easy substitution is the exception, and not the rule, for common items made from plastic.

/ / /

At our farm most of the water for gardening comes directly from rainwater, which we catch in barrels or ponds. We also have a number of trees beside the house who constantly shed leaves and twigs onto the roof and into the eaves trough. It's been a wet few months and we haven't needed much water, so it's been quite a while since the eaves troughs have been cleaned. However, one section became stopped up entirely, and it was my job to climb a ladder and clean it out.

When I reached the roof I looked into the eaves trough and found a forest in miniature. Leaves and other organic matter had filled the trough completely. The wet lower layers had decomposed into a rich organic humus. The upper, more recent layers had stayed mostly intact and acted as a mulch, keeping the bottom layers from drying out, and also allowing air to reach into the mix.

Black locust and maple seeds had germinated in the humus and the trough was filled with tiny trees. I dug a little underneath the surface and found sow bugs and millipedes, indicators of other soil life thriving there.

A soil science textbook will tell you that soil is made up of air and water, the organisms who live in the soil, and a substrate made of mineral particles and decaying organic matter. Because of the leaf litter that had collected there, we had all but one of those. The basis of that tiny forest was trash by the original definition (fallen leaves and twigs). The only soil component we were missing was mineral particles like sand or clay—the tiny broken bits of stone or rubble that gave us the word *rubbish*.

Soil itself is essentially trash and rubbish, but the living creatures in it turn it into *soil*, the foundation of terrestrial life. Just as it is true that industrial civilization turns land into waste, life turns trash into land.

/ / /

When I said earlier that waste is a matter of perspective, I was kind of lying. That's true so far as some waste materials. But some other waste materials are simply waste materials. Trash from trees and trash from a computer factory are two very different things. Garbage from a dead animal is much different than garbage from a nuclear power plant. Radioactive waste from a nuclear power plant is dangerous to life for hundreds of thousands of years or longer. In contrast, the organic "waste" scraps from my dinner are turned into compost for gardening. Not only is that compost not harmful, it's actively helpful both to the plants in the garden (weeds included) and to me.

Both materials are "waste" by the dictionary definition. But it seems silly, if not outright misleading, to use the same word to refer to something that kills all life, and then to turn around and use it to describe something highly beneficial to all life. We need to define exactly what we're talking about when we say "waste."

/ / /

When I defecate outside, not only shit is left behind. There's also toilet paper, or whatever else I use to clean myself up afterwards. The toilet paper almost invariably lasts longer on the ground than does the shit. During the winter the toilet paper gets broken into pieces by rain and pounded into the ground or carried off in rivulets. In the summer small mounds of off-white paper remain for months, or until I get sick of seeing them and toss them into the forest where they decompose out of my sight.

I sometimes wonder why it bothers me to see the toilet paper: by the time toilet paper hits the ground, it's really just parts of trees ready to fold back into the forest, ready to come home. Although it looks like trash it's really just litter. But that doesn't alter the fact that it *looks* like waste.

I think a big part of the problem is not what *I* think and feel about seeing the toilet paper, but what I fear other people might think and feel about seeing it. Not many people come to my home, and those who do generally know me. I'm not worried about what they think. But sometimes—very rarely—someone I don't know shows up: kids cutting through the forest, the meter reader, a botanist checking out plant habitat. And then I always wonder if they're thinking, *What the hell is this?* (Note that I'm not worried about them noticing the piles of shit: they'd probably just think they came from dogs or bears, and in any case the brown of the shit blends well with the ground. The papers, on the other hand, stand out, and one of my fears, I suppose, is that if the stranger stares at them long enough, the stranger might figure out what they are. . . .)

For this reason and because I wanted to see if I could, at least in this little way, stop supporting the paper industry, I soon began experimenting with other materials. At first, inspired by tales of old Montgomery Wards catalogs in outhouses, I tried glossy junk mail. It was too slick and worked for only the most rudimentary cleaning, and seemed to take forever to decompose. I'm not sure how the old-timers put up with it. Then I tried newspapers, which, while they decomposed at about two-thirds the rate of toilet paper, cleaned only slightly better than the catalogs. I also wasn't comfortable with the unsightly ink stains this might leave on my behind, and I especially wasn't comfortable having all those ink toxins so close to an orifice. After that I tried those cotton balls that come in the tops of vitamin bottles, which work great. They're soft and luxurious, and when I use them the only thing holding me back from feeling a sort of emperor-like decadence would be someone fanning me with a palm frond while discreetly looking away. But cotton balls are unfortunately in short supply, so I save them for my birthday and other special occasions. Orange peels work well, so long as I use the inside, not the outside, which is a bit tangy for my taste (and probably covered with pesticides). At first I was concerned about using orange peels, since I've read that in many places they can take decades to decay, but at least here that's not true. Slugs love them, and the peels disappear within days or weeks. Banana peels work okay too, so long as, once again, you use the inside. But they, like cotton balls and orange peels, are in short supply. Nowadays I save all my used paper napkins from restaurants—they work great. And there's the paper wrapping from recycled toilet paper rolls.

I've heard that some indigenous peoples use stones to clean themselves, but I haven't figured out how to make that work. And of course people the world over use fingers, but I'm nowhere near that hardcore. Nor am I yet hardcore enough to use and (after washing) reuse cloth, although that would finally give me a use for my worn-out tube socks.

I know what you may be thinking: *You live in a rainforest; why don't you just use leaves?* I have, and some work well. Thimbleberry leaves work great. The leaves are big, soft, and slightly fuzzy. My concern is that although there are a fair number of thimbleberry plants, each plant doesn't contain a huge number of leaves, meaning each lost leaf would be at least somewhat significant to the plant. I tried cascara leaves, which are plentiful enough on those trees, but they're a tad too slick. For obvious reasons, I'm not going to try pine, fir, hemlock, or redwood needles.

I wish blackberries didn't have thorns, because they're invasive, plentiful, and have tons of leaves. But the thorns aren't limited to the canes, and also run down the spines of the leaves, which means the leaves' use as toilet paper is at best suboptimal.

/ / /

Here's something else about toilet paper. The toilet paper I leave in a clearing remains relatively intact through the summer, not breaking down until the rains come. If, however, I leave toilet paper in the forest and return just one or two days later, I see that the stained part of the paper is gone, eaten by slugs, who clearly prefer to live in moist forests over open sunlight where trees have been removed.

I mention this for a few reasons. The first is that, if I got desperate, I suppose I could use slugs as toilet paper cleaners and reuse the toilet paper (er, maybe not). The second and far more important reason is that it returns us to the question of how long it takes for something to break down. Now, it may not matter much whether my toilet paper breaks down in six days or six months, but when you have a society producing toxic wastes lasting thousands, tens of thousands, or in some cases hundreds of thousands of years, the question of how long it takes for something to break down becomes far more important. But there's an even more important reason, which is that it leads to the question of whether and how our "garbage," our "refuse," our "waste," is useful or harmful to others in our landbase, and to our landbase in general. And this, of course, leads to the most important question of all: does our presence and do our actions help or harm the land who supports us, the land who is our home, the land on whom our own survival ultimately depends? It's very simple: any way of living that doesn't help the landbase—any way of living that doesn't feed the soil what it needs to survive and thrive—will not last. This means that any way of living that produces waste that doesn't help—or worse, harms—the plants and animals and fungi and bacteria and land and water and air upon whom that way of living—in fact, the species' survival—depends, will not last.

GARBAGE

Our willingness to part with something before it is completely worn out is a phenomenon noticeable in no other society in history. . . . It is soundly based on our economy of abundance. It must be further nurtured even though it runs contrary to one of the oldest inbred laws of humanity—the law of thrift. —J. GORDON LIPPINCOTT, INDUSTRIAL DESIGNER

SINCE THE UNITED STATES IS the world's largest generator of waste, we can learn a lot from looking into this country's history of waste, and into the history of waste in general.

It may or may not surprise people to learn that centralized garbage collection systems were rare in most American cities until the early twentieth century. But what is more likely to be surprising is the reason: people did not want to give up their "waste." The garbage that was collected was often collected informally, by people like the so-called "swill children" who once went from house to house in American cities to gather food refuse to sell as fertilizer or for hog food. In the mid-1870s, the city of Milwaukee created an early garbage collection system, in which contractors with carts were sent around the town to collect garbage for disposal. To the surprise of the municipal government, many residents "refused to give their refuse to the city collector." Accustomed to giving their trash to the "swill children," residents insisted that their garbage belonged to them, and they could "give it to whom they please." Milwaukee's early garbage collection system was abolished within three years.[16]

The attitude of the residents may seem almost incomprehensible from a modern perspective. They wanted to keep their garbage? They refused to give their garbage even to municipal collectors? But what seems like strange behavior can illuminate a critical point: they wouldn't simply give their garbage to the collectors because *their garbage wasn't waste.* It had value, and it was useful. And it had, in their minds, a future. It wasn't simply loaded into a garbage truck by nameless workers to disappear into oblivion. Their garbage was part of a relationship with the "swill children" that they valued and wanted to continue. It's a marked contrast to the present day, when most people are happy to be rid of their garbage in the most convenient way possible, so that they simply don't have to think about it.

/ / /

So how did things change? How did garbage go from trash to true waste? When did that happen, and why?

/ / /

As we look into the long history of garbage, we can see a number of different trends, trends that we can also see in civilization at large. First is increasing scale. Civilizations tend to increase scale as a matter of course—if those in power have a problem, there's a good chance the solution involves making something bigger, or making more of things.

A second trend is increasingly complex technology. The earliest recorded trash was comprised of things like wood ash, organic waste, and rags. But technological "progress" has brought us a refuse stream that includes many different kinds of metal and alloys, toxic byproducts of manufacturing, radioactive waste, and thousands of different kinds of plastics and polymers.

A third trend is decreasing community autonomy and control over garbage. This is very much a result of the first two trends. Denser and more wasteful populations, especially urban populations, made it increasingly difficult for communities to cope with their own waste locally. And as new industrial technology produced new kinds of waste, it became impossible for small communities to maintain the infrastructure necessary to deal effectively with that waste. Larger and more complicated waste management systems meant that dealing with rubbish was no longer a community or household activity—it was the specialized domain of technical experts and engineers. As a result, the ability to make decisions about waste was often lost by communities.

To a person in power, all of these trends are markers of progress. More production means more garbage, and more profit and power. More technology means more things to make and sell, and that means more profit and power. And control over garbage also means more profit and power. (And vice versa: poor neighborhoods are a favored location for incinerators and other polluting and unpleasant waste management facilities.) This last point may be a difficult one for us to understand, because of our place in history during a period of immense wastefulness, in which garbage has no value. But as we shall see, the value of garbage was something our ancestors understood well, and any descendents we have will come to understand it as well.

///

If you have an object that you don't want or find useful, you have some general options. You can store it until later, in the hope that you or someone else might find it useful then. You can give it, trade it, or sell it to someone else who does find it useful. You can get rid of it by sending it away so it isn't a problem (for you) anymore. You can try to break it down into smaller parts that are more useful or easier to deal with. Conversely, you can try to combine it with something else to produce something that is more useful or easier to deal with. Historically, strategies for dealing with garbage have been combinations of these basic acts.

///

Garbage wasn't really an issue for human beings until the advent of cities. Prior to cities and civilization, any trash produced by humans was either biodegradable and beneficial to the land (like shit or food scraps) or non-biodegradable but inert and harmless (like pottery). In either case, it was small in quantity.

When the first cities were built, garbage was still mostly harmless in nature. But the density of cities, their growing populations, and the large quantities of materials they brought in made the accumulation of garbage a problem. Instead of spreading small amounts of waste over large areas with a diversity of scavenging species above and below the soil, urban humans were now accumulating large amounts of waste in tight spaces where most nonhuman species had been wiped out. Piles of decomposing organic matter offered a wonderful environment for generalized scavengers like rats, flies, and cockroaches, who accompanied civilized humans as they spread around the world. The dense populations of humans and domestic animals allowed for the incubation of plagues and infections, many of which were communicated by those rats and insects.

While germ theory and public sanitation did not exist until much later, the residents of early cities sometimes realized that the garbage was making people sick, although they attributed this to gases emitted by them. Certainly the garbage smelled bad and looked unattractive in the streets.

So the question emerged: what to do with all this garbage?

/ / /

Garbage disposal strategies fall into several main categories: landfills and dumping; feeding to animals and composting; burning and incineration; and recycling. Each of these techniques has varied widely in popularity at different points in history, and in different locations. Different climates and terrains are often associated with different waste disposal techniques. For example, cities in colder climates tended to produce more ash than cities in warmer climates because of heating needs. And cities in warmer climates tended to produce more organic waste because of their longer growing season.

Landfills and dumping are related and overlapping strategies that differ in intent and approach. A "sanitary landfill," currently the most popular approach to waste disposal in the industrialized nations, is a site (usually lined with plastic) where garbage is spread, buried in layers, and capped to minimize the leaching of toxins from water flowing through. (This does not usually work as well in practice as in theory.)

Dumping is a far older, simpler, and more haphazard act. In fact, dumping is so old, and its consequences so severe and enduring, that a 1912 statement to the American Public Health Association even compared dumping to Original Sin: "In its simplicity and carelessness the dump probably dates back to the discarding of the first apple core in the Garden of Eden, and its subsequent train of evils is ample testimony of the Eternal Wrath elicited by that act."[17] Living as we do in an age of immortal plastics, persistent organic pollutants, and radioactive waste, we can chuckle at the author's quaint concern about an apple core. But the author is likely right about one thing: the modern act of dumping could be considered a toxic and grossly amplified version of the ancient act of discarding harmless—or more accurately, beneficial—organic material.

In a hunter-gatherer society, like the kind that has made up the vast majority of cultures for the vast majority of human existence, "carelessly" tossing away organic matter like food scraps or fruit cores is indeed a helpful act for the land. The food is quickly broken down and the nutrients reabsorbed into the environment. Scavengers like ants carry pieces of the food back to their nests and so incorporate the nutrients into the soil. The seeds of fruits are spread and can take root in new locales. In that context, discarding food into the environment is a universally helpful act. Only in recent history, as garbage became much greater in quantity and less digestible in composition, did this change from a helpful to a harmful act.

Dumping is kind of a combination of two of the basic acts we talked about earlier—sending things away, and storing them. Here's what I mean: the strategy of choice for dealing with garbage in early cities was to have it dumped on the outskirts of town, ideally in a ravine or small valley, for the sake of human convenience. The first municipal dump in the Western world was created around 500 B.C.E., in Athens, where citizens were required to dispose of their garbage at least one mile from the city walls.[18]

But over time, as the world became more crowded with humans, sending garbage away often meant sending it into someone else's backyard. By the late 1800s, some municipal officials in the US were already concerned about the long-term implications of the steady stream of waste. In 1889, the health officer for Washington, DC, complained that good sites for waste disposal were "becoming scarcer year by year," and that "inhabitants in proximity to the public dumps are beginning to complain."[19]

As people grew more discontented with the flow of garbage through or into the places where they lived, they sometimes turned to direct action.

For example, people living in Alexandria, Virginia, at the close of the 1800s were appalled by the passage of garbage barges floating down the Potomac River from Washington, DC. So they began to sink those barges upriver of Alexandria, before they could pass by.[20]

As a result of more crowded conditions, centralized dumps gained popularity. Organic matter would eventually decompose in these dumps, but metals, and later, plastics, would not. Those materials would essentially be stored indefinitely in the ground. Perhaps we can think of it as a variation on the act of sending things away: instead of being sent to neighbors, the garbage is being sent into the future, where it will become the problem of the descendents of the original dumpers.

Dumping has long been a popular strategy for cities on or near oceans and large waterways, and continued to be until quite recently. In the United States, the 1899 Rivers and Harbors Act banned the dumping of garbage in navigable rivers. This was not motivated by a concern for aquatic ecology—rather, legislators were concerned that continued dumping would obstruct rivers and interfere with shipping.[21] Ocean dumping wasn't brought under environmental regulation until 1972, and dumping of sewage sludge and industrial waste wasn't made illegal until 1992.[22] The perceived convenience of dumping into water comes from one of our other basic acts: combining garbage with something else to make it easier to deal with. Superficially, this is an effective solution for the dumper, because moving water will carry garbage out of sight, or allow it to sink into the depths. However, the consequence is that control of the garbage is lost— it can now spread through the water and may come back as a problem later, especially in the case of pesticides and other toxins. All of this makes a mockery of the slogan so popular among producers of waste and the governmental agencies at least nominally charged with their oversight: *the solution to pollution is dilution.*

This is a strategy we'll see again and again in civilized methods of dealing with waste (and with other problems). Emphasis is put on the quick fix, on a short term gain, even if the resulting problems are much greater in the long term.

One of the oldest and most sustainable methods of dealing with trash has historically been to feed organic garbage to (nonhuman) animals. Certainly, in hunter-gatherer societies, discarded organic material would be eaten by animals, fungi, bacteria, and so on. And the humans would benefit, most often indirectly, from having those nutrients returned to the broader living community. But agricultural societies with domesticated animals had a different dynamic. By feeding scraps to farm animals, especially more generalized eaters like pigs and chickens, agriculturalists received direct benefits from their shared garbage in the form of meat and eggs.

This sharing of waste didn't need to be especially organized, but it did require more sorting than simply tossing any and all organic matter aside. An 1835 edition of *The American Frugal Housewife* describes multiple levels of home "recycling." Food scraps inedible to humans were put in slop pails to be fed to pigs. But fatty food wastes were kept in the "grease-pot" where they could be used for cooking and soap-making.[23] In rural areas, farm animals sometimes grazed close enough to the house that slop buckets were not needed; waste would simply be tossed out the kitchen window and eaten.

Even in cities, livestock often roamed the streets until the twentieth century, eating garbage that people discarded from doors or windows. In 1834 in Charleston, West Virginia, killing vultures was outlawed because the birds kept the streets clean by eating garbage.[24] And in 1842 it was estimated that something like ten thousand pigs wandered the streets of New York. When municipal governments in New York and other cities attempted to remove the pigs they were met with political resistance because of the street-cleaning activities of the pigs, and the food they provided for the poor.[25]

Organic waste was also collected in a more deliberate fashion to be fed to animals in urban areas. In the early 1900s, many small and medium-sized towns built facilities called "piggeries," where pigs were fed raw or cooked garbage. It took about seventy-five pigs to eat one ton of garbage per day.[26] But life in a piggery was not like life on a farm. These pigs often died from eating foreign matter like broken glass, sharp oyster shells, or lye-based soap. A government pamphlet at the time pled with householders to sort out dangerous foreign matter, and to "imagine the tortures suffered by the unfortunate animals."[27]

In an industrial version of the farmstead's retention of grease and fats for candle- and soapmaking, many "Reduction Plants" were built in the US to extract greases and oils from organic waste. However, the growing extraction of fossil fuel oils meant that such efforts became uneconomical, and the last Reduction Plant closed in 1959.

/ / /

Feeding waste to nonhumans does not have to be a relic of the past, even in the United States. All through my twenties I'd wanted to get chickens, and I still remember the exact moment, when I was thirty, that I finally decided to do it. It was a summer day. I sat on the couch, looking out the window and half-heartedly reading an account of the lives of white settlers on the Great Plains. Two sentences stopped me. I read them again: "If these early settlers could travel forward in time they would probably be surprised and perplexed to learn that their descendants actually *purchased* special food for chickens, instead of feeding them scraps and letting them forage. If you have to feed the chickens special food, what's the point?"

I laughed, put down the book, and began dreaming of chickens. The next spring I built a little shelter and got some chicks. I quickly grew to love the chickens for themselves, for doing the things chickens do—dusting themselves, dancing in the sun, squabbling, grooming themselves—but I have to admit that I derived great satisfaction, even joy, from the knowledge that much of what they were eating was what most people would consider "waste."

I never bought them any chicken food. They mainly foraged, which is a polite way of saying they killed and ate everything that moved—insects, spiders, worms, mice, and snakes; I'm sure that had the chickens been large enough they would have been delighted to eat me—and they also ate just about everything that *didn't* move.

I did buy *some* food, but it wasn't just for them. A pasta factory bagged the noodles that fell on the floor and sold them for ten cents a pound. I bought them by the hundred-pound bag. Not only did this provide chicken food, but it allowed me to gorge on pasta, and one trip to the factory completed all my Christmas shopping. The only problem was figuring out how to fit hundred-pound bags under Christmas trees (and you should have seen the intendeds try to shake their gifts). The good news is that for the first year, at least, nobody—and I mean nobody—was able to guess what I was getting them.

I also bought great globs of congealed pasta that had been extruded but never formed into noodles. These were obviously not sellable, except to me for two cents a pound. As hard as I tried, I never was able to figure out how to make these work as a Mother's Day gift.

But the birds loved picking at them.

In addition, I dumpster dug. Oh, how I dumpster dug. As much as I loved the chickens, I may have loved dumpster digging even more. I loved it not only because of all the good free food I got for the chickens, and not only because of all the good free food I got for me and most everyone I knew, and not only because it often acted as a substitute for pesky Christmas, birthday, and other shopping ("Oh, how sweet! You brought me a bag full of, what's this, several hundred doughnuts! And what's over here? A whole crate of artichokes! For Valentine's Day, no less! Derrick, you really shouldn't have"), and not only because it so often made for intriguing second or third dates ("I'm sorry, Derrick, I don't think I heard you: what was it you said you wanted to do? Dive . . . in a what? A dumpster? Oh, umm, that sounds, well, umm, lovely! I'll just have to check my schedule and . . . oh, shoot, it looks like I'm busy all this week . . . Oh, you'll be going again next week? Unfortunately that's looking pretty bad, too"), but also because it kept a great deal of food out of the waste stream.

In Spokane, where I lived at the time, this meant keeping all of this food from being burned in an incinerator. (What sense does it make to burn watermelons? What sense does it make to throw away perfectly good watermelons simply because the store received a shipment of some other fruit and needs the space?) I got some tremendous hauls: a big black grocery bag full of cashews.[28] An even larger bag of coffee (chickens, by the way, don't like coffee beans). Gallon after gallon of slightly post-dated ice cream. (Grocery store employees waited for me to show up before bringing these out, since they, too, hated to see it go to waste. I had so much ice cream I ended up feeding that, too, to the cats, dogs, chickens, other birds, and anyone who would eat it: poultry, by the way, love ice cream.) Truckloads (literally truckloads) of perfectly fine watermelons (I felt as decadent as a Roman emperor: I had so many watermelons that neither I nor everyone I knew could eat them all, so I'd split them open with an axe, take two bites of the heart, and throw the rest to my humble subjects, er, the birds), cantaloupes, peaches (the birds didn't like peaches), tomatoes (they loved tomatoes, and would run around with red all over their breasts, as though they'd been wounded in battle), onions (they didn't much care for onions, and two of the worst smells I've ever encountered are deeply rotten onions and, oddly enough, deeply rotten honeydews), and so on. It's fun to get free food, and it's even more fun to share it with your (human and non-human) friends.

I dumpster dug a few times a week from probably 1990 to 1999, and over that time I witnessed the disturbing trend of grocery stores increasingly locking their dumpsters, or worse, installing huge and expensive enclosed systems that open only on top of the garbage truck. That seemed mean-spirited and unnecessary. When I asked managers of big stores why they'd done this (I always established relationships with store managers or owners where I dumpster dug: I learned to do that after the owner of a neighborhood grocery store once pulled a gun on me. I was there after-hours, and that day his store had been robbed; he thought I was the robber's accomplice coming to pull the loot out of the dumpster. When I told him I was there for another kind of green—lettuce—he smiled and requested I come back during the day. Soon after, he and his employees were setting huge piles of food aside for me, and then for all sorts of other people. He did this slightly out of generosity, but mainly because it allowed him to cut his garbage bill by two-thirds), they uniformly told me it was for "insurance purposes." Whether or not I believe them—and I'm not sure I do—the decision still seems wasteful and stupid.

Let's follow the chain. Subsistence farmers in Mexico are forced off their land so a transnational agricorporation can grow tomatoes on the land that used to be theirs. The soil of this land quickly becomes toxified by the use of pesticides. The local river—which has forever been the source of water for local villages—is ruined by the pesticide and fertilizer runoff from the new tomato fields. Or it would be ruined if it hadn't already been diverted for irrigation.

Now utterly impoverished, one of the men who used to live on this land (and whose wife is now dead from cancer and whose children are now brain-damaged from pesticides) leaves these children with his mother and crosses the (heavily guarded) border into the United States, which happens to be the destination for the tomatoes grown on land that used to be his. He works his way toward Spokane. Along the way he is robbed at gunpoint twice, beaten thrice, and raped once. Of course employers steal his labor any number of times. One night in Spokane he is hungry. He has no money. He goes to a dumpster. It is locked. Inside are a couple of dozen crates of tomatoes—coincidentally tomatoes that were grown on land that used to be his. He can't get at them. Tomorrow morning these tomatoes will be taken to the incinerator and burned (insofar as tomatoes can burn at all). Whatever ash is left from the tomatoes will not even nourish any soil, because the incinerator is at the same time burning all sorts of other wastes, including plastic containers that formerly held toxic chemicals used to manufacture pesticides that were sprayed on the land where the man used to live.

This makes no sense.

This is how the system works.

/ / /

In a strange way, forming that relationship with the neighborhood grocer, and finding that because of that relationship he set aside part of his parking lot specifically for people to come by and pick up the old food to feed chickens, pigs, or humans, made me more proud than just about anything else I did in Spokane, including writing *A Language Older Than Words.* The pride came from the knowledge that this accomplishment was tangible: because of me, those particular watermelons remained in the food stream, those particular tomatoes got eaten by ducks, those particular artichokes got eaten by me.

Writing is good work, but it's abstract. By itself, it accomplishes nothing directly; as Jung put it, "Philosophy butters no parsnips." Nor does philosophy reduce waste, make Styrofoam break down, make dioxin nontoxic, nor bring down civilization.

It felt good to do something positive—however small—in the real physical world.

/ / /

For me, dumpster diving wasn't particularly political. It was simply something I did for fun. I enjoyed the physical act of dumpster diving, and I'm also more or less a cheapskate, so I was delighted to get free things. I'm also by nature pretty conservative, so I derived great satisfaction from saving food from being wasted, in the worst sense of that term, and I also derived satisfaction from protecting the natural world by my actions, however slight the protection might be from those particular actions.

Over the years I've met a fair number of people who saw dumpster diving differently: they believed it to be a profoundly political act, a way of sticking it to The Man. There is some truth in that: we have intentionally and systematically been made dependent for our very lives upon those who control the economy. Elsewhere, I've written extensively about how it's almost impossible to get people to slave away for you unless you deprive them of access to land. Because access to land provides access to food, clothing, and shelter—which means access to land provides the possibility of self-sufficiency—if you want to maintain a dependent (and therefore somewhat dependable) workforce, it's crucial for you to sever their access to land. It's also crucial that you destroy wild foodstocks: why would I buy salmon from the grocery store if I could catch my dinner from the river? Now, with people having been effectively denied access to free food, clothing, and shelter, which means having been effectively denied access to self-sufficiency, if they are going to eat, they're going to have to buy their food, which means they're going to have to go to work to get the cash to buy what they need to survive. If you're a corporation, you've got them where you want them.

Unless, of course, they're diving into your dumpsters, parasitizing of your waste products. Then they're not paying you in order to eat. And if they're also squatting, they aren't paying you for shelter. And if they're also shoplifting clothing (or getting their clothes from dumpsters: I've known divers who specialize in food, clothes, furniture, metals from industrial dumpsters, and so on), then these people have no need for cash, and thus no need to become wage slaves. That's bad for a corporation's bottom line.

Which means it's good for everyone else.

So far, so good. But the problem I have with calling dumpster diving a powerful political tool is the same problem I have with calling any other form of lifestyle activism a powerful political tool. The problem is that any form of lifestyle activism (including dumpster diving, shitting in the forest, and so on) is an extremely short lever with which to increase your power to bring about social change. There are many actions we can take, as I've explored in *Endgame* and elsewhere, that are much more effective at bringing down The Man than simply living off the scraps he throws away.

I've also heard some dumpster divers say they're living sustainably. Again, that is only partly true. As with me and my watermelons, they can consume decadently without this consumption causing direct physical harm. But ultimately my watermelon extravaganza and other divers' dumpster meals and wardrobes are not really sustainable, because they all still depend on the larger, wholly unsustainable system. In the end, whether the tomatoes grown on land where the former subsistence farmer used to live are purchased, plucked from a dumpster, or burned alongside plastic pesticide containers, the truth is that all of these actions take place *after* the man was dispossessed, his wife killed, their children's health impaired, the land poisoned, the river dewatered, and his village destroyed. For an action to be sustainable, it can't merely parasitize off an unsustainable system. It must impede the system, and ultimately stop it from being unsustainable at all.

/ / /

I get giddy when I see chicks. All spring I have to stay away from feed stores, or I'll want to take home every chick in sight. It's like going to the pound (where I don't get giddy, but sad), where I want to give every dog and cat a good home.

So one spring I bought too many chicks, ducklings, goslings, and baby turkeys. Way too many. And then my chickens started raising babies. And then I was able to get a bunch of hens who'd been rescued from a factory farm (when I got them their combs were wilted and pale, but after they'd been at my place for a while their combs became firm and bright red).

I ended up with way more birds than the land could have supported had I not been dumpster diving.

I tried to garden at the same time I had the chickens, but the birds ate almost everything. For the longest time I felt bad for what I was doing to the land. It looked like a moonscape for quite a ways in all directions from the chicken coop. But one day I picked up a shovel and dug beneath the thick dried crust of chicken manure. The soil was rich and black, with a higher worm density than I've ever seen outside a worm bin. The land only looked dead on top, but underneath that skin, the chicken shit was working its magic.

/ / /

I kept chickens for about ten years, and quit only when I moved away. One of my great pleasures of that time (in addition to getting giddy over chicks, and giving away big bags of pasta) was watching the soil slowly crawl up the skirt of my double-wide mobile home. The chickens (along with the peaches they wouldn't eat, the watermelon rinds they left behind, the onions they despised, and the potatoes they ignored) were making soil. Rich, black, healthy soil. And this soil—this living and breathing soil, this carnivorous soil who eats us all in the end—was growing so fast—maybe two inches in the ten years I lived there—that I could see the changes.

/ / /

The particular method of waste disposal any historical town or city chose depended greatly on the composition of its garbage, and the makeup of human garbage has changed dramatically in the past century. A more detailed look at that change is illuminating and can help to answer some of our questions about how society got to be the way it is.

At the close of the nineteenth century, more than 80 percent of household waste was comprised of dust and cinders from stoves.[29] The use of woodstoves and coal for heating and cooking meant ash would make up the majority of household waste until the middle of the twentieth century, when new infrastructures allowed oil, gas, and electricity to become the primary sources of energy for heating and cooking. Of course, wood ash in the present day is effectively absent from household waste.

The remaining portion of late-nineteenth-century waste was made up of kitchen and organic scraps, as well as paper, glass, and a small amount of metal. Modern materials like plastic had not yet entered the picture. In urban areas, dead animals made up a significant portion of refuse tossed into the streets. In 1866, New York City's Metropolitan Board of Health forbade the "throwing of dead animals, garbage or ashes into the streets."

Despite this, in 1880 alone, some fifteen thousand dead horses were removed from the streets of New York City, more than forty a day.[30]

Household waste at the time was dealt with more effectively than one might imagine. The fact that household garbage was mostly two rather homogenous materials—and organics—meant that it was comparatively easy to deal with. Although movies have given us images of people dumping chamber pots willy-nilly out of their windows, early municipal garbage collection systems were well-organized and based on a system of "primary separation." That is, citizens were required to sort their garbage into distinct categories when they put it out for collection, much like (but significantly predating) modern recycling systems. In 1896, New York City already had a system for separating different kinds of household waste, dividing it into food waste, ash, and dry trash.[31] This system was enforced not just by garbage collectors but also by the police who would lay fines or even arrest people who failed to sort their garbage. And by 1898, the Street Cleaning Commissioner of New York City had organized a sorting plant for recyclables.[32] Primary separation was not limited to New York, though. Rather, it was commonplace during the first few decades of municipal garbage collection across the United States. In 1902, 80 percent of cities with populations larger than twenty-five thousand required some degree of primary separation.[33] New York's rather heavy-handed use of police to enforce primary separation lasted only a few years, with cities generally finding the most success for primary separation when it was endorsed by civic groups who helped educate and motivate the public.[34]

Partly as a result of primary separation, turn-of-the-century society had specialized occupations for dealing with different kinds of waste. Although the jobs were often very dirty, labor was cheap because of high levels of poverty in the industrialized nations. Coal dust was collected from homes and factories by "dustmen"[35] and taken to "dust-yards." There, workers (men, women, and children) would use sieves to separate the dust out into materials used as soil conditioners or for making bricks or pavement.[36] By 1939, coal and wood ash made up only 43 percent of garbage in major cities like New York, the proportion having dropped by almost half in only a quarter century.[37] The spread of natural gas infrastructures for cooking and heating was one cause. The effects of this were not entirely positive in terms of dealing with household waste. The use of gas was first available in wealthier homes, and since the use of gas produced no ashes, those more affluent city dwellers were even more apathetic about primary separation than they would normally be.[38]

In fact, ashes are the only category of waste which has decreased in municipal waste collection in the past century, to the point where ashes are now a negligible component of household waste. It would be nice to believe that this is due to a dramatic decrease in coal burning. Unfortunately, it merely reflects a shift in the *location* of coal burning. During the twentieth century the burning of coal became centralized to large coal burning plants, and the ash by-product became treated as an industrial waste, rather than a municipal one. The effect was to move coal burning out of local control, and largely out of the public mind.

An American living in a major city around the turn of the twentieth century produced somewhere between 1,000 and 1,750 pounds of waste per year.[39] But only a tiny fraction of that was actually heterogeneous, which is the sticking point in terms of effective disposal.

A lot has changed in a century. Ash is gone, but there've been increases in organic waste and dramatic increases in "product" waste, which includes containers, packaging, and disposable, broken, or unwanted goods.

The increase in organic waste is not due to an increase in kitchen scraps. Of course, in the past century there has been a great shift towards pre-processed and pre-packaged foods, so scraps such as vegetable peelings make up a smaller proportion of waste. Instead, the increase in the volume of organic waste is due to the incorporation of "yard trimmings" from mowing the grass. There's a paradoxical irony here, and it's one of many we'll see. It was the inexpensive availability of curbside waste collection that caused grass clippings—a source of mulch and valuable nutrients for lawns when left in place—to become a waste product, a form of garbage that could be disposed of simply out of convenience.

Of course, the disposal of "product" wastes has seen the greatest increase in quantity in the past century. In a hundred years they've grown from only ninety-two pounds per person per year, to a staggering 1,242 pounds per person per year. Most of the increase in waste generation happened in the later twentieth century, with per capita waste generation increasing from 2.7 to 4.4 pounds per person per day.

A 1905 study identified common kinds of product waste as glass bottles, paper, pasteboard, rags, mattresses, old clothes, old shoes, leather and scrap leather, carpets, and straw.[40] As we will discuss shortly, those were products that were commonly recycled until the twentieth century. In contrast, modern waste is comprised of a very wide variety of unsorted products, each of which may be comprised of many different materials joined together in many different ways, making these products far more difficult to deal with intelligently than the products of a hundred years ago.

The current municipal garbage collection systems—called "Municipal Solid Waste Management" (MSWM) systems in the industry vernacular—emerged from the context of that turn-of-the-century waste composition. They were designed to deal mainly two with homogeneous categories of waste: ash and organics. The basic structure of those systems, combined with the inherent difficulty of dealing with a varied mixture of waste, doomed them to a feeble compromise. These systems are good at getting garbage away from the curbside and to a disposal site. But they're not very good at recovering materials from waste (less than a third of material is recycled or recovered in some way[41]), and they're inherently ineffectual at reducing the creation of waste at the source.

Any system, or any community, must have a method for dealing with waste with an intelligence and complexity that mirrors the diversity of its wastes. If this is not that case—if a community can't sort and deal with its waste intelligently—then their waste truly *becomes* waste, because they can't close the loop back to a useable substance. This is what has occurred to municipal waste management systems in the industrial nations, and though they're effective at moving extremely large quantities of waste out of sight and mind of urban dwellers, they're not effective at sorting or reprocessing it. As a result, countries like the United States now dispose of more than two-thirds of their household waste into landfills and incinerators. The reason these disposal methods now predominate is simple: they require no sorting, and they displace the ultimate responsibility to dealing with waste to other places and to future generations.

/ / /

Burning waste isn't new. If you've ever gone camping in the backwoods, perhaps you've burned your food scraps to avoid attracting bears or other wild animals (including ants). Practices like this are probably as old as campfires, and essentially harmless. What's new, and not nearly so harmless, are chlorinated plastics, enormous municipal and industrial waste streams, and large-scale incinerators.

Though burning of refuse has been a common method for a long time, large public incinerators weren't common in the United States until the 1920s, when they were called cremators. It was around this time, you may recall, that growing urbanization and a trend towards disposability in consumer items kicked off a municipal garbage problem that seemed serious at the time, but pales in comparison to modern garbage production. Through the middle of the twentieth century, landfills and incinerators vied for the position of "preferred waste disposal method," although landfills generally predominated. Into the 1970s, the aim of incineration was still to reduce the volume of solid waste, but more recently incinerators have been built with the intent of capturing the energy released by burning garbage either as heat or by converting it to electricity.

Here's how a typical modern incinerator works. First, garbage trucks bring in loads of mixed municipal waste. In theory, hazardous materials should not be included in this garbage, but of course, the municipal waste stream is contaminated by a variety of items like batteries and electronics waste that should officially be disposed of in other ways. The trucks dump their loads into a waste bunker that ensures a constant supply of waste is available to be burned in the incinerator. Next, overhead cranes pick up loads of waste and drop it into a feed chute supplying the incinerator. The garbage is then pushed onto the incinerator grate, which moves the garbage through the combustion chamber over a period of several hours. Air to support combustion is commonly drawn through the waste bunker to reduce odors outside the incinerator.

Incineration reduces the mass of solid waste by about 75 percent, and the volume by close to 90 percent.[42] The flue gases heat a boiler to produce steam which generates electricity. The incinerator also produces "fly ash," the light ash that goes up the chimney, which contains various hazards and heavy metals including vanadium, manganese, chromium, nickel, arsenic, mercury, and lead. Although the emissions are filtered, many particles produced are simply too fine for the incinerator's filters to capture. The majority of the ash produced, however, is "bottom ash," which is enriched with heavy metals. Although the fly ash should theoretically be disposed of as hazardous waste, bottom ash isn't likely to get such cautious treatment. In some areas, bottom ash is even mixed in with construction materials as a form of "recycling."

"Waste to energy" schemes are commonly used to try to justify the construction of new incinerators, with advocates insisting that burning garbage represents a source of "green energy." As a desirable method of producing energy, garbage incinerators fall short, emitting more carbon dioxide (CO_2), watt for watt, than, say, natural gas power plants.[43] Furthermore, most of the embodied energy in garbage is derived directly or indirectly from fossil fuels, meaning that any energy gleaned from incineration is far from green, but simply a roundabout (and more polluting) way of burning fossil fuels. In fact, some researchers believe incineration to be the least energy efficient way of dealing with mixed waste.[44]

In addition, the construction of new incinerators essentially encourages the production of more garbage, by further displacing the consequences of wastefulness. "Building incinerators lets industry off the hook," says Dr. Paul Connett, professor emeritus of chemistry at St. Lawrence University. Instead, says Connett, industry should not be permitted to make materials that can't be disposed of safely. "The message to industry is this: if we can't reuse it, if we can't recycle it, if we can't compost it, you shouldn't be making it."[45]

There are economic implications as well. Building a large incinerator costs hundreds of millions of dollars, and can even exceed a billion. This money doesn't come from industries producing garbage, of course. Instead, it comes from taxpayers via governments. Again, the cost of producing garbage is externalized. And the high costs mean that incinerators often take decades to pay off, locking municipalities into that method of waste disposal for a long time to come.

There is some good news—to use the phrase rather loosely. The incinerators currently being built burn garbage at much higher temperatures than did historical incinerators. At these temperatures, chlorine atoms are unable to combine with carbon atoms to make dioxin molecules. According to the Environmental Protection Agency (EPA), backyard "burn barrels" used to dispose of household waste (including plastics) are now a significantly larger source of dioxin and furans than municipal incinerators.[46] Of course, dioxins and furans are still produced in large amounts by older incinerators, and newer incinerators during start-up and shutdown. And some critics of incinerators point out that when waste-to-energy schemes are used, flue gases are cooled as their heat is used to boil water, bringing the gas back into a temperature range where dioxins can form.

Advocates of incinerators often point out that burning garbage does save some greenhouse emissions in the long term. Although incinerators immediately release large amounts of carbon as CO_2, landfilled garbage may release carbon more slowly in the form of methane, a more potent greenhouse gas. But it's a Hobson's choice; instead of trying to decide whether to build more landfills or more incinerators, industry should stop producing wastes that need to be landfilled or incinerated.

/ / /

The development of urban garbage collection systems led to an unexpected, even paradoxical, result. As garbage collection became more regular and convenient, penalties for wastefulness were gradually removed. Free garbage collection led to an *increase* in the amount of garbage produced, because it was so easy to throw things away.

In a brief called "Unintended Consequences: Municipal Solid Waste Management and the Throwaway Society," Helen Spiegelman and Bill Sheehan of the nonprofit Product Policy Institute examine this exact point. They note that one of the primary effects of MSWM has been to provide a public subsidy to the manufacture of disposable products, the cost of which is borne by taxpayers instead of the manufacturing companies. This marginal advantage to disposable products created market conditions which encouraged the creation of disposable products.

They also noted another contributing factor to the development of a throwaway society. Although used bottles were once commonly collected, washed, and refilled in the US (as they still are in many nonindustrialized nations), the "public investment in the national highway system . . . made it more economical to ship one-way from distant production facilities than to operate local bottling plants."[47] Again, public monies acted as an incentive for manufacturers to make more and more disposable products.

/ / /

If you've gotten this far you've probably realized, if you didn't know already, that recycling isn't a method devised by earnest environmentalists in the 1970s. Rather, the recycling of human-made materials (as opposed to the organic "recycling" that is implicit in nature) dates back many thousands of years.

Recycling is often viewed as a very new and somewhat complicated procedure. In fact, recycling—collecting used materials to remake something useful—is very literally an ancient procedure. Notably, historical recycling was in many cases *more* effective than its modern, high-tech incarnations.

In its earliest forms, what we might call recycling was something so fundamental that the name hardly applies. The "recycling" of nutrients and elements by living organisms is an intrinsic process that dates back to the beginning of life. To equate the carbon cycle with the collection and recycling of plastic grocery bags would be more than a little misleading, because there are some very major differences which I will discuss in greater detail soon. I will use the term recycling to refer specifically to cases in which human-made materials that would otherwise go to waste are collected and processed into new materials.

Processing, also, is an important aspect of our definition. If I, for example, took an old, torn-up sweater and sewed it into a scarf, this would be an example of "reuse" instead of recycling because I'm not making any changes to the material itself. Melting a glass bottle down to create a new bottle would be an example of changing that material. The distinction is important because recycling requires more energy and a larger infrastructure than does reuse.

Recycling practices that we might find recognizable as such date from early civilizations. (It makes sense that this is the case, since non-civilized cultures produced garbage that was dealt with through natural practices like biological decomposition.) Often the recycled material was a kind of metal. Because of its rarity in early civilizations, and because of the immense amount of energy and labor required to mine, smelt, and shape metal, it made economic sense to recycle whenever possible. This was especially the case in places where the raw materials to create an item anew where not available. For example, bronze is an alloy of both copper and (usually) tin. Settlements located where those two metals were not both present had to rely on imports and recycling.

Recycling was not always a gradual process where used or damaged items were reformed continually. Oftentimes in history large-scale recycling would occur after a civilization had collapsed. Those who lived in the wake of those collapses would often make thorough use of the refined materials concentrated in urban centers. For example, the famous Roman Coliseum was used as a stone quarry for centuries after the collapse of the Roman Empire, with the marble façade either reused in other buildings or burned to make quicklime.[48] (This went on until 1749, when it was banned by the Pope). The Great Pyramids of Egypt were also used as stone quarries by those who lived after the collapse of the civilization that built them. The pyramids were covered in a polished casing of white limestone, and this limestone (and later the softer stone core underneath) was stripped for construction in adjacent Cairo.

/ / /

As manufacturing businesses proliferated in the US during the eighteenth and nineteenth centuries, and as the population continued to grow dramatically, the demand for certain manufacturing materials began to grow. This was especially true for rags and cloth scraps, since at the time paper was made solely out of cloth fibers and not out of wood. It's around this time that we find systems we can parallel with modern recycling systems. Reuse of materials within households and homesteads was common, as was the reuse of materials in workshops and even factories. However, the 1800s were full of historically documented and well-organized systems for bringing used materials from households back to urban scrap dealers and manufacturers, where these materials were then reprocessed into consumer goods.

This occurred largely through surprisingly sophisticated and widespread systems of barter. Shopkeepers and roaming peddlers would exchange manufactured goods from afar for quantities of used rags, bones, or certain metals and other recyclable materials. Some peddlers were employed by dealers and provided with equipment, and others worked more independently. The most reliable sources of information we have come from these dealers, who as businessmen kept detailed records of what was bought and sold. In some instances they set precise prices and terms for barter, so that the peddlers in their employ would not attempt to negotiate. The peddlers would travel from home to home, mostly in rural areas, where they would gather materials before heading back to their urban or railroad depots to drop off their collections and restock.

This system inherently placed limitations on the scale of collection and accumulation. Itinerant peddlers could only travel with whatever goods they could carry in backpacks, or sometimes on horses or in horse-drawn wagons. This meant that the recycling and exchange system was necessarily kept near human scale. It also meant that human relationships played an important role in dealing with rubbish.[49]

Lists kept by materials dealers show a wide variety of materials accepted for barter. Some were materials we'd expect to see, like metals, rubber, glass, and most importantly rags. However, items listed were not limited strictly to conventional recyclables. They also included organic materials or goods such as "fruit, flax, mustard seed, woolen yarn, beeswax, butter, eggs, feathers, bristles, hair, horns, bones, and the skins of deer, sheep, calves, bear, mink, raccoon, and even house cats."[50] Perishables like eggs and butter were resold at retail, such as in the general stores which acted as relay points for the dealers.

Many of the other materials were sold to industrial manufacturers. Paper manufacturers required a material to fill in the gaps between fibers in their porous paper. This material, called sizing, was made by cooking down animal products like horns, hoofs, and scraps of hide (hoofs and bones, of course, were also used to make glue). Fats were of great value for lighting and lubrication in any factory. These were eventually replaced by petroleum derivatives, but the first oil well wasn't even drilled in the US until more than a century after these peddlers were well established.[51] Fats were also processed to extract glycerine, which was used to manufacture explosives like nitroglycerine and dynamite.

But from the beginning, rag collection drove the development of these recycling systems. Rags and cloth scraps were needed to make paper, since wood pulp wasn't used to manufacture paper in the US until the twentieth century. Recycled rags were *the* fundamental requirement for paper manufacturing since the first paper mill was opened in the US in 1690. That mill was the Rittenhouse Mill, near Philadelphia, which made new paper from waste linen, cotton, and old paper.[52]

Rags were collected by storekeepers and peddlers, who relayed them to paper mills through "rag routes." Newspapers and printers also collected rags, and used the bundles of rags to pay their paper suppliers.[53] Indeed, most of these transactions were pure barter, with no money involved.

In the 1800s, clothing was still produced by tailors and in households, so there were very few large-scale sources of scrap fabric. To deal with constant shortages of rags, paper mills embarked on large-scale propaganda campaigns to convince women to save their rags and send them to the mills. These campaigns often included newspaper advertisements and witty or poetic slogans. For example, take this piece of verse from 1807: *"The scraps, which you reject, unfit / To clothe the tenant of a hovel, / May shine in sentiment and wit, / And help to make a charming novel."*[54] This and other pieces of propaganda at the time emphasized the benefits of recycling in class terms—from hovel to novel. This contrasts starkly to twentieth century propaganda, in which being of a higher class became identified with wastefulness rather than thrift.

Patriotism, also, was commonly identified with conservation. A 1734 advertisement requesting rags claimed that it was "the Duty of every Person . . . to help forward so useful to a Manufactory; *Therefore I intreat all those that are Lovers of their Country, to be very careful of their Linnen Rags,*" and gave instructions on where to send them.[55] The coming of the American Revolution dramatically increased local demand for rags and patriotic calls for rag saving. Paper products from England were banned, and Americans urgently wanted paper for printing propaganda, for printing money, and for the manufacture of cartridge cases.

/ / /

Rag and rag paper were in comparatively short supply until the twentieth century, with the onset of large-scale mechanization and the use of wood pulp for paper manufacturing. The difference was critical. To get a supply of rags, a society first had to clear and plow land, plant cotton or flax (for linen), harvest and process the fibers, weave cloth and sew clothing, and then collect scraps and used-up articles. This required large amounts of labor (often provided through slavery) and months or years of lead time. The use of wood pulp introduced a tremendous change. Instead of having to plant and harvest, papermakers could now cut down forests, liquidating fiber sources that took centuries to accumulate. In the short term, paper manufacturers were freed of the irritation of having to collect rags and could scale up production to whatever society demanded. In the long term, the change would doom many of the planet's remaining forests.

/ / /

The effects of rag and other recycling systems of the 1800s were somewhat counter-intuitive. Rather than solving rag shortages, recycling actually led to an increase in rag demand by fueling the growth of an embryonic industrial system. The availability of paper (with its importance for keeping records, communication, and advertising) facilitated a global trend of growth for business and an increase in economic activity. And this increase in economic activity and industrial production further drove the demand for paper.

It's a phenomenon often seen in the history of waste. Take computers. They were touted as tools to reduce waste, fostering ideas like "the paperless office." But one of the primarily real-world effects of computers was to increase the number of communications that took place and to vastly increase the amount of data that could be collected on any given subject. Since much of this information is printed out, computers ended up *increasing* the amount of paper used, rather than decreasing it. And through a system of planned obsolescence, the computer age also ushered a new era of electronic waste.

/ / /

If we can learn a lot about a society from the way it deals with garbage, we can also learn much about it from the way it treats the *people* who deal with garbage. Sadly but not surprisingly, people who work with waste are often socially stigmatized, or, in extreme cases, even labeled "untouchables."

Another way to say this is that the attitudes a society has toward trash closely mirror its attitudes toward trash workers and "garbage pickers." In situations where garbage is viewed as valuable, trash workers themselves often have greater social status or potential for "upward mobility." For example, the ragmen and itinerant peddlers of the 1800s, whose salvaged product was viewed as valuable and worth trading for, would often accumulate wealth and eventually open their own stores or materials dealerships. This makes sense—in a society where sorted and recovered trash has worth, its collection and trading can be viewed as a form of economically gainful production.

Class and caste attitudes are also deeply related. Members of highly stratified societies—and most especially members of their upper classes—tend to have more disdain towards trash workers than members of more egalitarian societies. A few examples of people who work with garbage may help clarify our understanding of these social and economic relationships at work.

After the economic collapse of Argentina in 2001, the country saw a sharp devaluation of its currency (the Argentine peso). As a result, common consumable materials like paper, cardboard, and copper became too expensive for most businesses to import, creating a larger market for local recycling. At the same time, there was a massive increase in unemployment. These factors combined to cause a burst in the number of *cartoneros*—poor people who collect and sort recyclables like cardboard ("carton") to sell to dealers for a small amount of money.[56] Each evening *cartoneros* take a government-supplied train from poor neighborhoods to wealthy areas of downtown Buenos Aires, where they search through the day's trash for recyclable materials of value. The train they take is stripped-down, without lights and without heating or air-conditioning. The seats have been removed to make room for shopping carts. Instead of glass, the windows are often covered in rusty wire.[57] The *cartoneros* call this *El Tren Blanco*, the white train. Because the train arrives at night and leaves downtown Buenos Aires before dawn, the wealthy Argentines almost never see the *cartoneros* at work. In a pointed commentary on their social invisibility, many *cartoneros* have dubbed the train *El Tren del Fantasma*—the ghost train.[58]

The International Red Cross estimated that in 2002 about two thousand *cartoneros* used the train each day.[59] In 2003, anthropologist Francisco Suarez estimated there were between thirty and forty thousand *cartoneros* working in the greater metropolitan area of Buenos Aires. In contrast to the behavior of many wealthier cities around the world, the municipal government of Buenos Aires has chosen to work with *cartoneros*, although the situation is still far from desirable. They've legalized the practice, which had been illegal for decades, and launched a campaign for wealthy residents to separate their recyclables in green bags, so that *cartoneros* will not have to sort through rotten food or hazardous or sharp objects. They've also begun to offer *cartoneros* vaccinations against tetanus.[60] And the *cartoneros* themselves have also become more organized, as the closure of factories brought an influx of laborers with union experience into the *cartonero* demographic.

Now, let's contrast that with the municipal dump for the city of Tijuana. A popular city for tourists, Tijuana grew rapidly as an industrial center following the ratification of the North American Free Trade Agreement (NAFTA) in 1994. The easy movement of products across the border to consumers in the US combined with barriers to the movement of poor Latin Americans across this same border provided the cheap labor and economic circumstances that allowed a boom of labor-intensive factories called *maquiladoras*. Often, these were simply sweatshops. Continuing American tourism and (often wasteful) local manufacturing helped to produce a steady stream of many different kinds of garbage.

Luis Alberto Urrea tells many stories about the Tijuana dump in his excellent, moving, and sometimes horrifying book *The Lake of Sleeping Children*. He writes that although the dump was once "a gaping Grand Canyon, it gradually filled with the endless glacier of trash until it rose, rose, swelling like a filling belly. The canyon filled and formed a flat plain, and the plain began to grow in bulldozed ramps, layers, sections, battlements."[61]

But Urrea's main focus is not on the landscape. It's on the people who live in the dump. The conditions of widespread poverty that allowed the *maquiladoras* to proliferate also meant that scavenging in the dump to find materials for personal use or to sell was an economically viable option for many people—if not a necessity out of desperation. Urrea observed that "poor Mexicans, transformed now by NAFTA into a kind of squadron of human tractors, made their way through the dump, lifting, sifting, bagging, hauling, carting, plucking, cutting, recycling."[62]

The decomposing garbage in the dump produces highly-flammable methane, which flows out of openings and crevices in the dump. Sometimes the residents will deliberately ignite methane-emitting openings to cook over. And sometimes the methane will flow out unexpectedly, bursting into flame and even burning down entire neighborhoods.[63] Toxic gases fill the air, not just from the mountains of garbage but from the places where garbage is burned for fuel, or to extract the metal from radial-belt automobile tires or insulated copper wires. Fires under the surface may smolder for years.

Often, families will eat food that comes directly from the garbage. And a local health clinic reports that those living in the dump have abnormally high incidences of "skin rashes, throat ailments and cancers."[64] And problems like these, awful as they are, are small compared to the ever-present risk of being crushed by one of the many huge garbage trucks.

The trash workers of the dump were not driven by the same goals of self-interest and personal profit which drove the architects of international trade arrangements. Instead, they developed community norms of mutual aid and egalitarianism. "The original *dompe* rules, a set of ordinances that sprang up organically from the people who have to work the garbage, prevailed. A set of rules, by the way, that are extraordinarily humane and sane."[65]

Just as the ethics and norms of indigenous groups were partly derived from their particular landbases, the rules of the Tijuana dump were partly determined by its structure. The middle of the *dompe* is where the trucks disgorge their loads. This is also the most promising site in the dump where canned food, (relatively) fresh produce, and working machines may be found. Because of the hazards of the trucks and bulldozers, the "youngest and the strongest" work here. When garbage-pickers drop into holes or fissures for choice finds, they erect poles with rags on the end to warn tractor drivers not to drive into them, crush them, or bury them in garbage. Despite precautions like this, garbage-pickers are still often killed in the Tijuana dump, and in other dumps throughout the world.

Urrea describes five rules that garbage-pickers follow. The first is to watch for dangerous heavy machinery. The second is that children are not allowed in "the trash," the main area of the dump piled with garbage. The third has to do with gender equity: men and women are equal in the trash. The fourth is that children and older people are allowed to work on the fringe of the trash, away from the dangerous machines, where garbage pushed by tractors down the slope of the expanding mountain is sifted by gravity so that some of the best finds roll out to the pickers. And the fifth rule is that a special "safe area" is set up by the healthy workers for the very old and for people with disabilities. Universally honored, this area is avoided by tractors and trucks, though the occasional load is dumped or carried there. This way, the old and people with disabilities can still work safely. Urrea observes: "There is no welfare in the dump, but there is work, care, sweat, and dignity."

In 2007, the city of Tijuana began to shut down the municipal dump Urrea wrote about. This is not necessarily good for the planet or the pickers. As far as the land around the dump is concerned, little has changed. Visiting the area of the dump, journalist Kinsee Morlan wrote, "When it rains, the hillsides of Fausto Gonzales bleed trash. Jagged bed springs pop out of the soft mud like broken bones from flesh. Plastic bags, rotted wood, pieces of Barbie dolls, bottle caps, cigarette butts and other debris join a muddy waterfall that spills into the canyon below. The smell is putrid." And of course, Tijuana has not stopped producing garbage, or even tried to staunch the flow. Instead, the garbage goes to new dumps to drown even more land. Many of the pickers still live there, though others have moved on to other dumps. Strange as it may seem, many prefer working in the dump to working in sweatshops, where the daily wage may be even less than one can gain by selling their best finds from the dump.

At a new landfill location outside of Tijuana, history is repeating itself. Simple shacks and shanties are being built from scavenged materials, and individuals and families are setting up shop, despite government efforts to keep them out. Alfonso López Posada, codirector of Tijuana's Municipal Cleaning Services, is resigned to the situation. "As long as there is trash, there will be people who work in the dump."[66]

SUSTAINABILITY

This culture—devouring, degrading, and insane—cannot continue. For sustainable to mean anything, we must embrace and then defend the bare truth: the planet is primary. The life-producing work of a million species are literally the earth, air, and water we depend on. No human activity—not the vacuous, not the sublime—is worth more than life on this planet. Neither, in the end, is any human life. If we use the word sustainable and don't mean that, then we are liars of the worst sort: the kind who let atrocities happen while we stand by and do nothing. —LIERRE KEITH

I am in love with this world. It has been my home. It has been my point of outlook into the universe. I have never bruised myself against it nor tried to use it ignobly. —JOHN BURROUGHS

I'VE WRITTEN ELSEWHERE THAT the predator-prey relationship is characterized by the following deal: if I consume the flesh of another (or otherwise kill this other) I now take responsibility for the continuation of that other's community. This deal holds morally, it holds spiritually, and it certainly holds physically. Those who do not know this—those who do not live it—do not survive. They destroy their own habitat, and in doing so, destroy themselves. It may take a while for those circles to close, to become self-made nooses around their own necks, but it happens. Every time.

I would say that the same holds true not only for what we take into our bodies, but for what comes out of them. If we leave something behind— whether it is shit or our bodies or our clothes or our shelters—we now are responsible for the well-being of this community.

The good news is that this is how humans have lived in place for most of our existence. The bad news is that almost no one in this culture lives this way now.

/ / /

The questions we need to ask ourselves about every action—as we live in the midst of a culture killing the planet—are these: Is this action sustainable? Why or why not? How would this action need to be different for it to be sustainable?

Before we can answer these questions, however, we have to define *sustainable*. Many politicians, business people, "green" architects, land managers, foresters and other resource specialists, and so on throw that word around a lot in meaningless or deceptive ways, labeling as "sustainable" many manifestly unsustainable actions that most often make a lot of money for them or for the corporations to whom they are beholden. We hear about sustainable buildings, sustainable agriculture, sustainable forestry, sustainable this and sustainable that, and of course, within this culture, little or none of it is even remotely sustainable.

For an action to be sustainable you must be able to perform it indefinitely. This means that the action must either help or at the very least not materially harm the landbase. If an action materially harms the landbase, it cannot be performed indefinitely: any line sloping downward eventually reaches zero.

Central to sustainability is the landbase itself. What may be beneficial to one landbase may be harmful or lethal to another. I feel good shitting outside and dropping pieces of toilet paper willy-nilly across the rainforest floor, comfortable in the knowledge that it will all break down at most within a year. Would this be appropriate behavior in a desert? Certainly beings in deserts still have to defecate, and certainly deserts have developed ways to turn shit into something they can use. But in a desert, I might have to spread my shit and paper over a larger area, and maybe not use paper at all, to ensure I won't negatively impact the land. A nonhuman example may help make this a little more clear. Cows did not evolve in a desert. Their poop makes big patties, which in moist climates break down into potent packages of food for scavengers and soil alike. In dry climates, however, these patties can ossify, turning into a sort of fecal asphalt that smothers and harms the soil. Antelope, bighorn sheep, and others who evolved in deserts do not poop in big mounds, but rather in tiny pellets that are more easily convertible to food the desert can use.

Here's another way to look at sustainability's dependence on context. I live on Tolowa land. Prior to conquest, the Tolowa lived here without materially harming the place for at least 12,500 years. By any reasonable definition they lived here sustainably. Their homes were made of wood. This means that, here in their rainforest, this particular use of wood was sustainable. But people who live where trees are sparse may not be able to sustainably use wood as a building material.

Any working definition of sustainability must emerge from and conform to a particular landbase—to what that landbase can freely give forever—and not be an abstract set of principles, or rationalizations, imposed upon the landbase. The landbase is primary, and what we do to it (or far more appropriately, with and for it) must always follow the landbase's lead.

What actions are sustainable is determined not only by context, but also by scale. One human shitting in the forest near my home is probably a good thing. Two, three, or four might make an even bigger bonanza for slugs. But let a thousand people shit right here and the slugs will quickly say *no más*. Likewise, the fact that the Tolowa took out a few trees does not mean this forest (or any forest) can survive industrial (or even extractive) forestry (which should more accurately be called deforestation). And the fact that the Tolowa took salmon to eat doesn't mean salmon can survive industrial (or even extractive) fishing (which once again should really be called de-fishing).

Similarly, I don't shit in the pond by my home. I don't believe the pond is big enough to take in my shit without materially harming it. If I lived near the Amazon River, on the other hand, and were the Amazon not horribly stressed by the various activities and wastes of this culture, I'd gladly defecate into it, knowing that I was helping the river. I remember years ago reading a story by Herbert R. Axelrod, an expert on tropical fish, in which he was wading waist-deep in a sluggish backwater of the Amazon when a big fish started nudging him in the butt. His native guides told him the fish was begging for food, hoping it could induce him to drop his pants and defecate.

At least my dogs have the patience to not begin their begging till I've removed my pants.

The land is a living entity who like any other living entity requires certain foods to survive. Certain other foods can be toxic. But even nourishing foods can be toxic out of scale.

My own shit is just one small bite for this land. But just as I might find one apple delectable and conducive to good health, and just as ten apples at once might make me sick to my stomach and give me the runs, and just as dropping ten tons of apples on my head will kill me, so too the land could find one person's shit delicious and beneficial, more shit harmful, and more shit than that all at once lethal.

Time is as important as scale to an action's sustainability. Indeed, they are related. The Tolowa, Yurok, Karuk, and other tribes probably killed more salmon in this region over the last 12,500 years than this culture has in the past 180. But they did it over 12,500 years (they also didn't dam or poison rivers, deforest hillsides, murder oceans, change the climate, and so on). Killing that many salmon to eat (and dying yourself on this same land, feeding the land and thus eventually the salmon in the same manner as they are feeding you) could have continued in human terms forever. But killing as many salmon as the dominant culture has in such a short time has not been a mutual feeding, but instead a slaughter, and has decimated—and in many cases extirpated—salmon.

/ / /

I asked my friend Terry Shistar for help understanding time and sustainability. She said, "I often think about why and how this culture values permanence. So often we hear this culture described as a throw-away culture, and that's certainly true. But there's a larger sense in which permanence is valued above all else. People want to make a mark. They want to build some monument that will last. They want to stop decay. They want to cheat death. It's all based on that disrespect for nature you've written about, and worse, a fear and hatred of nature and of death."

"Others may die, but I shall live on. Or at least what I create."

"Yes. And this striving for permanence—or at least a form of it—happens on a less grand scale, too. Even in conversations about 'green' products, permanence is valued. Disposable pens/razors/diapers/cameras/whatever are not considered as good as durable products. Well, those disposable products are generally not very biodegradable, so they're generally only less permanent in their usefulness to us while remaining waste for just as long. Nevertheless, it leads me to think about how much that 'green' preference for durable products affects my thinking in general. In reality, I should prefer structures that will be eaten by termites and fungi (and lived in by birds and mice along the way), and valuing that permanence interferes with my attempts to live sustainably. I think part of the problem is that civilization constantly tries to turn circles into straight lines. Ultimately, the circles *will* close, but it can take a long time."

/ / /

An action's sustainability is also determined in part by what other stressors the area is suffering. Massive fish kills by dams, logging, and so on make it so far fewer fish can be sustainably caught. Likewise, if someday this culture kills all the slugs in this forest—as will undoubtedly happen if this culture keeps going the way it is—then even one person shitting here may be too much. That's one of the great costs of living in this culture. Simple acts that previously would have been perfectly sustainable—and perfectly natural—now are not.

It's a pretty obvious point. If you are healthy, you may be able to give a lot to others, emotionally, physically, spiritually. If you are sick, you may have less to give, and what you do give may make you sicker. If you are grievously wounded or seriously ill, any additional injuries or insults could kill you.

/ / /

But of course me shitting here isn't really sustainable anyway, because no action takes place in a vacuum. The food that I ultimately shit has to come from somewhere, and if the hunting, gathering, harvesting and/or production as well as transportation of that food—the bringing of that food to my very lips—isn't sustainable, then my shitting isn't sustainable, either, because if I'm going to shit I need food. No food, no shit.

No shit.

It's a pretty basic point that's perhaps intentionally missed by almost everyone in this culture who claims to participate in sustainable activities: an action is sustainable if and only if all necessary associated actions are sustainable. To make this more clear, let's talk about so-called sustainable architecture.

SUSTAINABILITY™

The word sustainable *has been reduced to being the* Praise, Jesus! *of the eco-earnest. It's a word where corporate marketers, with their mediated upswell of green sentiment, mesh perfectly with the relentless denial of the privileged. It's a word I can barely stand to use because it's been so exsanguinated by the morally limp, the personally smug, the cheerleaders for the technotopic, consumer kingdom come. To doubt the vague promise now firmly embedded in the word—that we can have our cars, our corporations, our consumption and our planet, too—is both treason and heresy to the emotional well-being of most progressives. But here's the question: Do we want to feel better or do we want to be effective? Are we sentimentalists or are we warriors?* —LIERRE KEITH

WILLIAM MCDONOUGH IS A WORLD-RENOWNED architect who has been labeled a "priest" of "sustainable development."[67] He recognizes—as does anyone possessing both intelligence and integrity—that this culture is highly destructive, and that part of its destructiveness comes from the fact that its waste products do not help the natural world, but rather poison it. He rightly states that in the natural world, "the waste of one organism provides nourishment for another—waste equals food."[68] He follows this by saying that he wants to change industry so that, among other things, its wastes will be useful: the industrial equivalent of that same principle of waste equaling food. Central to his philosophy, he says [and throughout the rest of this discussion, you can see some of my line-by-line responses to the more absurd of his statements by looking in the endnotes: other analyses follow his text], are "fundamental design principles" that "yield products that are composed of materials that biodegrade and become food for biological cycles, or of synthetic materials that stay in closed-loop technical cycles, where they continually circulate as valuable nutrients for industry.[69] They yield buildings designed to accrue solar energy, sequester carbon, filter water, create habitat, and provide safe, healthy, delightful places to work.[70] Designs such as these aren't damage management strategies. They don't seek to retrofit a destructive system.[71] Instead, they aim to eliminate the very concept of waste[72] while providing goods and services that restore and support nature and human society.[73] They are built on the conviction that design can celebrate positive aspirations and create a wholly positive human footprint."[74] He gets right to the point: "Long-term prosperity depends not on making a fundamentally destructive system more efficient, but on transforming the system so that all of its products and processes are safe, healthful and regenerative."[75]

He writes, "Imagine a building, enmeshed in the landscape, that harvests the energy of the sun, sequesters carbon and makes oxygen. Imagine on-site wetlands and botanical gardens recovering nutrients from circulating water. Fresh air, flowering plants, and daylight everywhere. Beauty and comfort for every inhabitant. A roof covered in soil and sedum to absorb the falling rain. Birds nesting and feeding in the building's verdant footprint. In short, a life-support system in harmony with energy flows, human souls, and other living things."[76]

His rhetoric is grand, and the culture has rewarded him well for his work. He has an eponymous architectural firm with thirty members. He is the winner of three US presidential awards: the Presidential Award for Sustainable Development (1996), the Presidential Green Chemistry Challenge Award (2003), and the National Design Award (2004). In 1999, *Time* magazine recognized him as a "Hero for the Planet," stating that "his utopianism is grounded in a unified philosophy that—in demonstrable and practical ways—is changing the design of the world."[77]

What does this mean on the ground, where it really matters?

Let's look briefly at a few of his projects.

First, there's the Ford Rouge Dearborn Truck Plant, in Dearborn, Michigan. McDonough's website describes his work on this factory in language that might almost make us forget that this is a factory at all, and certainly has a shot at making us forget what is manufactured there: "Lying at the center of the Ford Rouge revitalization project, this new assembly plant represents the client's bold efforts to rethink the ecological footprint of a large manufacturing facility. The design synthesizes an emphasis on a safe and healthy workplace with an approach that optimizes the impact of industrial activity on the external environment."[78]

He continues, "The keystone of the site stormwater management system is the plant's 10-acre (454,000 sf) 'living' roof—the largest in the world. This green roof is expected to retain half the annual rainfall that falls on its surface. The roof will also provide habitat . . . With the sound of nesting songbirds chirping over factory workers' heads, the new Dearborn Truck Plant offers a glimpse of the transformative possibilities suggested by this new model for sustaining industry."[79]

The factory, this "new model for sustaining industry," manufactures trucks. Trucks. That's right: trucks. McDonough is correct about one thing: this model is all about "sustaining industry." It's certainly not about sustaining the natural world.

It is both absurd and obscene to suggest that just because he puts plants on the roof that this factory is even remotely sustainable. Where does he think the steel comes from for the trucks (or for that matter, for the factory)? The bauxite for the aluminum? The rubber for the tires? Is the production of each of these materials also sustainable? And what happens with the trucks after they're manufactured? What damage do they cause? How much do trucks contribute to global warming? Is a culture based on the routine transportation over great distances of people, raw materials, and finished products sustainable? Is it possible for a culture with trucks to be sustainable at all?

It's easy to break down any process and say that this or that particular part of the process is sustainable. If, for example, I were going to "push the button" and blow up the world with nuclear weapons—which by any definition is not a sustainable activity—I could describe my thought processes as sustainable: my thought processes aren't causing any damage at all. I could describe walking over to the button as sustainable: me walking from here to there doesn't cause any damage, especially if I'm wearing shoes made of recycled materials. I could describe pushing the button itself as sustainable: I'm just pushing a button, and how much harm can there be in that? It's only what happens after—for which I need take no responsibility—that is not sustainable. I am as green as could be. I am a Hero for the Planet™.

Would we consider a nuclear bomb factory to be sustainable because we put native grasses on the roof?

And for those of you who think my jump from a truck factory to nuclear weapons spurious, ask yourself: given global warming, and given the other effects of car culture (and given the implications of a culture based on the long distance transportation of people, raw materials, and finished products), which have caused more damage to the biosphere: nuclear weapons or trucks?

The answer of course is both, and more particularly the mindset that leads to them.

Next, let's read McDonough's description of his work on Nike's European headquarters: "Nike's business revolves around world-class athletic performance,[80] so the design of its European Headquarters aspires to equivalent levels of building and human performance. The campus creates an active habitat that promotes physical, social, and cultural health in the broadest possible senses.[81] Indoor environments contain virtually no PVC, sustainably harvested wood,[82] and abundant daylight and fresh air. One of Europe's largest geothermal installations provides safe space conditioning, contributing to Nike EHQ's place as the most energy-effective office of its size in the Netherlands. The design encourages strong local connections by evoking the regional landscape of water and native plants and embracing a rich architectural context that includes an adjacent grandstand designed by Hilversum architect Willem Dudok for the 1928 Olympics equestrian events."[83]

In an article extolling his own work on this "campus," and extolling also, as the article's subtitle states, "Nike's Giant Steps Toward Sustainability," McDonough writes, "What, they [Nike] wondered, are the long-term environmental and social impacts of the athletic footwear industry? How does a company with annual revenues in the billions (over $9 billion in 2001) and more than 700 contract factories worldwide profitably integrate ecology and social equity into the way it does business, every day at every level of operation?"[84]

He answers the questions: "Rather than trying to limit the impact of industry through the management of harmful emissions, cradle-to-cradle thinking posits that intelligent design can eliminate the concept of waste,[85] resolving the conflict between nature and commerce.[86] By modeling indus-

trial systems on nature's nutrient flows, designers can create highly productive facilities that have positive effects on their surroundings,[87] and
completely healthful products that are either returned to the soil or flow
back to industry forever.[88] It's a life-affirming strategy that celebrates
human creativity and the abundance of nature—a perfect fit with Nike's
positive, innovative culture."[89]

McDonough quotes Nike's director of corporate sustainable development (a triple oxymoron if ever I've heard one), as saying that
McDonough's philosophy "meshes very well with the culture here. And
it's an exciting message. If you talk about environmental management systems and eco-efficiency, people just roll their eyes. But if you talk about
innovation and abundance, it's inspirational. People get very, very
excited."[90]

Of course people at the upper levels of a corporation get excited at
McDonough's message: nothing here suggests they must fundamentally
change the way they do business, a way of doing business that makes them
rich. Ah, so now we finally get to the point: they can continue to exploit
workers and destroy the planet, with all their niggling fears erased because
they are now participating in a life-affirming process, a "sustainable" (or
rather, "sustainable™") process.

He continues, "And get excited they did. . . . In 1996 . . . Nike contracted William McDonough + Partners to design a new, state-of-the-art
campus for its European headquarters in The Netherlands. A complex of
five new buildings, the campus was designed to integrate the indoors with
the surrounding environment, tapping into local energy flows to create
healthy, beneficial relationships between nature and human culture.[91]

But McDonough and his philosophy did not merely lead to a beautiful
"campus" in Europe where beautiful European executives can work, play,
eat at a bistro, all the while managing this company that profits from the
near-slave labor of brown people—mainly young brown women—in factories where they're forced to work sixty-five hours per week at starvation
wages to make luxury athletic shoes; in factories where chemicals causing
liver, kidney, and brain damage can be 177 times the legal limit (even for
countries with such lax standards as Vietnam and China); in factories
where 77 percent of workers suffer respiratory problems;[92] in factories
where nearly all of those who work there would have neither the strength

nor the time to play tennis. No, McDonough's philosophy has accomplished far more than this; as McDonough proudly notes, "By 2010, Nike plans to use a minimum of 5% organically grown cotton in all cotton apparel."

McDonough concludes, in language that makes *me* want to go out and buy some Nike shoes, "We agree. And whether the once-sleeping giant is now striding along in Swoosh Slides, Air Jordans, or organic cotton socks, we've been delighted to see it rise to its feet."[93]

This company that McDonough describes so glowingly, so lovingly, this company that turns "inspiration into fruitful action" and that "will do the same as it takes on each new challenge on the path to sustainability," is Nike. Nike. Sweatshop Nike. Exploitative Nike. Nike that makes luxury athletic shoes—a product that is absolutely unnecessary to a good life—in filthy factories filled with young women who are not paid enough to eat, much less raise their families, young women who are often subject to sexual harassment.[94] Nike that allows (poor brown) workers five minutes per day for bathroom breaks and forces women to show bloody underwear to prove they're menstruating.[95] Nike, which often fires workers who take even one day of sick leave.[96] Nike, where in Indonesia 30 percent of the company's total business costs are payoffs to "Indonesian generals, government officials, and cronies."[97] Nike, with contract factories that routinely burn rubber, then deny this was ever done.

I'm glad Nike is moving toward using 5 percent organic cotton. Using "only" 95 percent pesticide-laden cotton is certainly better than using 100 percent pesticide-laden cotton. But we should never forget that the slave-based cotton industry of the Antebellum American South was 100 percent organic. Nike has 95 percent more to go before it reaches even that wretched standard.

Nike. Would McDonough have applauded Ford Motor Company's World War II German subsidiary (Ford Werke A.G., with between 55 and 90 percent of stock owned by Ford USA from 1933 to 1945)[98] for making a small percentage of its raw materials slightly less toxic in its Nazi slave factories? Would he have designed Bayer's headquarters, then written encomia about Bayer improving material usage in *its* slave factories? What about Daimler-Benz? What about the Japanese companies such as Mitsubishi and Kawasaki that used slave laborers during World War II? Would

McDonough have sung their praises if they'd hired him to build a lovely headquarters for them and then provided recycled (or 5 percent non-toxic) materials for the slaves to assemble?[99]

A slave camp is a slave camp, no matter the rhetoric.

And for those of you who think my jump from a Nike factory to a World War II slave camp spurious, consider the words of 1996 Nobel Peace Prize winner José Ramos-Horta: "Nike should be treated as enemies, in the same manner we view armies and governments that perpetrate human rights violations. What is the difference between the behavior of Nike in Indonesia and elsewhere, and the Japanese imperial army during WWII?"[100]

Finally, a brief look at one more of McDonough's projects. This is a "corporate flight center" in Detroit, Michigan. McDonough writes, "This design takes the passenger experience as the point of departure, seeking to restore the wonder of flight to both commuters and flight staff. A new entry sequence begins beneath the shelter of a winglike canopy and leads to a central atrium that more clearly defines the 'placeness' of the terminal for both groups.[101] Extensive glazing, skylights, and a dramatic balcony open to expansive views of the runway and the sky. By creating a more vibrant and people-centered environment, the new center enhances the flight community's workplace and provides travelers with a more inspiring gathering place."[102]

He's describing a corporate airport. Airport. Where corporate jets land and take off. What an airport—even one that's "a vibrant and people-centered environment"—has to do with sustainability entirely escapes me. Airports are not sustainable. They cannot be sustainable. They will never be sustainable. A culture based on the sort of movement of people and materials implied by an airport can never be sustainable. A culture that has the physical, social, political, and economic infrastructure necessary to support air travel can never be sustainable. A culture with the physical, social, political, and economic infrastructure made possible and reinforced by air travel can never be sustainable. The only sustainable culture is a local culture, deriving its support from and giving support back to a specific place. The only sustainable culture is one in which its wastes are not industrial and do not serve industry, but rather are organic and serve the local landbase.

A few pages ago I derisively referenced *Time* magazine's characterization of McDonough as a "Hero for the Planet." But the truth is that if all McDonough did was put native grasses on top of truck factories, try to get transnational corporations to recycle more, and make "vibrant" airports, he would, for what it's worth, be a minor hero to me.

I've written extensively about how this culture—civilization—is irredeemable, and is based on its systematic and functionally necessary[103] exploitation and destruction of the natural world. I've written that civilization needs to be brought down before it destroys life on this planet.[104]

But I've written as well that we need it all. We not only need people doing whatever it takes to protect the places they live (and/or love)—to protect life itself from its destruction by this culture and its members (I just today read that the baiji, or white dolphin, which survived for 20 million years, has been declared extinct)—but we also need people trying in the meantime to make this culture incrementally less destructive. Neither alone is sufficient. If we all work and wait for the great glorious uprising that will bring down this culture of death, there will be nothing left when it finally comes crashing down. So to the degree that McDonough is responsible for getting Nike to reduce its use of pesticide-cotton by 5 percent, I am grateful, and I gladly acknowledge and celebrate the importance of his work. But likewise, if we all merely tinker, merely mitigate—and rhetoric aside, mitigation is at best what McDonough is doing—this culture will continue to grind away at all life until, once again, there is nothing left to save anyway.

Unfortunately, as we saw, McDonough claims his mitigation and tinkering are far more than they are. Imagine how different McDonough's work and rhetoric would be if he said things like, "It's great we're putting native grasses on the roof of this truck factory, but it's the tiniest of tiny steps, and given the rapidity with which industrial civilization is dismantling the natural world, these transitional steps need to be much bigger, much faster. Today we put grasses on the roof of a truck factory. Tomorrow we do away with trucks. The day after, we do away with factories, as we throttle down our way toward some semblance of sustainability. The design of this factory is in no way sustainable, and while I am flattered that some may call it that, I cannot accept that praise, since we do not have time to delude ourselves as to the magnitude of what must be changed, what must be unmade, what must be rejected outright. But at least this design is movement, and we desperately need movement—any movement—in the right direction." If McDonough said anything even remotely as honest as this, he would (once again, for what it's worth) be one of my major heroes.

Of course he says nothing of this sort. If he did, Ford, Nike, and so on would never hire him, and certainly Presidents Clinton and Bush would never give him awards for sustainability™. *Time* would not call him a "Hero for the Planet."

Unfortunately this sort of social reinforcement of pseudo-solutions is a big (and necessary, and routine) part of the problem. We could perform the same sort of analysis for any of the other "green capitalists" or "green foresters" or "green entrepreneurs," from Paul Hawken to Amory Lovins to Al Gore, who try to tell us, against our evidence, that nonsystemic changes can transform a fundamentally unjust and unsustainable system into something it clearly is not. Industrial civilization is killing life on Earth. As I mentioned before, pretty much anyone possessing both intelligence and integrity knows this (even if they don't speak it aloud or consciously acknowledge it even to themselves). But to recognize that industrial civilization is killing the planet can be incredibly scary and threatening, especially if one's lifestyle, job, work, fame, fortune, power, *identity*, are all dependent upon the continuation of the industrial economy, and more fundamentally upon the giant Ponzi scheme known as industrial civilization.[105] And if this realization is scary, if it threatens our identity, if it threatens many of those *things* we hold dear, if this realization threatens *us* by threatening that upon which we have been made dependent (Could you survive without industrial civilization? Where would you get your water? Your food? Your athletic shoes?), we will so often eagerly seek out any excuse to dismiss this realization; we will cling tight to any flimsy rationalization that anyone (especially some authority figure) cobbles together that will allow us to maintain, in the face of all evidence, all logic, all intuition, our old lifestyle, no matter how destructive. It's as I wrote on the first page of *A Language Older Than Words*, "In order to maintain our way of living, we must tell lies to each other, and especially to ourselves. It's not necessary that the lies be particularly believable. The lies act as barriers to truth. These barriers to truth are necessary because without them, many deplorable acts would become impossibilities. Truth must at all costs be avoided."[106]

And that is precisely what McDonough (and Hawkin, Levins, Gore, and so on) does. He feeds us sweet and soothing story after sweet and soothing story, each with the same destructive message, which boils down to: *Industrial civilization can continue, if only we make minor changes and call them great transformations.*

We all know this way of "life" isn't working. We all know that something must change. We know this change must be dramatic. McDonough co-opts this understanding of the necessity of dramatic change and puts that energy back into the service of the very same system that is destroying life. He effectively puts us back to sleep: *Don't you worry. Things are going to be okay; Ford now has plants on top of its factory. A revolution is underway.*

I can already see the commercials.

I don't think McDonough is intentionally attempting to mislead us. I think he's doing something many of us often do, something that Robert Jay Lifton describes so well in his crucial book *The Nazi Doctors.* I've written about this in other of my books, and I write about it again here because what Lifton realizes is so important, and the actions he describes happen so frequently.

Lifton wanted to know how it was that doctors—people who had taken the Hippocratic Oath—could work in Nazi concentration and death camps. He wasn't so much talking about Mengele or others like him (although he did discuss Mengele), but rather run-of-the-mill Good German doctors. He found something quite extraordinary, which is that many of the doctors actually cared about the inmates and did everything within their power to make the inmates' condition ever-so-slightly less intolerable.[107] So they might give sick inmates an aspirin to lick. They might put them to bed. They might give them a tiny bit of extra food. The doctors, once again, did everything they could. Everything, that is, except the most important thing of all: they did not question the existence of the camps themselves. They did not question working the inmates to death. They did not question starving them to death. They did not question gassing them to death. They did not question the hubris, bigotry, perceived entitlement to exploit, and utilitarianism that led to the camps. They did not question the function of the camps. They did not hinder the operation of the camps. Within these constraints they mitigated the best they could.

What, precisely, does McDonough do? He seems to be doing everything within his power to make factories just the eensiest bit less destructive (all the while, of course, being sure to never hinder corporate functioning: he is quite insistent that his work increase profits for the corporations that hire him), that is, he is doing everything but the most important thing of all: questioning the existence of factories in the first place. He does everything but question the hubris, bigotry, perceived entitlement to exploit, and utilitarianism that is leading to the entire planet being turned into first a work camp, then a death camp (just ask the oceans, grasslands, or forests whether the world is being turned into a death camp). He doesn't question working the planet to death. He doesn't question starving it to death. He doesn't question poisoning it to death. He, like the Nazi doctors, does what pathetically little is left to him given the constraints he refuses to question.

One of the problems is that McDonough, like so many in this culture, pretends that the culture is primary and the world is secondary (or more precisely, that this culture and its perceived right to exploit and destroy is the only thing that really exists, and everything else must conform to this "reality"). McDonough's work, like the work of so many others in this culture, does not, cannot, and will not question this culture's perceived right to exploit and destroy. He, like so many others in this culture, takes the culture as a given, to which everyone else must adapt, or, like the baiji and so many others, die. He, like so many in this culture, confuses independent and dependent variables. He forgets that this culture—any culture—is dependent upon the health of the land. No health of the land, no culture. In fact, no health of the land, no life. The first principle of sustainability is and must be that the health of the landbase is primary, and everything else—and I mean everything else—is subordinated to that. To ask, "How can this factory be made sustainable?" is to ask the wrong question, and is to guarantee unsustainability, because the question presumes factories can be sustainable, and takes the existence of factories as a given. It is precisely equivalent to asking, "How can we make concentration/death camps more sustainable for the inmates (the feedstock for the camps' productivity)?" The question in each case should be, "What will it take for the land (or the inmates) to be healthy?" The most obvious and most important answer to this question is the destruction of that larger superstructure which is causing the harm: factories and in a larger sense industrial civilization that gives rise to them in the first case, camps and in a larger sense the Nazi government that gave rise to them in the second.

This is a lot easier to see at a safe historical distance.

I'll be even more explicit, since so many of us, myself certainly included, have been systematically made insane and selectively stupid (or, as R. D. Laing might have put it, have been made into imbeciles with high IQs). This inducement of stupidity is necessary because otherwise, of course, we would tear down this wretched system that is killing the planet instead of merely tweaking factories, calling our work sustainable, and continuing to expect the natural world to accommodate itself to whatever we wish to impose (e.g., truck factories, athletic shoe manufacturers, airports, transnational corporations, mass-produced books (in my case), industry, civilization). If we wish to survive, *we* need to accommodate ourselves to the land, to give it what *it* needs and wants, to accept from it only what it needs and wants for us to have.

It's really not so hard. It's how humans have lived for most of our existence, and it's how nonhumans live. It is the *only* way to survive. It's the only way to live sustainably. If you don't, if you harm your landbase, your habitat,[108] you don't survive.

Our stupidity and our denial make McDonough's rhetoric all the more dangerous. Instead of us recognizing factories for what they are—machine-structures that turn members of the natural world into *things*, that is, convert the living to the dead (forests into two-by-fours, mountains into truck frames, and so on)—these sustainable™ factories become modern-day Potemkin villages.

Potemkin villages, if you recall, were façade-villages supposedly constructed by Russian minister Grigori Aleksandrovich Potemkin along the banks of the Dnieper River in order to impress his lover the Empress Catherine the Great. That the story of him creating these fake villages is itself fake hasn't stopped the phrase from entering common usage, meaning something that looks impressive but is hollow, a fundamental deception.

I had a friend who never picked up trash by the side of the road. I asked her why. She said she wanted people to see this culture in all its ugliness and wastefulness, and she didn't want to superficially clean up a roadside—a roadside!—and facilitate the delusion that this way of life is anything other than filthy and wasteful.[109]

I respect her position, and although I do pick up trash I can't say I disagree with her.

I think the same, once again, about McDonough's work. If we recognize it for what it is—a superficial cleaning up—then it's good work to do. But the degree to which these factories become "Potemkin factories"—to the degree that his work masks the inescapable fact that factories, and more broadly this way of life, are filthy, destructive, exploitative, and unsustainable—is the degree to which this work does more harm than good.

/ / /

Throughout the last few weeks it's taken me to write this analysis of McDonough, I've kept thinking about something I learned when I first started writing for publication. I began with book reviews. I was told to ask three questions about any book: What was the author trying to do? How well did the author do it? And, was it worth doing?

Let's take *The Nazi Doctors*. I answered these questions in my paragraph introducing that book. What was the author trying to do? He was exploring how people who had taken the Hippocratic Oath—people who were presumably of good heart—could participate in such an evil and destructive enterprise. How well did he do it? The book is crucial. Was it worth doing? Once again, the book is crucial.

We could do the same exercise for any book, and in fact for any action. What was George Bush attempting to do by ordering the invasion of Iraq? Certainly attempting to gain access to Iraq's oil fields. Also, he was attempting to protect Israel and attempting to solidify US control of the Middle East.[110] How well did he do it? Pretty damn bad. Was it worth doing? I think not. Or maybe this is all wrong, and the purpose was to increase the power of the US government over its own citizens, and Bush and his allies needed a war as the excuse for it. How well did he do that? So far it's worked. Was it worth doing? He hasn't yet been put on trial and hanged for shredding the US Constitution, ordering the torture or illegal detention of thousands, and causing the deaths of hundreds of thousands, so maybe from his perspective and the perspective of his autocratic buddies, yes. Or maybe this is all wrong, too, and the truth is that George Bush is a closet revolutionary intent on destroying the US empire through wasteful military spending and fiscal, domestic, and foreign policies guaranteed to run the US economy into the ground. How well has he done that? Better than the worst enemy of the US could have hoped. Was it worth doing? Good question.

So, what is McDonough trying to do? One answer is that he's trying to make factories less destructive. How well is he doing it? McDonough *did* plant plants on a Ford truck factory, and he *was* associated with getting Nike to use only 95-percent pesticide cotton. Is it worth doing? McDonough should be asking this same question: Is it worthwhile to build a truck manufacturing facility? Is building a truck factory—no matter how well done—conducive to sustainability, indeed human (and much nonhuman) survival? It doesn't matter so much how well you do something that shouldn't be done: it still shouldn't be done.

Or maybe here is what McDonough is doing: he is deluding us into believing that industrial civilization can be sustainable. How well is he doing that? According to Ford, Nike, two presidents, a lot of liberals—just today the *San Francisco Chronicle* called him "the star of the sustainability movement"—and many mainstream environmentalists, he's succeeding extremely well: they seem to think the world of him, or more precisely, they think more of him than they think of the real world, which is being destroyed by the very industrial civilization he is supposedly trying to make more sustainable.

All that said, I still didn't really want to write this section. I hate saying negative things about people who are going even remotely in the right direction. For example, while there are many activists I like very much, there are also many I don't care for or actively dislike (sometimes because I have significant concerns about their work, and sometimes because even though I think their work is good or important I know them and can't stand them personally: they're jerks) I don't speak ill of them in public.[III] I try to save my venom for those who are my *real* enemies. I generally think it's harmful to spend a lot of time attacking possible allies.

This is an important enough point that I want to tell two stories about it. The first is that soon after *A Language Older Than Words* came out I was asked to participate in a conference of children's health advocates. The organizer told me she wanted the perspective of an environmental activist, and wanted me to "stir things up." The conference ended up being, at least for me, two days of hell. Whenever I brought up anything about the destruction of the natural world, I was in one way or another told to shut up: there was to the minds of many of these children's health advocates no correlation between sick baby salmon and sick baby humans. One woman approached me during a break and said, "I wish you'd stop wasting the time of all these brilliant people with your talk of some apocalypse." During a breakout session devoted to environmental issues, I talked in my group about the destruction of the biosphere and its effects on human and nonhuman children (as well as, of course, all of the human and non-human children who will come after, and who will inherit an unlivable world). Our group secretary dutifully wrote it all down on the whiteboard. Afterwards, as I was leaving the room, I saw someone who had not said much during the session approach the secretary, point at the section of the whiteboard where my comments were written, and begin to speak earnestly. When a few moments later the secretary brought out the white-board to make a presentation to the whole conference about our group's discussion, I saw that all of my comments had been erased. I later learned that the person who had approached the secretary was a lobbyist by pro-fession. This helped me better understand the slimy processes by which laws are made. But the worst part happened toward the conference's end. Someone happened to mention PETA (People for the Ethical Treatment of Animals), and the room erupted into hisses and catcalls, because PETA opposes vivisection. I told these children's health advocates that while I sometimes find PETA silly and off-putting, I needed to comment that the people at this conference were expressing far more hostility toward PETA than I'd heard any of them express toward Monsanto. I asked why that might be.

Someone said, "Because PETA hates children."

I blinked.

He continued, "They kill children by opposing animal testing of toxic chemicals."

"No," I said. "*Monsanto* kills children by *producing* toxic chemicals."

"You must hate children, too," he said.

No, I'm not making that up.

Someone else said, tired, somewhat condescending, "Without animal testing, how will we regulate the production of these toxic chemicals?"

"I have no interest in regulating the production of toxic chemicals," I responded.

The room exploded. The first person said, "Ha! I *knew* he hated children. I knew that all along."

I said, "I don't want toxic chemicals produced at all."

Red faces. Balled fists. No, I'm still not making this up. I was glad that one man in particular had neither a gun nor a length of rope.

I finished, "I don't think we should countenance the poisoning of *any* children—human or nonhuman—by Monsanto or any other corporation."

The red-faced man with neither gun nor rope started shouting at me. The lobbyist broke with vocational pattern and confronted me directly: "You need to spend some time in the real world."

Disgusted, I disengaged. I was disgusted not merely at their conflation of the current political situation and more broadly civilization with "the real world" (while all the while in the real world, real breast milk has been made toxic with real industrial chemicals). And not merely at them falling into the same trap as McDonough and the Nazi doctors—doing what little is left to them after they refuse to question the constraints this death camp culture places on them, indeed after they refuse to question the death camp culture at all. And not merely at them falling into yet another trap that so many of the powerless do, which is to act out their anger at others of the relatively powerless instead of expressing it at those who are really harming them—in this case Monsanto and other producers of bulk industrial chemicals, many of them toxic. No, I was mainly disgusted because I'm so tired of the infighting and petty attacks that characterize so much of our so-called resistance. Their attacks on PETA and later on me didn't make sense, yet unfortunately also didn't really surprise me, because that's what those of us who at least pretend to oppose the system so often *do*—attack each other.

Which brings me to my second story. Over the past several years I've received about 700 pieces of hate email. Only two of these were from right wingers. (One was a death threat because I shared the stage with Ward Churchill; the writer didn't even have the courtesy to threaten me because of my own work, but rather because of someone else's. The other was from someone who objected to me using quantitative analysis to prove one of my points; I'm still not sure what his problem was.) All of the others were from those whom I would have thought would have been allies. Some vegetarians and vegans write me hate mail because I eat meat. Some permaculture activists write me hate mail because I don't think that gardening will by itself stop this culture from destroying the planet. Pacifists write me hate mail because I say that sometimes it's okay to fight back. Hitchhikers have written me hate mail because I fly to give talks. People who don't do much of anything have written me hate mail because my books are printed on the flesh of trees. A Trotskyist (I didn't know there were any left) wrote me a note that began, "I hate you. I hate you. You are an anarchist and so I hate you." Anarchists have written me hate mail because they say I'm not enough of an anarchist. And so it goes.

It's all a monumental waste, and I wish these people would devote this time, this energy, this emotion, to stopping the culture that's killing this planet.

It happens often enough to have a name: horizontal hostility. It has destroyed many movements for resistance against this culture, and driven people away from these movements in hordes. It's much easier to attack our allies for their minor failings rather than take on Monsanto, Wal-Mart, Ford, Nike, Weyerhaeuser, and so on.

Intellectually, I understand why it happens, and even understand its roots in abusive power dynamics. Children who were abused by one parent often as adults end up hating the non-offending parent more than the offending parent, because the non-offending parent is safer. This safety allows those feelings to come out. This happened to me some in my thirties. I dated some women who had been treated very poorly by men (as many women have) and was the first nice guy they'd known. All of their anger toward men finally had a safe place to come roaring out. It wasn't fun, and it's not an experience I would repeat. The same thing is true on the larger scale: it can be so much easier and safer to get angry at the activist next door because that activist has a car, or eats foods you don't approve of, and so on, than at the corporations and governments that are killing the planet, and the police who support them.

That's why I hesitated to write this section. Would I be doing the same thing? I wasn't sure. I wrote to ask my friend, the activist Lierre Keith, whether I should write a section criticizing McDonough. He is, after all, at least heading in the right direction.

She responded, "But in the end, McDonough *isn't* heading in the right direction. He's heading in exactly the same direction—complete drawdown of planetary reserves of metal, oil, water, whatever—but we'll get there a bit slower on his plan. Industrialization is still industrialization. This way of life is over. It has to change.

"And people who would otherwise have to *face facts* have an emotional/intellectual out: *Look! The Rocky Mountain Institute has a car that can get 100 miles to the gallon!* Yeah? So fucking what? Where will the steel and plastic come from? What about the asphalt? And the main problem is that if we're building for cars—or truck factories and so on—we can't build for human and other biotic communities: cars require the exact opposite.

"There's no way trucks and an economy based on those kinds of distances are ever going to be sustainable. Why are we as a culture wasting our time and resources building one more fucking truck?

"So I think their project is corrupt and it's only prolonging the inevitable. They're still fighting for a way of life that necessitates destroying the planet."

She's right. I wrote the section.

/ / /

In his powerful book *Overshoot: The Ecological Basis for Revolutionary Change*, William R. Catton defined a term I haven't seen elsewhere. It is *Cosmeticism:* "faith that relatively superficial adjustments in our activities will keep the New World new and will perpetuate the Age of Exuberance."[112]

This is what William McDonough does.

/ / /

Just today I received a mass email from an organization called the Crisis Coalition. It gives dreadful details about glaciers melting, methane burps giving positive feedback to global warming, and so on: the murder of the planet.

And how does this passionate email end? By saying: "We can transform our life on this planet, and maintain our lifestyles. We can do both—if we start NOW."

❧ ❧ ❧

How does someone who understands that this lifestyle is killing the planet still insist that we can maintain this lifestyle and not kill the planet?

❧ ❧ ❧

Our denial makes us really stupid.

❧ ❧ ❧

No, I mean *really* stupid.

❧ ❧ ❧

Does anyone besides me experience deep sorrow that someone called a "Hero for the Planet" and a "star of the sustainability movement" is designing truck factories and Nike headquarters? Ninety percent of the large fish in the oceans are gone. Ninety-seven percent of the world's native forests have been cut. There are 2 million dams just in the United States. Once-mighty flocks of passenger pigeons are gone. Islands full of great auks, gone. Rich runs of salmon, gone. Gone. Gone. Gone. The oceans are filled with plastic. Every stream in the United States is contaminated with carcinogens. The world is being killed, and this is the response? Not only am I angry, not only am I disgusted, I am also deeply, deeply sorrowful.

And I am deeply ashamed.

We need to act differently.

COMPARTMENTALIZATION
AND ITS OPPOSITE

Compartmentalization is the principal mechanism of evil. —M. SCOTT PECK

THIS CULTURE SPECIALIZES in compartmentalization. If people didn't blind themselves (and allow this culture to blind them) to the effects of their actions—in other words, if they didn't put their actions into one compartment and the harmful effects of these actions into another compartment that must never be examined—this culture and its members would not be able to continue a lifestyle based on systematic exploitation and theft. So walls must be erected and maintained. People must be trained to be selectively deaf and blind, to sensate only when necessary to perform the task at hand. Do you want to design a "sustainable" truck factory? No problem. Throw some plants on the roof, then just ignore the effects of car culture. Do you want to design an eco-groovy Nike headquarters? No problem. Just make sure you ignore the slave labor. Do you want to manifest your destiny? Great! Just make sure you eliminate everyone who already lives on the continent. Do you want to run an industrial economy? Wonderful! Just make sure you pay little or no heed to the fact that you're killing the planet you live on.

More or less all of us in this culture—and I am explicitly including myself—are adept at this sort of compartmentalization. It's what we *do*.

The training starts early. If familial abuse hasn't shattered our psyches—forcing us to compartmentalize our experiences, storing trauma in a compartment we will never allow ourselves to look into, and keeping happy feelings and memories where we can hold on to them—then school will surely teach us to compartmentalize. It does this by separating subjects, and even moreso by separating school from home, and school from land-base, and schooltime from playtime. School is school, and play is play, and never the twain shall meet.

The same is true philosophically. Science is science, ethics is ethics, and these two, also, shall never have more than a passing and uncomfortable acquaintance. We can say the same for politics and ethics, and we can say the same for economics and ethics. (Of course politics and economics, handily for those in power, share no such separation from each other, and in fact politics, economics, and science are all united by their mutual raison d'être, which is the raison d'être of this culture: the drive to accumulate and use power over others.)

Compartments. Men and women in different boxes. Humans and nonhumans in different boxes (except when it comes to vivisection: evidently we have enough in common for nonhumans to be models for how toxins will affect us, yet we have little enough in common that nonhumans can't feel the agony, frustration, despair, sorrow, rage, and helplessness that we would feel were we similarly tormented). Thought goes in one box. Emotion goes in another. One's job goes in one box. One's life goes in another.

In *The Nazi Doctors*, Lifton describes what he calls "doubling," by which guards in concentration camps would have one moral code at home, where they might be good parents, and so on; and another at work, where they might torture and murder inmates. In another book, he describes a similar doubling—which, in some ways, is just another word for compartmentalization—among scientists and technicians working on nuclear weapons. The same applies to all of us, really. We try to be good people while participating in this inherently destructive death-camp culture.

People like McDonough claim that actions like putting plants on truck factories—truck factories!—are significant steps toward sustainability, and it should come as no surprise that others in this culture laud this disturbing compartmentalized thinking. So long as you ignore everything but your own particular little action, so long as you ignore every preceding step and every inevitable consequence, that is, so long as you think in a linear, compartmentalized, rationalized[113] way, you, too, can proclaim every one of your actions to be sustainable. You, too, can be praised by presidents and CEOs alike. You, too, can with a clear conscience continue to assist in this culture's ultimate consumption of the planet.

/ / /

The real world does not resemble our compartmentalized version of it. Where do I end and you begin? Are you still in the air you just exhaled, the air I now take in, the air that carries with it the sweetness of your breath? Do you end where your fingers touch my skin, or do you follow these sensations into my body? Where do I end when I move inside of you, and where do you end when you move inside of me? And when you leave are you still inside of me? Am I still inside of you? Where do you end and where do I begin?

A week ago my dog Amaru died. He was fourteen. He had been sick for more than a year, from Cushing's, arthritis, and most of all age. He had lost about 30 percent of his weight, and even more of his strength. When he was younger he gladly chased bears, but those days were gone. He was losing his sight and hearing. And then, the night before he died, he began to throw up. I stroked him, spoke with him, tried to reassure him, but I did not know how to take away his pain and nausea.

He finally fell asleep around midnight, and I fell asleep sometime later. I awoke near dawn knowing something was wrong. I had to find him. He wasn't near my bed, wasn't in the house. He was outside, collapsed in shallow water at the edge of a pond. He had only been there a short while (I could tell because the water had not wicked up his body), but still he was very cold. I brought him in, warmed him up. He got up once, hours later, to stagger outside to relieve himself, taking care of me by not soiling the house even as he was dying. He couldn't make it back inside. I carried him in, laid him back down. He never got up again.

I held him as he died, as I'd held him all through the day, talking to him, rubbing him, not quite believing he was dying. But he did. His breathing grew forced, and then stopped altogether, although his heart still beat. I pushed on his chest, and pushed again, and again, but he was dead.

I did not bury him that night. He wasn't ready to go, and I wasn't ready to let him go. I let him sleep—or maybe he let me let him sleep—next to my bed one last time, near the head, where for so long I had reached down to pet him in the middle of the night.

The next day I buried him, where he will over time become more and more a part of the forest.

I miss him terribly, though he is even now only thirty feet from me.

I have always had difficulty falling asleep, often lying awake for hours. I've written elsewhere that it sometimes feels as though I've forgotten how to sleep, or like I never learned. In my late thirties I finally experimented with various herbs—valerian, chamomile, and so on—but none seemed to help, until finally I came upon a combination that put me to sleep: kava kava, 5-htp, and melatonin.

For years then I marveled at the deeply sensuous pleasure of slipping so seamlessly between waking and dreaming worlds. But a few months ago the herbs stopped working, and once again I lay awake each night till near dawn, thinking and not thinking, writing and not writing, meandering sometimes into shallow dreams but rarely going deeper, instead nearly always bubbling back a few moments later to and through that permeable surface, back to this side.

I tried different herbs, and I tried changing other variables in my life. Nothing worked.

Nothing worked, that is, until the night after I buried Amaru. Ever since then I've slept soundly, and deeply.

It took me only a couple of days to understand that Amaru had been keeping me awake. Not by snoring (my other dog Narcissus does that, sometimes driving me to wear earplugs or to carry him gently into another room) but because he was in so much pain and carrying so much psychic distress that it hung in the air, seeped through my skin.

Where does Amaru end, and where do I begin?

I often have dreams that are not my own, but rather they come from one friend or another. I may, for example, understand everything in a dream except a strange blue car that drives by too fast and scares me. I try to make some interpretation fit. I think about every blue car I can recall, think about their owners, try to assign some meaning. Nothing works. Then a friend calls, tells me she was nearly run over the day before by someone driving a blue car. This happens often enough my friends and I have a name for it: *leakage*, because their realities leak into my dreams.

Where do my friends end, and where do I begin?

You and I touch. We make love. I touch the middle of your chest with two fingers. You kiss my forehead, my lips, my shoulder.

One of us gets up, leaves. What is left behind in this room? Where do you end, and where do I begin?

I live in a forest. What is this forest? Is it redwood trees? Or is it soil? Or is it frogs? Slender salamanders? Huckleberries? Shrews? Steller's jays? Or is this forest all of these and so much more? Where do frogs end and where does water begin? And when the small streams taste like tea from flowing past and through and with all this vegetation—living, dead, whatever—where does the forest end, and where does the water begin?

/ / /

I'm not saying that we are all one. We aren't. I am me. You are you. Amaru was and is Amaru. This frog is this frog and that tree is that tree.

But just as we are not one, we are not entirely separable. We are not monads. We are not impermeable. Our boundaries are blurry, shifting, porous, and ultimately indefinable. Mysterious.

When you put us—and by "us" I mean anyone, from me to you to Amaru to the fly that's buzzing around my face to the frogs outside the window—into a conceptual box, you obviously miss all of this other's complexity that does not fit into your box. Worse, when you forget that the conceptual box is nothing more than a concept—a tool, a projection, a figment of your rational mind—and is not reality, and does not necessarily even come close to corresponding to reality, you have foreclosed all possibility of entering into any sort of meaningful relationship with this other. Of course you have: you no longer perceive this other; you perceive only your precious and projected box. You are at that point interacting only with yourself and your projected delusions.

Welcome to what passes for thinking in this culture of compartmentalization. Welcome to the wonderful world of rationalized exploitation.

/ / /

Have you wondered why, in this book ostensibly about shit and more broadly decay, I wrote about making love, followed by a description of my

Amaru's death, jumped from there to insomnia and then to the porous boundary between waking and sleeping, from there to dream leakage, and then back to making love? I do this sort of thing in all my books and talks.

Years ago after one of my talks, someone said to me, "At first I didn't understand your jumps, and then I thought you were crazy, but now I see the patterns, and I see how the different subjects come together, and I like it." I've gotten a fair number of letters over the years from people who say, "Thank god I've finally found someone who writes like I think."

I like these notes, but I don't like the notes in which people compliment me on what they call my "disorganized style," and for "writing whatever happens to come to mind." The reason I don't like them is that my style *isn't* disorganized, and I don't (usually) write whatever comes to mind. I try to make the organization as tight as I can. But my writing is organized along different principles than those that normally guide discourse and thought in this culture. I write this way to undercut or even destroy the monopoly, the stranglehold, that linear thinking has over our discourse, our thinking, our lives. This hegemony is incredibly dangerous.

It's dangerous in part because it doesn't match reality. The world is not organized linearly. A forest is not organized linearly. A river is not organized linearly. An intrahuman (or interspecies) relationship is not organized linearly. A human (or nonhuman) is not organized linearly. Life is not organized linearly, or parabolically, or sinusoidally, or any of the other mathematical models we can try to project onto it; these regressions may sometimes describe a portion of what we see, but they do not come close to defining reality. These regressions are not reality, and they are no substitute for reality. The combination of the near-ubiquitous cultural delusion that one's concepts are more important than the real and the near-absolute cultural narcissism where one perceives oneself as entitled to exploit everything and everyone else has led to a culture capable of destroying everything.

/ / /

But there are other ways to be, other ways to perceive, other ways to think. The Lakota man Brave Buffalo said, "I have noticed in my life that all men have a liking for some special animal, tree, plant, or spot of earth. If men would pay more attention to these preferences and seek what is best to do in order to make themselves worthy of that toward which they are attracted, they might have dreams which would purify their lives. Let a man decide upon his favorite animal and make a study of it, learning its innocent ways. Let him learn to understand its sounds and motions. The animals want to communicate with man, but Waka´Taka [the Great Spirit] does not intend they shall do so directly—man must do the greater part in securing an understanding."[114]

I asked the American Indian writer Vine Deloria what, in the Indian perspective, is the ultimate goal of life.

He said, "Maturity," by which he meant "the ability to reflect on the ordinary things of life and discover both their real meaning and the proper way to understand them when they appear in our lives. Now, I know this sounds as abstract as anything ever said by a Western scientist or philosopher, but within the context of Indian experience, it isn't abstract at all. Maturity in this context is a reflective situation that suggests a lifetime of experience, as a person travels from information to knowledge to wisdom. A person gathers information, and as it accumulates and achieves a sort of critical mass, patterns of interpretation and explanation begin to appear. This is where Western science aborts the process to derive its 'laws,' and assumes that the products of its own mind are inherent to the structure of the universe. But American Indians allow the process to continue, because premature analysis gives incomplete understanding. When we reach a very old age, or have the capacity to reflect and meditate on our experiences, or more often have the goal revealed to us in visions, we begin to understand how the intensity of experience, the particularity of individuality, and the rationality of the cycles of nature all relate to each other. That state is maturity, and seems to produce wisdom. Because Western society concentrates so heavily on information and theory, its product is youth, not maturity. The existence of thousands of plastic surgeons in America attests to the fact that we haven't crossed the emotional barriers that keep us from understanding and experiencing maturity."[115]

One of those barriers—and I want to smash this barrier—is the stranglehold that linear thought has on thought itself, such that any other thought is considered nonthought, any other organization is considered disorganization.

A forest, for example, is organized in a way we have been trained not to recognize (in fact, often to fear). Our culture has trained us not to understand a forest in even a shadow of its full complexity because understanding would impede the exploitation of that forest. Consider this: perceiving a forest only or primarily in terms of board feet facilitates the forest's exploitation. Perceiving the complex interrelationships that make up a forest does not. Likewise, perceiving a river only or primarily in terms of the amount of hydroelectricity it can produce facilitates that river's exploitation. Perceiving the complex interrelationships that make up a river does not. Likewise, perceiving a woman only or primarily in terms of her orifices and how you can gain sexual satisfaction from those orifices facilitates that woman's exploitation. Perceiving the complex interrelationships that make up a woman does not.

I want to be able to begin to recognize the organization of a forest, the organization of a stream, the organization of a woman.

No, I want to be able to understand what a forest, a stream, a woman may wish to communicate to me. Even more than that, I want to be able to understand a forest, stream, woman on each of their own terms, insofar as each of them may wish me to understand them.

To do so, I must begin to break the shackles of linear thought.

I'm not saying that my work is anywhere near so complex as a forest. Of course it's not. But I'm attempting to introduce—or rather reintroduce—myself and readers to a new—or rather old—way of seeing, thinking, being in the world, relating to the world, a way that does not rely on the particular and peculiar compartmentalization that characterizes this culture, a way that is far more complex than this culture's "normal" mode of thinking, a way that is, I believe, far more natural.

/ / /

Just today I did a radio interview that manifested compartmentalized thinking. The interviewer, named Andy, was a Bible-quoting, property-rights-über-alles antienvironmentalist. I don't know if he requested an interview with me because he wanted to bash an environmentalist, because he didn't like the sound of my name, because God told him to, or because environmentalism is, according to many Bible-quoting, property-rights-über-alles antienvironmentalists, a religion that must be stamped out using any means necessary. Most likely he requested an interview for reasons I'll never know and wouldn't understand anyway. I know it wasn't because he's read any of my work; he said so on air when I challenged him about the abysmal quality of the interview.

Normally people who interview me—and there've been hundreds of them, ranging from national radio interviewers to major print journalists to enthusiastic (and nervous) teenagers interviewing me for their photo-copied zines—do so because they're concerned about the murder of the planet, and they want to explore with me the deep underlying reasons for this culture's ubiquitous destructiveness, and far more importantly, how to stop it. This interviewer cared about none of this, and in fact did not seem to consider environmental problems "real": whenever I gave any—and I mean *any*—specifics about planetary collapse he said as much. I said that 90 percent of the large fish in the oceans are gone, and he interrupted me to say, "Let's keep this interview about the real world." I talked about dioxin in every mother's breast milk, and he cut me off to insist once again I talk about what he called "the real world," which evidently does not include the real world. Salmon are being exterminated, I said, but before I finished he cut me off with the same refrain. I told him that there's no world more real than the physical world of soil and trees and birds and insects. But he, like so many members of this delusional culture, believes that the real physical world is secondary, peripheral, and that industrial capitalism is primary—in his words, "the real world." If insanity can be defined as being out of touch with physical reality, this man is insane. Literally.

He was also extraordinarily hostile. His hostility made no sense to me. Why would he invite someone onto his radio program, then cut the person off, talk over the person, chide the person, ignore everything the person said except for one or two key words he would then use to launch an attack? What's the point? If he's not going to listen anyway, he should just get a recording he can turn off and on: the interview equivalent of masturbating to porn, as opposed to actually engaging another human being.

Early in the interview I spoke about about dams killing rivers. The inter-viewer said that we need dams for electricity: "Some environmentalists, especially radical environmentalists, say that air conditioners aren't sus-tainable. They say that dams aren't sustainable. They seem to say that everything isn't sustainable."

"I wouldn't call air conditioning and dams everything, but those two things aren't sustainable. Dams kill rivers by forcing—"

He cut me off: "Oh, come on. Get back in the real world. We need air conditioning. We need dams. We need all those things."

I said, "I live on Tolowa land. The Tolowa lived here for at least 12,500 years, and did so without air conditioning, and without dams. They lived here without materially harming the landb—"

"You're doing it again," he said. "Get back in the real world. I don't care about history."

Which means, I thought, *that you'll never learn anything from it.*

He continued, "I want you to stop being so ethereal."

He actually used the word *ethereal* (definition: characterized by lightness and insubstantiality; intangible; not of this world) to describe my reference to a culture that has lived on and with this same land for at least six hundred human generations. Six hundred generations of eating what the land gives willingly; six hundred generations of living, dying, being eaten by the soil, bones becoming food for trees, food for forbs eaten by elk then eaten by the human children of those whose bones fed the soil. Six hundred human generations is even long enough to witness several generations of redwoods. He called my reference to humans living in an intact native redwood forest filled with salmon, grizzly bears, and lamprey *ethereal*.

He continued, "I'm talking about people right now whose property would be destroyed by floods if it weren't for dams. What would you tell these people who've built their homes—their dream homes—right next to a river?"

"I would tell them that—"

This time he made no pretense of listening, but started up again, "And what about all the cropland that's near rivers, the cropland that would get flooded?"

He took a breath, and before he could start up again I said, "Flooding is crucial to the health of the soil. It's one of the ways that lands near rivers regenerate. That's one reason the Nile Delta, for example, was so extraordinarily fertile, and—"

"Derrick, Derrick, you're doing it again." Yes, he actually was that condescending. "We're not talking about history. Can you please stay in the real world? I'm not talking about what the soil needs over the long term . . ."

He actually said that.

He continued, "I'm talking about fields and homes getting flooded."

"Floods are necessary to the health not only of the river but of the whole river plain."

"You're avoiding my questions. What are you going to say to the people whose dream homes are next to a river?"

It occurred to me he might be talking about his own home. I said, "I would tell them that to force a river not to flood is to kill the river and to kill the surrounding lands."

He almost shouted, "Just answer the question!" He'd clearly watched too many bad courtroom dramas. But his demand was as rhetorical as his questions, as he didn't let me answer anyway. He continued, "And what about the tens of thousands of people in some poor country who just died in a flood? We need dams to protect us from the ravages of nature."

"How much does deforestation add to catastrophic—as opposed to normal—flooding?"

"Answer the question, Derrick. Floods kill people. Dams are necessary to protect us."

"It may or may not be the case that dams protect people from flooding, but how many people suffer mercury poisoning because of dams?"

"We're not talking about mercury, and we're not talking about deforestation. We're talking about dams saving people from floods."

I knew by now of course that he was locked into a compartmentalized way of thinking, and that his compartment was very small indeed. The only thing that mattered to him was the question of whether dams protect people's dream homes from floods. (I knew that his expressed concern about the lives of poor people was purely rhetorical, since he'd already said that the global poor are not that way because of the theft of their resources by the rich, but rather because corruption and local caste systems combine to drag them into poverty; and since he'd already told me to get back into the real world when I'd pointed out that a half a million human children die each year as a direct result of so-called debt repayment by nonindustrialized nations to industrialized nations: his real concern was property.) I said, "If the relationship between dams and human safety is important to

you, then surely you know about endemic onchocerciasis and schistosomiasis."

For once, he said nothing.

I said, "Do you know what they are?"

He still didn't say anything.

I said, "I'm surprised you'd talk about dams and safety if you don't know
what endemic onchocerciasis and schistosomiasis are. As a direct result
of the Akosombo Dam on the Volta River, in Ghana, which powers huge
bauxite smelters (and from which the displaced or otherwise affected
people don't even receive electricity), 100,000 people have contracted
endemic onchocerciasis, or river blindness, rendering 70,000 of them
totally sightless; and another 80,000 have been permanently disabled by
schistosomiasis."

I have to admit that this little checkmate wasn't as satisfying as I'd
hoped, because the next thing he said was, "You're changing the subject.
Floods kill people and they destroy property."

In his compartmentalized mind, the only victims who matter are those
killed by floods; the only damage that matters is that caused by floods; the only
effects of floods that matter are the ones that harm property (as opposed to
those that keep rivers and floodplains alive); the only floods that matter are
the ones dams prevent; and the only cause of floods is a lack of dams. Never
mind the permanent floods caused by dams that destroy tens of thousands of
villages. Never mind the great runs of salmon killed by dams. Never mind the
soil exhausted because dams don't allow flooding. Never mind the "dream
homes" that fall off sandy cliffs because dams rob beaches of sediment. Never
mind the cataclysmic floods caused by deforestation or climate collapse.

I kept trying to smash the compartment he was building, and he kept
trying to reconstruct it and force us both to climb back inside.

If you can make the compartment of discourse and thought small
enough, you can make sure that all questions you ask lead only to prede-
termined answers, which takes us right back to the rhetorical equivalent
of masturbation. This was clearly his (probably entirely unconscious)
intent. This sort of compartmentalized thinking is not thinking at all, and
this sort of compartmentalized discourse is not discourse at all. This sort
of compartmentalized thinking and discourse is fundamentally a lie. And
I was not willing to go along with his lie.

/ / /

I'm not nearly so smart as my books make me seem. One of the advantages of being a writer is that I get to win every argument. Most of the conversations in my books don't happen the way I represent them, where a friend says something really smart, and I say something smart back, and then the friend responds brilliantly, and so on. What really happens is that my friend says something really smart, and I say, "Gosh, that's really smart." A few weeks later I call up my friend and finally have a good response. My friend will immediately say something brilliant to me, and I then say, "I'll get back to you in a month."

That happened in this interview. I came up with the perfect response to his final comment. Unfortunately, my "perfect response" came about twelve hours after we'd hung up. Here's the story. Toward the end of the interview he finally came clear on perhaps the most fundamental premise underlying his opinions. He said, "You don't seem to have an appreciation for the fact that Genesis says God gave man dominion over all of the earth."

I stammered a response: "If you attempt to have dominion over or dominate something, you destroy all possibility of relationship with it, and if you attempt to have dominion over the world, you destroy it."

He didn't believe me, of course, and I had to walk him through—rather, drag him through—an analysis of the complexity of, for example, a forest, and how removing one component can lead to unintended consequences. This time he didn't tell me to return to the real world. He said, "Time's up. The interview's over." I didn't complain.

My response was okay as far as it went—or, to be honest, it was pretty tepid—but a few hours later this came to me: "Actually, Andy, I do have a great appreciation for God handing out dominion. In fact just last night I was talking to God, and God said to me, 'Derrick, when I gave man dominion over the earth, I was way too general. What I *really* meant was that Derrick has dominion over the earth, and especially over everyone named Andy. So you, Andy, need to give me, Derrick, everything you own—including your dream house on the river. If you don't, then God says I can bring in an army and take it."

He probably would have responded, "My Bible doesn't say that."

And I would have said, "I can't help it if you read the wrong Bible."

"That's absurd," he would have said.

And I would have responded, "But that's why I love this particular God so much. He always tells me exactly what I want to hear."

/ / /

Shit breaks down the barriers created by this culture of compartmental-ization. When it's in my intestines, is it part of me? Or is it something else, not me, simply held inside of me, in a tube made of intestines? Are they my intestines? Or just intestines that (who?) happen to work for me? Or are they somewhere in between, part me and part not me? Is that line between me and not me blurry? *When* is that line between me and not me blurry? When is that line *not* blurry?

I have Crohn's disease, an incurable progressive autoimmune disorder that (who?) can cause sores anywhere along the digestive tract from the lips to the anus. Is this Crohn's disease part of me, or is it something outside of me? Where is the boundary? Crohn's disease is a disease of civilization: it is very rare in nonindustrialized regions. As industry invades a region, so does Crohn's disease. There have been many theories as to why this is the case, ranging from increased pollution to changes in diet to changes in hygiene. A few years ago, researchers noticed that people whose intestines are pop-ulated by a certain sort of intestinal parasite called helminths, or whipworms—and approximately a billion humans share their intestines with whipworms (or, to use more common language, have whipworm infes-tations), and, as recently as the 1930s, essentially 100 percent of children in the southeastern United States had whipworms—generally do not have Crohn's disease (nor, for that matter, multiple sclerosis, asthma, diabetes, allergies, or hay fever, although they do sometimes have filariasis or schis-tosomiasis). For a time, researchers thought that perhaps people with Crohn's disease just had extra strong immune systems, and that immune systems without whipworms to fight off started attacking their own bodies instead ("a police riot" is how one doctor who, interestingly enough, didn't know my politics, explained it to me). But more recently the accepted theory is that whipworms somehow modulate the immune system, such that "our" immune system (is "my" immune system still part of me when it's attacking me?) can go a little crazy when whipworms are not present. This should probably have surprised me less than it did. Our bodies are natural communities, just as forests are natural communities. And when you remove members of these communities, it can be hard (especially for linear thinkers) to predict how the removal of those members will affect the rest of the communities. For example, Douglas firs require that a certain sort of fungus live in or on their root tips in order for the fir and the fungus to together metabolize food. Without the fungus, the firs die. Red-backed voles eat the fruits of this fungus and then distribute its spores around the forest by shitting them out in nice little pellets (pooparoonies, to use the language of one highly respected forest ecologist). If you get rid of the red-backed voles, the Douglas firs have a much harder time growing. Why should we expect any less complexity and interdependence from the natural communities that are our bodies? If we remove some of the members of this community—members with whom we have evolved—can we not expect for there to be repercussions? Now, if we can ask what is a forest, and ask if a red-backed vole is part of that forest along with the Douglas firs and that particular fungus, can we not ask the same of our intestines? Are the whipworms part of us? What about the others—worms, bacteria, viruses, and so on—who live in or on or with us, and with whom we also evolved?

Where do we stop and the whipworms begin?

Last year, I had the worst Crohn's flare-up of my life. Some modern medicines saved my life. But they didn't make me better. What ended that flare-up was that I started (with my doctor's enthusiastic approval) to ingest whipworm eggs. Unfortunately (or fortunately, I'm not 100 percent sure which) they're not human whipworms, but rather pig whipworms. If the doctor had given me human whipworms and the whipworms didn't stop the Crohn's, I would have had both Crohn's *and* a whipworm infestation. Not good.

I started worrying about the welfare of the whipworms I was ingesting, and the pigs who were providing the whipworms. After all, they were making me better; the least I could do was hope both the pigs and the worms had happy lives. I asked the supplier of the eggs. The news was bad: I found that the pigs are slaughtered to provide the worms. But that should change soon: it's technically feasible to harvest eggs without killing pigs. That latter method of collection is unfortunately not yet approved by either the Food and Drug Administration or its European equivalent. I also found that after the eggs hatch, the baby worms are unable to properly attach to my intestinal wall—I am, after all, not their proper host—so they lead short miserable lives before dying of starvation and being excreted. I felt bad about this—although not bad enough to quit taking the eggs who were making me well—and I brought up the subject with my doctor. He immediately asked if I eat meat.

I said yes.

"Consider them food, then, if that helps."

It helped my anxiety (though not the lives of the worms), but also takes us back to that question about whether the worms are part of me. If those worms who attach to my intestinal walls and live there for a year or more are a part of me, then what of these more transient worms?

And when I defecate, is this shit—which we can presume is no longer me—part of the soil? Or before it can be considered a part of the soil, does it first need to be digested by this soil, as I suppose the carrots, potatoes, and others I take into my body need to be digested by me before we consider them a part of me (unless we consider them a part of me as soon as I take them inside)?

If I shit in a forest, where, then, do I stop, and where does the forest begin?

/ / /

I go for a walk through a forest with a mushroom hunter. I know some of the basic mushrooms, but his eyes and his experience help me see this familiar forest anew. He teaches me about fungi. Their names. Their habits. Their preferences. He teaches me that fungi can live for a very long time, and can grow underground to almost unimaginable sizes: an *Armillaria ostoyae* mushroom in eastern Oregon may be 8500 years old and cover nearly four square miles. He tells me what he gives mushroom patches in return for their mushrooms: he makes sure no one overpicks, he protects the patches from loggers, he asks if it's okay with the mushrooms for him to take the fruit, he says please and thank you, and he gives the forest his shit.

He stops in his story to point out some chanterelles. Even I can recognize these. Then some turkey tails. They're easy to spot, too, since they look like, well, turkey tails. We keep walking. He looks at me out of the corners of his eyes. He looks ahead. Then back to me. Then back ahead. Finally he says, "And the biggest part of what I've promised them is my body after I'm dead, not pumped full of poison to keep me from decomposing, but rather a body that will feed them as they have fed me."

/ / /

Where does he stop, and where does the mushroom patch begin?

PLASTIC

To desire immortality is to desire the eternal perpetuation of a great mistake. —ARTHUR SCHOPENHAUER

HERE IS THE BOTTOM LINE: the world is being killed. It is being murdered. And one of the ways it is being murdered is that it is being poisoned: the waste products of this culture do not help landbases—as waste products are *supposed to do*, and *have always done*—but instead they harm and kill them. In nature there is no waste. Waste in this sense is a modern invention. And it's a rotten one.[116]

As epidemiologist Rosalie Bertel has pointed out, the probable fate of our species is extermination by poisoning. We could at this point add that this is the probable—in fact, looming—fate for the oceans, the air, the soil, the bodies of most every living being.

All of the fancy talk of sustainability—by us and others—is just dancing around the central issue: this culture is killing the planet.

This culture is killing the planet.

This culture is killing the planet.

This culture is killing the planet.

If we repeat this enough times, perhaps we will start to comprehend even the tiniest terrifying bit of *what this means*, and we will begin to act as if any of this matters to us.

I have in front of me a photograph. It is a photograph of a turtle. Or what would be a turtle. Or what could or should be a turtle, and of course still is a turtle, but is a turtle who got a plastic ring caught around the shell's middle.[117]

The turtle grew. The shell surrounding the ring did not. The turtle—and I wish I were making this up—looks like an hourglass. I first saw this photograph a few weeks ago, and have not been able to get it out of my head, nor my heart. But of course there is a difference between feeling empathy for another and actually having to live the life of that other. I can walk away. I can live my life pretending nothing is wrong. The turtle cannot do that. The plastic ring deformed the turtle, changed the turtle's life for much the worse.

Now I have another picture before me. It is of a river, or so I am told. I cannot tell, because there's too much trash. As the accompanying article states: "It was once a gently flowing river, where fishermen cast their nets, sea birds came to feed and natural beauty left visitors spellbound. Villagers collected water for their simple homes and rice paddies thrived on its irrigation channels. Today, the Citarum is a river in crisis, choked by the domestic waste of nine million people and thick with the cast-off from hundreds of factories. So dense is the carpet of refuse that the tiny wooden fishing craft which float through it are the only clue to the presence of water. Their occupants no longer try to fish. It is more profitable to forage for rubbish they can salvage and trade—plastic bottles, broken chair legs, rubber gloves—risking disease for one or two pounds a week if they are lucky."[118]

And now another picture. It is of the skeleton of a seagull. Inside the rib cage is a mound of plastic. There's a sense in which this picture is less horrible than the previous ones, since it could have been staged: someone could have placed the plastic inside the skeleton.[119] But I know that a study of fulmars—a type of seagull—in the North Sea revealed that the gulls had an average of forty-four pieces of plastic in their stomachs, weighing what in a human would be the equivalent of five pounds.[120] One animal had consumed and retained more than 1000 pieces of plastic.

Now I see a picture of a sea turtle with a plastic bag hanging out of its mouth; creatures in the ocean often mistake plastic bags for jellyfish. Sometimes they eat them. Sometimes they die.

Now I see a picture of a "ghost net," a plastic fishing net that was cut loose from a commercial fishing vessel. The net hangs in the water, fills with fish, turtles, sea mammals, seabirds—anyone captured by it—and eventually sinks to the bottom. When the bodies decompose sufficiently to fall apart, the net floats again toward the surface, where it begins this process anew.

I'm sure by now you know the numbers. Marine trash kills more than a million seabirds and 100,000 mammals and turtles each year, as well as unimaginable numbers of fish[121]—each and every one of these an individual worthy of consideration. There is at least six times more plastic in the middle of the Pacific Ocean than phytoplankton. (Imagine trying to eat and six out of every seven swallows bring only plastic; it is no wonder that so many sea creatures are starving to death, bellies full of plastic. Others die of constipation brought on by plastic blocking their intestines; having suffered a blocked intestine, with its pain at least an order of magnitude worse than a broken bone, I can tell you that I cannot imagine many more excruciating ways to die). This plastic is not degrading, but merely breaking into smaller and smaller pieces, until by now it is routinely found inside the cells of phytoplankton.

Plastic is everywhere in the oceans—and I mean everywhere—but it also accumulates where currents carry it. In these places the essence and endpoint of this culture could not be more clear. As one author wrote, "It began with a line of plastic bags ghosting the surface, followed by an ugly tangle of junk: nets and ropes and bottles, motor-oil jugs and cracked bath toys, a mangled tarp. Tires. A traffic cone. Moore could not believe his eyes. Out here in this desolate place, the water was a stew of plastic crap. It was as though someone had taken the pristine seascape of his youth and swapped it for a landfill."

The article continues, "How did all the plastic end up here? How did this trash tsunami begin? What did it mean? If the questions seemed overwhelming, Moore would soon learn that the answers were even more so, and that his discovery had dire implications for human—and planetary—health. As Alguita glided through the area that scientists now refer to as the 'Eastern Garbage Patch,' Moore realized that the trail of plastic went on for hundreds of miles. Depressed and stunned, he sailed for a week through bobbing, toxic debris trapped in a purgatory of circling currents. To his horror, he had stumbled across the 21st-century Leviathan. It had no head, no tail. Just an endless body."[122]

That particular "Garbage Patch" is nearly the size of Africa. And there are six others. Combined, they cover 40 percent of all of the oceans, or 25 percent of the entire planet.[123]

It's not merely river and ocean creatures—and rivers and oceans themselves—who are being murdered by plastic. So are land dwellers, including us. And frankly, although I love humans, at this point I feel even worse for those like the turtle, whose species have done nothing to deserve this, than I do for most, especially the rich humans whose lifestyles are causing these murders. At least rich humans get to drink from plastic cups and play with plastic Barbies and watch televisions housed in plastic—using electricity flowing through wires insulated with plastic—before these plastics poison and suffocate us all.

And plastics *are* poisoning and suffocating us all.

/ / /

What is plastic? We certainly feel and see a lot of it. I'm typing this on a plastic keyboard, sitting on a chair that has plastics in it, looking at a screen framed by plastic, and when I'm interrupted by a phone call, the phone is made of, you guessed it, plastic. Later, my clock (one of those very popular clocks that uses bird songs to note the hour, which means that even as birds are exterminated in the real world we can all still hear their voices— or rather reproductions of their voices—on the artificial hour, every artificial hour, until the batteries run down) marks eleven, so I look there, and, well, we all know what it's made of, too. We even ingest a lot of it, far more than most of us know. But I'm not sure a lot of us know what it is. I'm not sure *I* know what it is.

So I went to Google and typed in "What is plastic." Surprisingly— although I guess this shouldn't have surprised me—by far the most common result answered a somewhat different question: "What is plastic surgery?" Although it struck me as odd that in the marketplace of ideas more people might want to know about plastic surgery than the substance that they touch more than almost any other, more even than food or human flesh (I stroke a computer keyboard far more often than I do another human being, which is both pathetic and revolting; what is even more pathetic and revolting is that I never until this moment articulated that even to myself), I skipped over the sites about changing my own face and body and went straight to some detailing this substance that is changing the face and body of the earth.[124]

Here's what I learned: "Plastics are polymers, and are composed primarily of carbon, hydrogen and oxygen." I know about the carbon, hydrogen, and oxygen, but what the hell's a polymer? College was a long time ago, and at this point the only things I remember from my organic chemistry class are: a) *organic* in this context doesn't mean "organic" as in "no petroleum-based pesticides or fertilizers," but rather means "made of carbon," which means that nonorganic pesticides are in this sense organic, which sort of clarity might help explain the only other thing I remember, which is; b) I got a C-. So, what's a polymer? Oh, maybe I should quit worrying and read the next sentence: "Polymers are just very long chains of atoms which repeat over [and] over again."

Okay. That's what plastic is. Now how did it come to overrun the planet? Well, reading on, I learn that "Alexander Parkes, English inventor (1813–1890) created the earliest form of plastic in 1855. He mixed pyroxylin, a partially nitrated form of cellulose (cellulose is the major component of plant cell walls), with alcohol and camphor. This produced a hard but flexible transparent material." Can you guess what Parkes named the stuff? Why does it not surprise me that, in this supremely narcissistic culture, this material that was fabricated with absolutely no concern for its effects on others was named, narcissistically enough, "Parkesine"?

The article continues, "The first plastic based on a synthetic polymer was made from phenol and formaldehyde, with the first viable and cheap synthesis methods invented by Leo Hendrik Baekeland in 1909, the product being known as Bakelite."

Baekeland invented Bakelite as an insulator to coat wires in electric motors and generators. This sticky mix of phenol and formaldehyde became extremely hard if heated, then cooled and dried. By the 1920s, Bakelite was commonly used in consumer goods. The website goes on: "It was molded into thousands of forms, such as radios, telephones, clocks, and, of course, billiard balls. The US government even considered making one-cent coins out of it when World War II caused a copper shortage."[125] My favorite phrase in the previous paragraph is *of course*. "Of course, billiard balls." Where would this world be without plastic billiard balls?

Actually, without plastic billiard balls, there would undoubtedly be fewer elephants, and without billiard balls at all, there might be no plastic. You see, although wood and clay were sometimes used to make billiard balls, "No natural material other than elephant ivory," according to the Smithsonian Museum, "had the physical size, strength, and beauty to perform in the billiard room."[126] Given this culture, it was pretty much inevitable then that elephants would be slaughtered *en masse* to keep up with the demand for billiard balls. The demand for piano keys took its toll, too, as did the demand for ivory combs and doodads, but by far the main factor in the economically driven crash of elephant populations was billiard balls. Yes, elephants were nearly driven extinct for billiard balls. Yes, I hate this culture. One reason billiards drove the demand was that while you can make hundreds of piano key slips from a single tusk, that same tusk will yield only between four and eight billiard balls. That's one or two dead elephants per pool table.

Consequently, in the middle of the nineteenth century, the billiards industry offered a $10,000 prize to anyone who would invent an alternative to ivory for the balls. They did this not because they cared about elephants, of course, but rather because the slaughter was so great it was endangering their own industry.

To capture this fortune, the inventor John Wesley Hyatt began experimenting with Parkesine, and eventually invented what many call the first industrial plastic: cellulose nitrate. This compound worked great for billiard balls except for its rather unfortunate volatility: cellulose nitrate is also known as guncotton, and can be used as a propellant or explosive. It's six times more powerful than black powder, but its instability renders it useless (or at least dangerous) in firearms. The substance is also known as flash paper, and is used by magicians because it burns instantly and leaves no ash. Be all that as it may, cellulose nitrate and other plastics that followed began to be used for billiard balls.[127]

This brings us back to Bakelite. Although cellulose nitrate was useful for billiard balls (and might have been especially useful for James Bond-style stunts), and although cellulose nitrate is generally considered the first plastic, it's not considered the first *true* plastic. That distinction belongs to Bakelite, in the twentieth century, in that it was purely synthetic, based not on any molecule found in nature.[128]

This means that nobody—no animal, fungi, plant, bacteria: *nobody*—had ever eaten Bakelite before, which means nobody had ever broken it down into usable form for any landbase.

That's a problem.

That's a big problem.

No, that's a really big problem.

///

I am acutely aware of the relationship between this culture's fear of death and its fabrication of plastics and other materials that do not decompose. I am acutely aware of the relationship between this culture's complete fear of natural processes—processes outside its members' control; processes like aging, death, decomposition, consumption by others—and its fabrication of more or less permanent materials. I am *also* acutely aware of the relationship between the fabrication of these more or less permanent materials—which, because they are more or less permanent, do not help the land and air and water and the beings who live in these—and the ongoing murder of the planet.

I am further acutely aware of mortality— my own and others.

It is morning. My guts—remember the Crohn's disease, the incurable, progressive Crohn's disease—woke me from a dream in which my reading glasses, which age has forced me to start wearing these last few years, became increasingly insufficient as my vision became even more blurred.

But I awaken. I can see. I get up, joints complaining. My ankles, destroyed by one of the drugs I was prescribed to control the Crohn's disease, are especially loud in these complaints. Sometimes I can't walk.

Narcissus—my dog who did not die a few months ago, sixteen-year-old Narcissus, who no longer hears and whose eyes are cloudy and whose primary pleasures these days are getting pets, eating, sleeping, and dreaming whatever dreams old dogs dream—moans, then cries aloud from the pain and difficulty of merely trying to stand up. I help him. Once standing, he looks in my general direction and wags his tail. My heart breaks with love and joy and sorrow, as it does each time I look in his ancient face.

Yesterday I helped my mom in her flower garden. She, too, is getting older, and each day I see her I am acutely aware that both she and I are, as Pink Floyd sang, older, shorter of breath, and one day closer to death. These past few years she has begun speaking of who will get her handmade quilts after she dies.

She asked me to hack down or pull out dead lavender plants. She is getting too old, too tired, to weed quite as much as she used to, and grasses had choked out the lavender. I pulled out dead plants, acutely aware, as I am each time I help her here, that gardening is an attempt to stop succession, that it, too, can be and nearly always is an interruption of natural cycles.

Back to this morning. I pick up some old paper napkins from a visit to a restaurant, and I step outside. My ankles hurt. I hear songbirds calling back and forth,[129] and they remind me of the ongoing murder of the planet: yesterday I read a report of a forty-year study of songbird populations, many of which are collapsing, as are so many populations of so many wild beings. Bobwhites, down more than 80 percent these past forty years. Whippoorwills, down 70 percent. Boreal chickadees, down 60 percent. Rufous hummingbirds, down almost 60 percent. Creature after creature. And of course, because this is the mainstream press, they found a mainstream environmentalist—Carol Browner, former head of the EPA—to tell us that this is not an emergency. If human populations went down by 80 percent in forty years, the capitalist press would call this an emergency. (Or how's this: imagine how the capitalist press would bray "Emergency!" if the gross national product (GNP) declined 80 percent in forty years. Hell, imagine how it would bray "Emergency!" if the *growth* of the GNP declined 80 percent in forty years. It's a measure of the grotesque, irredeemable, and near-complete insanity of this culture that the GNP is deemed more important than life. I've written a dozen books on this insanity, and the horror and stupidity still stun me.) At this point, though, I'm not sure that most creatures would be sad to see us decline. And I couldn't blame them. There would also be creatures who would complain that the decline had come too late to save much of the planet. And I couldn't blame them either.

Part of my point is that there is of course a difference between death and death. There is a difference between the death of this chickadee and the death of chickadee communities. There is a difference between the death of an individual and the deaths of entire species, communities, the planet. Our fear of the former is a major cause of the latter, in part through our desperate attempts to create permanent—nonbiodegradable, in other words, inedible—materials, which is a subset of our desperate attempts to control, which is a subset of our desperate attempts to deny personal death (our own, as we kill those around us).

I know the altitude here, and I know that all of these beautiful trees, all of these beautiful creatures—at the moment I see two large banana slugs tangled together, and last night I carried a frog from the relatively sterile linoleum of my kitchen to the riot of life outside—will be doomed if global warming proceeds even much less fiercely than many predict.

I squat. My knees, ankles, and starting this past month even my elbows, creak. I let go. More food for the soil, the slugs, and ancient Narcissus, standing here wagging his tail.

*   *   *

I have seen the results of this culture's quest for immortality, and these results are not pretty. We have achieved a sort of immortality, and its name is plastic.

This quest for immortality is wrong, not only because life requires death, not only because we must be eaten just as we eat, not only because it manifests wrong thinking and wrong being in the world, and not only because it is killing the planet, but also because it stems from a fundamental confusion about what will continue. Like all toxic mimics, it takes a truth and distorts or perverts it.[130] In this case it confuses the continuation of the living landbase with our own personal (and cultural) continuation.[131]

Maybe, I think, it would help if we would drop our narcissism, drop our precious pretense that everything is here for us and that we are the point or even the star of this heartbreakingly and stunningly beautiful show— and somehow transfer our allegiance back to our natural communities, transform our longing to them, transfer our desire for continuation to them and to the ever-surprising processes of life itself.

The point is not to freeze ourselves or anyone else in time, like so many insects stuck in so many clear plastic paperweights, but to fall into the heartbreaking and joyful processes of living.

Before I get up to go back inside, I look once more at the beautiful redwoods and fir and spruce and alder that surround me. I look at wispy clouds forming and reforming before my eyes. I look at tiny flowers who bloom only a few days before they close, make their seeds, then die, and whose names I do not know. I look at the slugs. I look at the soil, made of

decomposed leaves and grass and wood and shit—my own and others—
and home to (and made up of) so many creatures living, eating, shitting,
loving, dying, decomposing. I look over at Amaru's grave, the grave that
like everything and everyone else here will be under water if or when the
ice caps melt. I look at the happy face of the ten-year-old border collie cross
I rescued from a pound—for his own sake, and to be a friend to Narcissus,
and to be a friend to me. I look at my own hands, see that they, like the rest
of me, are aging (my mother says she realized she was getting old when
she put her arm into the sleeve of a sweater, and saw her mother's hand
come out the other end). I am getting older. Someday I will die.

And then I look once more at Narcissus, at his sweet devoted face and
at his cloudy eyes, and once again my heart breaks with love and joy and
sorrow, as it does each time I look at his ancient face.

/ / /

By now plastic is almost everywhere. By *everywhere* I mean in a huge portion of consumer products, in food and packaging, in liquid containers and
the liquids they contain. By *everywhere* I mean in the oceans and in the air
and on the land. By *everywhere* I mean on Mount Everest and in the Mari-
anas Trench and in remote forests. By *everywhere* I mean inside every
mother's breast milk, inside polar bear fat, inside every fish, inside every
monkey, inside every songbird, inside every frog. And rest assured, it's
inside of you, too.

And that's a very bad thing.

Let's start with poly-brominated diphenyl ethers (PBDEs), used as flame
retardants in products as varied as computers, carpeting, and paint. They're
also used extensively in automobiles, and along with pthalates (more on
them in a moment) contribute to that "new car smell" that, while roman-
ticized by some, in truth signals the off-gassing of poisons. PDBEs have
been shown to cause liver and thyroid toxicity, reproductive problems, and
memory loss.[132] For the last three decades, the quantity of PDBEs dumped
into the environment has doubled every three to five years. Not surprisingly, the load of PDBEs in our bodies has also doubled every three to five
years (which means we have about 128 times as much of these PDBEs in
our bodies as we did only twenty-eight years ago).[133]

Pthalates, used to make plastic soft and pliable, are just as bad. Industrial civilization fabricates about a billion pounds of pthalates per year. Pthalates, used in millions of products from varnishes to cosmetics to the coatings of timed-release pharmaceuticals to packaged food, leach readily enough from those products so that by now they are found in our blood, urine, saliva, seminal fluid, breast milk, amniotic fluid. Pthalates are toxic to our reproductive systems.[134]

But the danger of pthalates doesn't stop there. In some food containers and plastic bottles, pthalates are used in tandem with bisphenol A (BPA). Industrial civilization pumps out about 6 billion pounds of BPA per year, and it's found in nearly every human being, and presumably a similar proportion of nonhumans (not that most of us particularly care about them). The effects on living beings are horrific. Exposure levels of only .025 micrograms per kilogram of body weight per day (at this point low levels of human exposure from diet are around 1.5 micrograms per kilogram of body weight per day, and relatively high levels are around 13; of course prior to the invention of plastic, everyone's ingestion was at precisely zero micrograms per kilogram of body weight per day) cause permanent changes to the genital tract, as well as changes in breast tissue that predispose cells to the effects of hormones and carcinogens. Two micrograms per kilogram of body weight per day lead to a 30 percent increase in prostate weight (have you wondered why there are so many ads for chemicals to reduce the effects of enlarged prostates?). At 2.4, the victims (for that's what they, or rather we, are) suffer early puberty and a decline in testicular testosterone. At 2.5, there is an increased risk of breast cancer (and you have noticed the explosion in breast cancer rates, have you not?). Doses of ten micrograms per kilogram per day lead to increased risk of prostate cancer (Do I need to keep putting in these parenthetical comments, or do you see this now in your own life and the lives of those you love?). That same dose leads to decreases in maternal behavior. Double it and you've got damage to eggs and chromosomes. Raise it up to thirty micrograms per kilogram per day and you've got hyperactivity, and also a reversal of normal sex differences in brain structure (where are those damn family values people when you need them?). Raise it all the way to fifty-one micrograms per kilogram and you've finally exceeded what the United States deems safe exposure.[135]

Of course industry liars, I mean, representatives, tell us repeatedly that bisphenol A is safe and that our exposure is tiny. But then again, industry liars—I mean, representatives—are paid to lie about—I mean, represent—the financial interests of the corporations, the rest of us (and the world) be damned. Literally. In this case, these lies take the form of doing tests on rats specially bred to be immune to the effects of this chemical, and then passing off these studies as meaningful; failing to disclose the fact that they're receiving money from chemical or plastics corporations; attempting to bribe scientists whose studies reveal bisphenol A to be dangerous; doing the sorts of statistical manipulations we've come to expect from corporate scientists; and of course good old-fashioned lying (as well as any number of forms of lying that would surprise even the most jaded).[136]

But bisphenol A is not the worst of the plastics. That honor probably belongs to polyvinylchloride, or PVC. PVC is also one of the most common plastics: over 14 billion pounds are manufactured each year just in North America. About 80 percent of this is used for construction, and a good portion of that is used for piping, and for vinyl siding and flooring. But it's also used to insulate wires and for carpet backing, window and door frames, shower curtains, furniture, gutters, weather stripping, moldings, and so on.[137] And of course it's also used extensively in hospitals, in everything from bedpans to catheters to enteral feeding devices to hemodialysis equipment to gloves to tubing. This gives people dioxins in two ways: through leaching, and through inhalation of smoke from burning this equipment later in the hospital incinerator. As Carolyn Raffensperger, executive director of the Science and Environmental Health Network, said to me, "What are we going to do with the irony that conventional western health care is such a toxic industry? They incinerate PVC medical devices that have been used to treat your cancer, sending the toxic residue out to cause someone else's disease. What sense does that make? They use mercury in thermometers in hospitals, then send *that* up the incinerator to be deposited in fish and eventually to give your child brain damage. Where does that make sense?"

Of course none of it makes sense.

Back to PVC. Part of the problem with PVC is that it's a product of the chlorine industry, and has by now become the single largest use for chlorine. We're all familiar with one form of chlorine: it's in salt (sodium chloride). But the chlorine industry uses vast amounts of (taxpayer-subsidized) electricity to break that chemical bond and release an extremely reactive and rare form of chlorine. Also released are dioxins—lots of them.

Dioxin is the general name for a whole class of chemicals (polychlorinated dibenzodioxins, for those of you whose grades in chemistry were higher than mine) that exist in trace amounts in the natural world, but are for all practical purposes creations of the chemical and plastics industries. Dioxins are some of the most toxic substances imaginable, dangerous at doses of several parts per trillion (the equivalent of a few drops contaminating a trainload of liquid many miles long). They are highly carcinogenic and poisonous, and they also bind to the hormone receptors of cells, modifying not only the cells' functioning (which would be bad enough) but also their actual genetic structure. Further, dioxins bind with fat, so they get stored there and when someone else eats this fat, they also eat the dioxins. Once they're in your body, you are, for the most part, stuck with them.[138] Most dioxins have half-lives of between four and twenty years in your body.[139]

Unfortunately, dioxins are not made only when PVC is manufactured. They can leach out from the PVC itself, and they're also produced when PVC is burned. There are many other industrial sources of dioxins as well, including the infamous Agent Orange, but also including the production and use of chemicals such as herbicides and wood preservatives, the refining and burning of oil, car and truck exhaust, and so on. Dioxins are, truly, an inevitable by-product of the chemical and plastics industries.

You'll be glad to know, however, that the PVC industry has your best interests at heart. In response to increasing outrage over the effects of dioxins on human health, in the 1990s the Vinyl Institute came up with a plan. The Institute stated, "The short-term objective of the plan is to mitigate the effects of negative press coverage by positioning the vinyl industry as a proactive and cooperating entity, working in tandem with EPA to characterize and minimize sources of dioxin." And how did this "cooperating" work? Well, the Vinyl Institute and the Environmental Protection [sic] Agency cooperated on a plan whereby the PVC industry would be the sole supplier of information about dioxin emissions: it would collect, analyze, and interpret all of the data, and hand these interpretations to a criminally credulous EPA, which would then put its stamp of approval on this "data."

These two members of this criminal conspiracy to poison us all agreed that this data could be peer reviewed, but because the collection processes themselves are confidential—invisible even to the peer review committee—there is no way for anyone outside of the industry to have any idea of the veracity of any part of this process.[140] Which of course is the whole point.

Don't you feel so much safer now?

Pthalates, bisphenol A, and dioxins are all in a class of chemicals known as endocrine disruptors, meaning that they can, as one writer put it, "disrupt the endocrine system—the delicately balanced set of hormones and glands that affect virtually every organ and cell—by mimicking the female hormone estrogen. In marine environments, excess estrogen has led to Twilight Zone-esque discoveries of male fish and seagulls that [sic] have sprouted female sex organs."[141]

It ends up that, as Marc Goldstein, M.D., director of the Cornell Institute for Reproductive Medicine, says of pregnant women, "Prenatal exposure, even in very low doses, can cause irreversible damage in an unborn baby's reproductive organs."[142]

Further, exposure, for example, to BPA also can make its victims fat (with both more and bigger fat cells). In the words of Dr. Frederick vom Saal, Ph.D., a professor at the University of Missouri at Columbia who specifically studies estrogenic chemicals in plastics (and whose research has caused him to remove every polycarbonate plastic item from his home, and to stop buying plastic-wrapped food and canned goods), "These findings suggest that developmental exposure to BPA is contributing to the obesity epidemic that has occurred during the last two decades in the developed world, associated with the dramatic increase in the amount of plastic being produced each year."[143] BPA also causes victims' insulin to surge and then crash. Is there a correlation between the rise of plastic and the 735 percent increase in diabetes in the United States since 1935?

No, of course not. There couldn't be.

Please.

Please.

And this culture isn't killing the world.

And the oceans aren't full of plastic.

And neither is my body.

/ / /

Somebody make it stop.

Somebody stop this insanity before it kills us all.

No, we can't wait for *somebody*. *We* have to do it.

/ / /

I sometimes picture the people who will come after the current planetary blowout (presuming humans survive, presuming any life survives). I go back and forth on what I think they will say about plastic. Sometimes I think they will stand in disgusted amazement and say, "You had light-weight waterproof containers, and you threw them away?" (Or maybe they'll still have them, since the damn things don't decompose). And sometimes I think they'll stand in disgusted amazement and say, "God damn you for making plastic in the first place."

More and more I suspect the latter.

/ / /

Gosh, would life be worth living without CDs, plastic pacifiers, plastic wrap, sandwich bags, syringes, bottled water (and soda bottles), single serving packets of potato chips, automobiles, straws (and crazy straws!), plastic grocery bags, freezer bags, ice cube trays, bubble wrap and packing peanuts, carpet-backing, Styrofoam life preservers and take-out trays, disposable pens, disposable diapers, hairspray and plastic hair brushes, plastic toothbrushes (and toothpaste!), milk crates, packing tape, plastic forks, telephones, computers, hair clips, billiard balls, shower curtains, beach balls, balloons, latex condoms, and polyester pants?

Surrounded by all of these necessary wonders, it can be easy to forget that humans lived without plastics for tens or hundreds of thousands of years—and to the best of our knowledge, they lived relatively cancer-free for tens or hundreds of thousands of years—and that plastics were invented only a century or so ago. By this point, plastic is central to modern life. How could we live without it? But if you think that life without plastic is unthinkable, the deeper truth is that life with plastic may very well be impossible.

/ / /

Part of the answer to living sustainably is not simply to reuse materials that should never have been made in the first place. It's to not make them at all.

/ / /

Just to drive the point home, here is another extremely incomplete list of some of the health effects of exposure to various forms of plastic: physical deformities, cancer (brain, breast, cervix, colon, testicular, prostate, and on and on), early puberty, immune deficiencies, endometriosis, behavioral problems, lowered intelligence, impaired memory, impaired sexuality, low sperm count, motor skill deficits, reduced eye-hand coordination, reduced physical stamina, and much more.[144]

Industry liars and their pet politicians will of course point to an inability to tell which poison (from which factory) caused which particular cancer. I don't disagree that it may sometimes be difficult to pin down the precise murder weapon. As Paul Goettlich wrote, "Since we do live in a sea of man-made toxicity, there is great difficulty in pinpointing exactly which chemical or combination of chemicals was the cause of a cancer or deformity."[145] But industry liars and their pet politicians will then say this uncertainty is reason enough for them to continue business as usual: *It would cause undue economic damage to remove this chemical—which makes all of your lives so much better and easier—from the free market without clear proof of the harm it is alleged to cause.* That's absurd, and it's murderous. Not knowing which specific carcinogen gave my grandfather cancer or my friend breast cancer or my other friend uterine cancer or my other friend's mother breast cancer and her father prostate cancer or my other friend leukemia (She has a t-shirt that reads, "My father dropped Agent Orange on Vietnam, and all I got was this lousy leukemia."), and so on, is like standing on a battle-field watching your friends die, yet not knowing precisely which gun fired which bullet that killed your grandfather, your lover, your friend, your child. The situation is not one that calls for bemused academic interest, it calls for action: we need to stop the bullets, and if those on the other side won't stop shooting at us, then we need to stop them, using any means necessary. Likewise, if those on the other side won't stop poisoning us—and I guarantee they're not going to stop because we ask nicely, or because our grand-parents die, or because our children die, or because we sign petitions, or because they're killing the world, or because they agree to closely cooperate with the EPA—then we need to stop them. Using any means necessary.

Or maybe not. Maybe it's all too big and too scary, and after all, we *don't* know precisely which poison killed your grandfather, made you sick, killed your dog, made it so you can't fish anymore at your favorite fishing spot because the fish all have tumors, killed your cousin, killed your mother, killed your niece, made your nephew fat, made your granddaughter develop pubic hair and breasts before she entered preschool, gave your sister asthma, gave frogs eight legs, fucked up the genitals of alligators and fish and seagulls, messed with your ability to remember, messed with your ability to think clearly, killed your best friend from childhood, and so on. And of course if we don't know precisely which poison did each of these—if we can't nail it down with 100 percent certainty—then fuck it, we should just keep studying—or rather let industry and government keep studying—until there is nothing left of the world. After all, it's only our lives that are at stake, and the lives of those we love, and the life of the planet.

Good luck.

/ / /

Things are actually far worse than I've described. To get the barest hint of how bad they are, just think for a moment about nurdles.

Nurdles are rabbit-poop-sized resin pellets not yet made into full-fledged plastic products. They make up about 10 percent of all plastic ocean trash. For obvious reasons, they're essentially impossible to clean up.

Part of the problem—apart from the fact that there are so damn many, and apart from the fact that the nurdles themselves are poisonous, as are all plastics—is that nurdles bond with other poisons such as dichloro-diphenyl-trichloroethane (DDT) and dioxins. In fact they *attract* these other poisons, so that concentrations of these toxins can be a million times higher than in the surrounding water.[146] You read that correctly. A million times.

It gets worse. Nurdles are the size of fish eggs, and often when they're not ingested incidentally they're ingested by mistake. But the reasons don't much matter, because in either case ingested along with the nurdles are other toxins, at rates up to a million times higher than in the already polluted oceans.

But who cares about a bunch of fish and whales and albatrosses, right? Clearly not many of us in this unremittingly narcissistic culture, or we wouldn't allow this culture to perpetrate these atrocities in the first place. But don't forget that these toxins are stored in fat, and ultimately they end up in you. Yes, you.

/ / /

The United States produces about 300 billion pounds of plastic per year. Billion. Not million. Billion.[147]

/ / /

As Captain Charles Moore, the discoverer of the "Garbage Patch" in the Pacific puts it, "If 'more is better' and that's the only mantra we have, we're doomed."[148]

/ / /

Here's another way to put it. Oceanographer Curtis Ebbesmeyer, Ph.D., an expert on marine debris, has said, "If you could fast-forward 10,000 years and do an archaeological dig . . . you'd find a little line of plastic. What happened to those people? Well, they ate their own plastic and disrupted their genetic structure and weren't able to reproduce. They didn't last very long because they killed themselves. . . . The ocean is warning us, and if we don't listen, it's very easy for her to get rid of us."[149]

/ / /

If we were only disrupting our own genetic structure and hindering our own ability to reproduce, there are many who would say good riddance to the members of this culture who are killing the planet.

But this culture is taking everyone else down with it.

Once again, there are those in the wild who I am sure would not be sad to see this culture go. The fulmars and albatrosses starving with bellies full of plastic. The alligators and fish whose genitals have been made ambiguous by this culture's poisons. The infant orcas dying because their mothers' breast milk has been rendered toxic by this culture's poisons. The turtles deformed by plastic rings, the fish and whales and birds caught in ghost nets. The ocean herself. The earth herself.

I would be hard pressed to blame a single one of them for their sorrow, their rage, their hatred.

I feel that sorrow, rage, and hatred myself.

Do you?

* * *

What are you going to do about it? What are you going to do to stop these atrocities? How far will you go? What will you do to make yourself worthy of the life that this planet has given you? Where is your allegiance? What will you do?

And why aren't you doing it?

* * *

Most especially given the near infinite depth of this culture's narcissism, abusiveness, and its perceived entitlement to exploit, I need to be clear. I do not hate human beings. I love humans, and I love humanity. But I love a living planet far more.

And that is how it should be.

* * *

I need to emend what I just wrote. I love humans, but I do not love what so many humans have become. I hate what this culture has done to us, and I hate that so many value profits and power over life. I hate the narcissists who are killing those I love. I hate those who are poisoning the planet, who are poisoning my landbase, who are poisoning members of my family, who are poisoning me. I hate them, and I will stop them from killing the planet.

Who will join me?

/ / /

I need to be clear about something else. Humans are not killing the planet. Industrial humans are killing the planet. Humans who identify more with this culture—this culture of plastic—than they do with life are killing the planet. Humans lived on this planet for tens or hundreds of thousands of years without destroying it. It is only in the last 6000 years or so that a culture toxic enough to kill the planet has emerged.

The choice we face is not between killing ourselves and killing the planet. The choice we face is continuing this way of life that is killing the planet (and us) or not. We can allow this culture to kill the planet, or we can destroy this culture. It really is that simple.

MINING

Some of us have become used to thinking that woman is the nigger of the world, that a person of color is the nigger of the world, that a poor person is the nigger of the world. . . . But, in truth, Earth itself has become the nigger of the world. —ALICE WALKER

PLASTICS ARE OF COURSE NOT the only persistent pollutants made by this culture. The list is long. There are plutonium-239 and other radioactive wastes, many of which will persist for the life of the planet (of course given the velocity and acceleration of this culture's destructiveness, the death of the planet will probably come sooner than we think, insofar as most of us think about it at all). There are heavy metals removed from the earth and put into living bodies. There is carbon dioxide, which used to simply be part of the air but is now a pollutant, significant quantities of which will not be metabolized by plants or otherwise removed from the biosphere for an extremely long time. Look around your own life—and get out into the land where you live—and find the long-term pollutants produced or released by this culture.

I grew up in Colorado and spent a lot of time in the mountains, exploring ghost towns and the fringes of old mines. I say fringes because I was forbidden by my mom from entering mines for fear they would collapse or I would fall down a shaft. So I spent those days wandering around the tailings: great mounds of crushed rock that a hundred years later still supported no life. In retrospect, the mineshafts may have been the least of my safety worries; the tailings are probably more dangerous than the mines themselves.

Mine tailings are what's left over after miners dig up the ground and extract whatever they're going to sell. These previously buried, now-exposed rocks have normally been broken, and range from much larger than a softball down to coarse sand or even fine powder. Some minerals commonly found in tailings include—but are certainly not restricted to—arsenic (especially in gold mine wastes), barite, calcite, fluorite, many radioactive materials (that the earth had previously stored where the earth wanted them: underground), sulfur (and many sulfide compounds), cadmium, zinc, lead, manganese, and so on.[150]

Many of these minerals are toxic. That explains the dead zones so familiar to anyone who has had the misfortune to go near a mine.

Many of these minerals don't stay in place, but weather away—their rapid weathering facilitated by their relatively small particle size, which results in greater surface area available for wind, air, and water to affect—into surrounding soil, and more ominously, water. This happens most frequently with the sulfide minerals—especially pyrite (iron disulfide: one molecule of iron and two of sulfur), the most common mineral in the majority of mine tailings[151]—which are oxidated by chemical processes and by being metabolized by certain bacteria.[152] These processes produce sulfuric acid, which often reduces the pH of afflicted streams and groundwater to less than 3.0. Of course, acidifying streams and groundwater kills the streams, as well as the plants and animals who live in them. As even one pro-mining website puts it: "The presence of these toxic heavy metal ions and acidic pH's has an adverse affect on every aquatic species found in the stream. In many instances, streams are almost completely void of life for many miles downstream of a mine drainage source."[153]

As the acidified water slowly gets diluted, the pH rises back above 3.0. You'd think this would be a good thing, but making the stream less acidic causes iron ions (from that original iron disulfide) to, as that same website says, "precipitate out of the water and coat the stream bottom with a slimy orange sludge (iron III hydroxide, $Fe(OH)_3$, and related compounds). This unsightly sludge, called 'Yellow Boy,' tints the stream water an unnatural reddish-orange color and smothers the organisms that thrive on the stream bottom."[154]

Great.

And it gets worse yet. Once miners have dug ore from the ground, they extract from it the gold, platinum, silver, lead, coal, or whatever else they want to sell. Frequently—and it gets ever more frequent with each passing year—this extraction (or "milling") uses toxic or otherwise harmful chemicals: in other words, the valued minerals are no longer simply separated mechanically, but are separated through processes often involving chemicals. For example, because these days gold remaining in its native state typically occurs at concentrations of less than a third of an ounce per ton— meaning mining corporations dig up and later dump three tons of rock and soil for every ounce of gold—it's not economically feasible to mechanically separate gold from ore. Instead, the gold is dissolved into a liquid, adsorbed from this liquid onto activated carbon, and then washed away from the carbon using solvents. Because gold isn't soluble in water, it requires both a complexant and an oxidant in order to dissolve. Many of the chemicals used in these extraction processes are toxic or otherwise harm the land and water, and of course also harm those who require land or water in order to live. In other words, they harm all of us.

Even though the mining industry claims these chemicals are recycled— and they often are recycled to the best of their technological capability, for what that's worth—the chemicals still often end up in mine tailings or tailings ponds, and from there the chemicals move into streams or groundwater. From there they end up in living beings.

You may recognize some of these chemicals. Let's start with cyanide. Yes, that cyanide. The one from those World War II movies where a character bites down on an ampoule of hydrogen cyanide and immediately goes into convulsions and dies. The one that, so we hear in all these movies, smells like bitter almonds. The one that prevents cellular respiration, and kills at concentrations of 100 to 300 parts per million (and kills fish and other aquatic life at concentrations in the range of parts per billion). The one that went by the name Zyklon B and was used by the Nazis as an efficient form of mass murder. The one used in gas chambers. The one responsible for many of the deaths at Bhopal.[155] The one commonly listed as a chemical warfare agent. The one allegedly used by Saddam Hussein's Iraq regime against both Iran and the Kurds (with ingredients provided with the help of the United States). The one the Aum Shinrikyo cult attempted to use to commit mass murder in a Tokyo subway in May 1995. The one al-Qaeda allegedly planned to use to commit mass murder in a New York subway in 2003. Yes, that cyanide. The same cyanide produced routinely—1.4 million tons per year—for use in the production of plastics, adhesives, cosmetics, pharmaceuticals, and so on. The same cyanide used in mass quantities—182,000 tons per year—to facilitate the extraction of gold from surrounding ores.[156] The same cyanide used to extract about 90 percent of the gold mined each year.

It seems that those who put small amounts of cyanide in subways are terrorists. But those who produce it in mass quantities and contaminate broad reaches of soil, water, and air, killing countless living beings, are not terrorists, but rather capitalists, and are counted among the finest and most powerful people on the planet.

Mining industry representatives—whose relationship to truth is as strained as that of representatives of other industries—like to tell us that using cyanide to extract gold from ore is reasonably safe. But at this stage in the corporate destruction of both the planet and honest discourse, it should not surprise us that they are lying.

Above, when I wrote that "gold is dissolved into some liquid, adsorbed from this liquid onto activated carbon, and then washed away from the carbon using solvents," and also wrote that since "gold isn't soluble in water, it requires both a complexant and an oxidant in order to dissolve," my description failed to get to the essence of the process: this articulation makes the process seem clean, even if that "complexant" is something as toxic as cyanide. If the miners are careful enough—and we know they will be—then this could and should be a safe process. Right?

Well, let's try this description. First, when we're talking about miners, we're not talking about Humphrey Bogart, Tim Holt, and Walter Huston protecting their hard-earned treasure by fighting off "cops" who have no stinking badges (nor, and this is more to the point, are we talking about miners who say, and more importantly actualize, the understanding that they've "wounded this mountain. It's our duty to close her wounds. It's the least we can do to show our gratitude for all the wealth she's given us").[157] Instead, we're talking about huge transnational mining corporations that often bribe local, regional, or national officials, politicians, dictators—whoever has the power to sign pieces of paper that legalize their activities—to give them permits to dig up entire mountains. They often buy off, bully, beat, capture, or kill those who oppose them; or they get their money's worth from the local, regional, national officials, politicians, or dictators who send in the military or police (with stinking badges) to bully, beat, capture, or kill those who oppose their mines. In the United States and Canada, the preferred tactic is to buy off the opposition (except especially in the case of those of the indigenous who cannot be bought off), whereas in the colonies it's often more cost-effective for these transnational corporations—and the military and police who serve them—to skip directly to the latter three options: in both the short- and long-run, bullets are cheaper for them than bread, and bullets are certainly (fiscally) cheaper for them than not digging up the earth.

Next, when these huge corporations mine and mill ore, these processes are not quite so spick and span as the technical descriptions make them seem. First, picture a living mountain, its base covered in trees, its top above the tree line. Now, picture this mountain flattened. Picture its guts removed. Picture pits so large that, as one pamphlet puts it, "they could swallow cities."[158] Picture heaps of extracted ore several hundred feet high and several times larger than a football field. Now, picture spraying a solution containing cyanide over those heaps. Picture the cyanide trickling down, chemically bonding with microscopic bits of gold. Picture this solution draining to a huge rubber blanket beneath this heap. Picture this blanket channeling the cyanide solution toward a large holding pond, where the gold is stripped away, and as much of the cyanide as possible is recovered to be reused.[159]

Picture birds landing in this pond. Picture birds dying. Picture every living being who comes in contact with this pond dying.

Picture these heaps being dumped on plains. Picture them being dumped in valleys. Picture them filling these valleys, until the valleys no longer exist. Picture these heaps being dumped anywhere these transnational mining corporations have permits to dump them. Picture governments handing out these permits to transnational corporations. Picture transnational corporations handing money to local, regional, or national officials, politicians, dictators—whoever has the power to sign pieces of paper that legalize these activities. Picture streams below turning to sulfuric acid. Picture all creatures dying. Picture the bottoms of these streams coated with Yellow Boy, and picture the water itself a sickly orange. Picture those humans who live by these streams dying. Picture paid representatives of these mines telling us that these processes are safe. Picture newspapers repeating these claims. Picture paid local, regional, or national officials, politicians, dictators repeating these claims as well. Picture paid local, regional, or national officials, politicians, dictators passing laws making it illegal or impossible (at least through "proper channels") to effectively stop the gutting of these mountains, the poisoning of these streams, the poisoning of this water, this land, this air, these nonhumans, these humans. Picture paid local, regional, or national officials, politicians, dictators buying off some of those who resist, and bullying, beating, capturing, or killing the rest. Picture the owners and CEOs of these corporations living nowhere near the tailings piles or the acidic orange streams.

Picture this cycle being repeated.

Picture the planet being killed.

Now picture a catastrophe. Picture the Baia Mare gold mine in northwest Romania. Picture a dam holding back a tailings pond. A large tailings pond. Full of water polluted with cyanide (120 tons just of cyanide) and other toxins. Now, picture this dam breaking. Picture a twenty-five-foot tall wave of toxic mud and water rushing into the Lapus River, and from there into the Somes and Tisza rivers. Picture this wave crossing the border into Hungary, hitting the Danube, and crossing another border, this time into Yugoslavia. Picture this wave killing nearly everything in its path. Picture the Tisza River being habitat and home to nineteen of Romania's twenty-nine protected fish species. Picture one hundred tons of dead fish being collected from the surface of this river. Picture otters on the Tisza and Somes rivers eating cyanide-laced fish. Picture these otters disappearing, presumed dead: casualties gone MIA in this culture's war on the world. Picture the Tisza and Somes rivers without otters. Now picture another spill higher on the Tisza, and this time picture 20,000 tons of sediment toxified by lead, zinc, copper, aluminum, and cyanide. Picture a river being systematically murdered.[160]

Picture the owners living elsewhere, and not being poisoned by the effects of their decisions.

Now, picture the Zortman-Landusky mine in Zortman, Montana. Picture spill after spill of cyanide-laced water. Picture once again a tailings pond. Picture a heavy rain. Picture a tailings pond in danger of overflowing. Picture mine operators deciding to pump this cyanide solution out of the tailings pond and onto another plot of land. Picture them not telling anyone about this. Now, picture living in Zortman. Pretend you are thirsty. You walk to your tap. You turn it on. You smell bitter almonds. You've seen enough World War II movies to know what this means. You turn off the tap. Your water has been poisoned with cyanide,[161] and not by Aum Shinrikyo or al-Qaeda. Instead by capitalists (who, by the way, kill a hell of a lot more people than terrorists). By a transnational mining corporation. Of course.

Now, picture these mine owners. Pretend that one of them is thirsty. He gets a drink of water. Do you think it smells like bitter almonds?

Of course not.

Now, pretend you are a rancher in southern Colorado. Pretend your cattle are on the range. You check on them. You notice they will not drink. You look more closely at the Alamosa River, and more closely still. You notice something different. You can't quite puzzle it out. And then you understand. You see far less aquatic life than normal.

Life goes on. You keep checking the river. And over time a pattern becomes clear to you. You realize you are seeing a river being murdered. You see no aquatic life: no fish, no insects, nothing.

Somehow this does not surprise you, because you know about the Summitville Mine, a nearly two-square-mile open pit mine: nearly two square miles of toxified landscape high in the mountains. You know that this mine tore open the mountainside, and you know that the absentee mine owners—mainly European investors, although Bank of America invested $20 million when it learned that the huge corporation Bechtel would be involved—did not follow the ethic laid out in *The Treasure of the Sierra Madre*, but rather walked away after wounding the mountain. You know that this landscape will remain toxified long after you are dead, long after your own bones have become no longer yours but have rejoined the earth, have become soil, become trees, become those others who live on this land (presuming they still live).

You want to know how severely these absentee mine owners have wounded not only the mountain—which you know they have wounded gravely—but the river. Studies are done. The government—always a better friend to the absentee mine owners than to you (after all, who pays the policy makers to make their policies?)—declares that all aquatic life was killed in a seventeen-mile stretch of the Alamosa River. You know that is not correct. You believe the local alfalfa farmer—who is in no way beholden to the mining corporation—who says, "There were fifty-five miles killed. Usually, the papers just mention the top of the watershed and not the residential areas that were contaminated. It affected our entire watershed including the river, its laterals, and the stock ponds where all life was killed."

You know, don't you, that none of the absentee mineowners poisoned. None of them were killed.

On the contrary, they, as always seems to happen, increased their wealth, at the expense of the mountain, the river, you, and your community.

Oh, and by the way, the total value of gold and silver taken from this mine is less than half of what the cleanup has cost *so far*.

Somehow this doesn't surprise you.[162]

/ / /

Did I mention that by far the most common use—85 percent—of gold is jewelry?

These mountains are being killed, these rivers are being killed, these fish and otters and all these other animals are being exterminated, for earrings, pendants, wedding bands, and wristwatches.

/ / /

Not that it would be okay to kill mountains and rivers, to extirpate plants, animals, and so on, for "more important purposes" like cell phones, airplanes, aluminum cans, coins, automobiles, solar panels, electric wiring, televisions, computers, and so on: there is *no* reason good enough to destroy a landbase.

None.

None at all.

/ / /

Cyanide is not the only toxic chemical used to process ore. For example, mine owners often use sodium ethyl xanthate (called SEX by the mining industry; having graduated from the Colorado School of Mines I can attest that this is the only sex that many of these miners will routinely encounter). Sodium ethyl xanthate is toxic in its own right, and under normal conditions it will form carbon disulphide, a toxic and highly flammable gas that is easily absorbed through the skin. It is even more toxic in water. In this case, even the most ardent hedonist will agree that not all SEX is good.

Another would be potassium amyl xanthate, or PAX, which is highly toxic to trout. Yet another is sulfuric acid, which is used an large quantities in some forms of leaching. Since sulfuric acid is created anyway by the weathering of tailings, additional sulfuric acid is the last thing local rivers need.

And so it goes.

MEDICINE

What we really need to do is rethink medicine with the understanding that we are the air and the water. —CAROLYN RAFFENSPERGER

WHEN I'M NOT WRITING, doing activism, or working on the farm, I work part time on an ambulance—I'm a practicing paramedic. Medical waste is an issue that's close to me, something I deal with every day.

Here are the disposable items I used and threw away today at work. Eleven pairs of synthetic nitrile gloves. Two used syringes. Two empty glass drug ampoules. Three plastic lancets for checking blood sugar. Two plastic blood glucometry test strips. Six feet of electrocardiogram printout paper. Twenty conductive electrocardiogram (ECG) stickers. Four oxygen masks, two pairs of nasal prongs for oxygen delivery, and their associated packaging. An empty IV bag. Two feet of medical tape. Three blood-soaked four-by-four dressings. A plastic cervical collar used to protect from neck injuries. And a styrofoam clamshell that contained the takeout food I bought on our sixteenth hour of work, after we transferred a patient to an emergency center a few hundred kilometers from our home area.

It wasn't an unusually wasteful day. Well, except for the takeout: I usually bring my own food that lasts the twelve hours of our official shift.

Medical waste generally includes pretty much everything discarded from health care facilities like hospitals, doctors', or dentistry practices, veterinary clinics, and medical labs or research facilities. Typical wastes would have a lot in common with what I used today, things like used gloves or masks, needles and syringes used to take blood or give drugs, blood-soaked dressings or bandages, used surgical instruments, or disposable portions of diagnostic equipment. Medical waste can even include removed bodily fluids or organs, or amputated body parts. Medical waste may be hazardous or infectious, or even radioactive, since radioisotopes are used for some diagnostic procedures and medical treatments.

Technically, most medical waste, by volume, isn't actually that different from waste created by homes or businesses. A doctor's office will dispose of a lot more used paper than used needles, pound for pound. According to the EPA, infectious waste only comprises about 10 to 15 percent of the medical waste stream.[163]

But shredded paper medical records and empty cartons of milk from the hospital cafeteria aren't what most people are talking about when they use the words *medical waste*. They're thinking of the stuff that's unique to medical waste. Maybe they're thinking about the "Syringe Tide."

In 1987 and 1988, large amounts of medical waste and other floating garbage washed up on beaches in New Jersey. In just one incident in 1988, a garbage slick close to a mile long washed up on the shore. This wasn't the first or last time this happened on New Jersey shores, but this event was different in that images of syringes on the beach hit newspapers and television during a time of growing public fears about AIDS and the growing popularity of environmentalism in the mainstream. And if all that wasn't enough to spur public officials to action, the reports of washed-up medical waste and sewage drove away billions of dollars in tourism: now *that* is something public officials will pay attention to. Exactly where the waste came from is still a matter of debate, but New York City ended up paying a million dollars for the cleanup, after some of the waste was blamed on the Fresh Kills Landfill on Staten Island.

The event has been called a turning point in modern environmentalism. It's been argued that people finally started to realize that the oceans aren't bottomless disposal pits. Of course, rafts of plastic currently suffocating oceans are powerful counterarguments to that, but in any case, government regulations did change. New Jersey initiated a program to clean up "floatable" garbage, and now specifies that medical waste must be "ground or minced into small unrecognizable pieces . . . prior to disposal."[164] I leave it to readers to decide whether such acts have reduced the amount of dumped waste, or simply decreased that waste's visibility.

In part because of events like those in New Jersey, medical waste occupies a special place in our hierarchy of revulsion. Sure, obsolete cell phones and computers also comprise hazardous waste, but we touch them all the time, at least before we dispose of them. People eat or drink from what later become discarded fast-food wrappers. Fertilizer runoff and phosphate-rich detergent we flush down the drain are more destructive of aquatic ecosystems. Even sewage can seem pretty disgusting, but how can that really compete with used scalpels, vomit- and blood-soaked bedsheets, or even, gods forbid, an amputated human foot?

Medical waste is also a more difficult problem, morally speaking. Everyone knows it's not *good* to throw away that styrofoam coffee cup, or the multi-coat packaging from some processed meal. People still do it, but more out of convenience or habit. It's easy to say that those wastes should be eliminated, and it's not a technically difficult problem. Medical waste is a different story. If you're a health care worker like a nurse or a para-medic, would you turn down disposable gloves when you're called to treat that bloody car accident survivor? Without gloves, not only would that put you at risk of contracting blood-borne diseases carried by the patient, but you could then pass on those diseases to future patients. Health care workers who refuse to wear gloves would be in violation of their organi-zation's health and safety rules, could be sued for malpractice, and could even be criminally negligent.

Or imagine you're a patient. Would you want medical staff to reuse items that had been used on another patient, and perhaps cleaned in between? Gowns? Gloves? Thermometers? Needles? In wealthier parts of the world, hypodermic needles have been single-use and disposable for decades, because they're so difficult to clean. In the rest of the world—and everywhere not so long ago—hypodermic needles and syringes are washed, sharpened, and reused. It's not a safe practice, and I'm not by any means suggesting it be emulated to reduce waste, but it happens. There's a deep irony to the fact that so many hospitals or clinics in Africa and other parts of the world are too poor to afford inexpensive disposable needles when resources from their countries are being exploited, exported, and turned into garbage for rich people in other parts of the world.

How dangerous is medical waste? Well, the answer, as for many kinds of waste, depends on how it's disposed of. Most medical waste isn't so dif-ferent from household waste—not tremendously dangerous on a small scale in the short term, but very destructive on a global scale over the very long term—and is disposed of in the same fashion. In the US, it's not known exactly how much medical waste is produced, though estimates are in the millions of tons.[165] Some of that waste is considered infectious, and more than 90 percent of that waste is incinerated, commonly using on-site incinerators at hospitals.[166] Incineration of medical waste brings up the same issues that incineration in general does: the creation of dioxin, the emission of mercury and other hazards, and the conversion of local-ized solid waste problems into globalized air pollution problems, with all the associated health and environmental risks.

/ / /

Of course, the most harmful medical waste isn't sterilized, incinerated, or put in landfills. It isn't collected in bright yellow biohazard bags or sharp containers. It doesn't wash up on tourist beaches or make graphic head-lines on front pages of national newspapers. It's disposed of through our very bodies.

When we take a medication, it doesn't stay in our body. Some of it may be processed by our body into byproducts called metabolites. Those metabolites, as well as the unmetabolized portion of the dose, are excreted from our bodies when we urinate, defecate, sweat, spit, and so on. And society consumes absolutely massive amounts of pharmaceuticals. For example, take the drug acetaminophen, the active ingredient in Tylenol and Panadol. All those tiny pills add up—North Americans consume more than 100 million pounds of acetaminophen each year.[167] That quantity of acetaminophen is worth about $200 million, only one product among a panoply of drugs.[168] The global consumption of pharmaceuticals is worth close to $700 *billion.*[169]

A 2002 survey by the US Geological Survey (USGS) Toxic Substances Hydrology Program found that steroids and non-prescription drugs were the chemicals most commonly found downstream from areas of intense urbanization and animal production (though detergent metabolites, plas-ticizers, and steroids were usually found in the highest concentrations). Of the ninety-five chemicals sampled, "33 are known or suspected to be hormonally active; 46 are pharmaceutically active."[170]

To be clear, those commonly found and relatively concentrated steroids mostly aren't from bull-necked jocks or major league baseball players shooting up in the locker room. The word "steroids" doesn't refer just to muscle-boosting anabolic steroids, but a broad chemical family defined by its four joined carbon rings. Steroids are special because they're commonly found in hormones—specifically, sex hormones like estrogens and prog-estagens. These are the hormones found in birth control pills.

More than 100 million women around the world use some form of pharmaceutical contraception, about one tenth of those in the United States.[171] I don't need to tell you it's a commonly prescribed drug—it's called *the* pill, after all. But liberating and empowering as it may be, when the drugs in contraceptive pills find their way into water they can very damaging to aquatic communities.

Despite looking very different on the outside, all vertebrate animals are based on much of the same genes and biochemistry. More than two thirds of our genes have a counterpart in the puffer fish, for example.[172] Hor-mones can work in species other than those they originated in. This includes the hormones in the pill, which are now coursing through water-ways around the world. The bottom line? Fish are being forced to change sex.

Actually, that's too cut and dry. Many fish can and do change sex them-selves. Clown fish—like the hero of the movie *Finding Nemo*—possess two pairs of fully functional gonads and can change sex based on social con-text. That's not what we're talking about. Change induced by external hormones tends to be much messier. Rather than changing completely from a male fish to a female fish, a male fish exposed to high levels of estrogens may become reproductively "feminized." Which is to say, he still has male fish parts, they're just structurally and functionally more like female fish parts. Which means he may be less fertile, or unable to repro-duce at all.[173] Considering the damage aquatic communities have already experienced over the last century, that's pretty bad news. Actually, it's really fucking bad news. Heck, how many salmon actually make it back to their spawning grounds in a good year? Imagine a coho salmon fighting his way up a river to the place he was spawned. He has to dodge commercial fishing nets, garbage dumped in the water, waterfront construction that has obliterated the shoreline. The river level is the lowest it's ever been because of water diversion for irrigation. Our coho is lucky, because his home hasn't been blocked by a dam. He smells his way to his spawning grounds, exhausted and battered, his body literally falling apart. But when he arrives, with only one goal left in his entire life, he is infertile. The contraceptive pill has done its job in a tragically unintended way.

Drugs other than contraceptives can have adverse effects, too, but the number of pharmaceuticals being dumped into the water (tens of thousands), and the number of species living in freshwater ecosystems (close to a million), mean that pinning down pharmaceutical impact definitively is almost impossible.[174] And the adverse effects of any given drug can vary greatly across species. Small doses of common analgesics like acetaminophen can be lethal to cats, for example. Conversely, some antiarthritis drugs that can cause kidney damage in humans seem harmless in mice.[175] It's extremely difficult to predict the effects.

Now, not every single drug molecule found in waterways goes directly through human beings, of course. Drugs can also be introduced into the water when people flush their expired or unused medications down the toilet, a practice long recommended and only recently discouraged. Pharmaceuticals are also put into products not traditionally recognized as medication. Antibacterial soap often contains the drug Triclosan, which has been found to disrupt the development of frogs living downstream.[176] Pharmaceuticals can also come from intensive animal agriculture, where animals are commonly given growth hormones or antibiotics. And it's possible to reduce those releases, certainly—but a lot of people are taking medications, and pretty much all of them are using sewage systems that will eventually run into waterways.

/ / /

I keep thinking back to the medical waste I threw away today. We spent about thirty minutes on the side of the highway extricating a patient from a smashed car. I used, among other things, a plastic oxygen mask, a pair of nitrile rubber gloves, and several sterile cotton gauze dressings, which became permeated with blood and were later discarded "safely"—by which I mean they either ended up in an incinerator or landfill. If you were to simply leave a blood-soaked dressing beside a road, it would be relatively harmless. Though it's unlikely that our patient had an infectious bloodborne illness, even if he did the vast majority of pathogens survive only briefly outside of a human host, and cannot be spread to wild animals who might encounter the dressing. Ultraviolet light from the sun would kill pathogens and start the breakdown of the thin cotton fibers. Within a few months of warm and sunny weather there would be nothing recognizable left of the dressing. Some of the nitrogen, carbon, and hydrogen in the cotton would be released as gas, and some of it would be gobbled up by soil microbes.

The nitrile rubber gloves and oxygen mask are a different story. The oxygen mask and its several feet of tubing are made mostly of transparent polyvinyl chloride (PVC) softened with phthalate plasticizing agents. This and similar plastics, which are made with chlorine and release dioxin both when manufactured and disposed of, are the base material for many common single-use medical devices, including IV bags and tubing. Pediatric studies have shown that intravenous bags and tubing leach their plasticizers into IV fluid as the pharmaceuticals are administered to patients.[177]

If you recall, plasticizers like phthalates make plastic soft by acting as a kind of lubricant between the plastic molecules. Most plastics are made of long strings of repeating molecular units—think of them as long strings, like spaghetti. A plasticizer is like a sauce that allows the strands of spaghetti to move against each other. Over time, plasticizers can gradually migrate out of the spaghetti into the surrounding environment, whether into the interior of a car (remember, plasticizers escaping from the dash and upholstery make that "new car smell") or the blood being transfused into a premature baby in an intensive car unit.[178] As plasticizers bleed out of the plastic, the material loses its softness—think of that leftover bowl of spaghetti in the fridge, that seems to solidify into a single bowl-shaped blob of starch overnight. When this happens, the aging plastic becomes brittle and is more likely to snap, crack, or break.

If left by the side of the road, the oxygen masks would slowly leach their plasticizers into the environment. At the same time, ultraviolet light would break the plastic molecules, our spaghetti, into smaller and smaller pieces. Both trends would make the plastic more brittle. The UV light would also cause the mask and tubing to lose transparency and turn an opaque whitish colour. Over time, the brittle plastic would break into smaller and smaller pieces, but the lack of living creatures capable of metabolizing organochloride polymers means that it would not be reincorporated into the soil. If the mask were washed into the ditch by a rainstorm, the mask might find its way into a nearby river, even all the way to the ocean. Resembling a jellyfish or other appealing snack for some predators, it or pieces of it may be gobbled up by fish or marine birds. The fish or marine birds, of course, would be completely unable to digest it, and if they ate enough of these materials their stomachs would eventually become clogged and engorged with indigestible plastics until they starved to death.

I just read an article saying that a juvenile sperm whale washed up on the shore in Northern California. The whale starved to death with a belly full of plastics. Far from leading to a call to ban plastics or to clean up the oceans, the discovery of this whale led to all sorts of other excitement. For example, someone learned that the teeth are valuable to some indigenous cultures, and so rushed to the beach to cut them out for souvenirs herself. This person was caught when she brought the teeth to the junior high so her daughter could use them for show and tell. A teacher called the cops, who confiscated the teeth: evidently destroying the oceans with plastic is not a crime, but making souvenirs of the teeth of creatures starved—which of course is insensitive and reprehensible—because of this plastic is. Further, there have been calls for the elementary schools to hold "name the whale" contests. Not exactly the sort of excitement that inspires us to believe this culture will any time soon change its attitudes and stop killing the planet.

My gloves aren't much better for the environment than are the mask and tubing. For a long time, latex gloves were the default hand protection. Biological latex, coming as it does from rubber trees, is at least biodegradable under the right circumstances. But largely because of growing concerns about potentially lethal allergies to latex, our gloves are made from the more puncture-resistant nitrile rubber, also called acrylonitrile butadiene rubber. Acrylonitrile butadiene is a polymer made of alternating units of acrylonitrile and butadiene (a simple molecular unit made up of carbon and hydrogen).

The acrylonitrile is a greater cause for concern. Even if you've never heard of it, you've probably encountered it many times in different polymers. Acrylonitrile is used to make the acrylic fibers in synthetic "wool" sweaters and other clothing, as well as being an ingredient in plexiglas.

Designated a "toxin" and probable human carcinogen, acrylonitrile was historically mixed with carbon tetrachloride and used as a pesticide. In Ontario, where I work as a paramedic, the government has identified acrylonitrile as one of only a dozen "designated substances"—dangerous materials including arsenic, asbestos, lead, and mercury—that require special regulation and safety protection for workers.

Fortunately for me and my patients, acrylonitrile is bound up in my gloves in a way that (we all hope) eliminates the risk for us. It's not part of the sauce, it's part of the spaghetti. It's all part of the polymer, so it shouldn't be able to leach out like plasticizers into my hands, or my patients' skin. That's the good news. The bad news is that it's still a plastic. It's unknown exactly how long nitrile rubber takes to break down, but we can expect that it decomposes in a similar way to other plastics, by slowly breaking into smaller and smaller pieces.

In the long run, the blood-soaked cotton dressings I threw away are probably less dangerous than the synthetic gloves I was wearing at the time. It's ironic that to deal with frightening dangers like infectious waste, the medical system has employed methods that create even longer-lasting hazards.

TOXIC GIFTS

EMBALM, v.i. To cheat vegetation by locking up the gases upon which it feeds. By embalming their dead and thereby deranging the natural balance between animal and vegetable life, the Egyptians made their once fertile and populous country barren and incapable of supporting more than a meagre crew. The modern metallic burial casket is a step in the same direction, and many a dead man who ought now to be ornamenting his neighbor's lawn as a tree, or enriching his table as a bunch of radishes, is doomed to a long inutility. We shall get him after awhile if we are spared, but in the meantime the violet and rose are languishing for a nibble at his glutoeus maximus. —AMBROSE BIERCE

OF COURSE, THESE HAZARDS ARE by-products. The goal of the health industry isn't to create enduring hazards. That isn't true for the death industry. I'm talking about embalming.

Early embalming was most famously performed by ancient Egyptians when they mummified royalty with the aim of benefitting the soul in the afterlife. For all of its intricate ritual complexity, ancient Egyptian mummification was chemically simple by modern embalming standards. Mummified bodies were preserved essentially by rapid drying—the desiccation made the mummy an inhospitable environment for microbial agents of decay. Sometimes the process was hastened by immersing the bodies in natron, a naturally occurring mixture of salts found in saline lake beds. (Natron, from the Arabic *natrun*, has given us the modern symbol for the element sodium, Na.) Ancient Egyptian mummification was ecologically benign (unlike their large burial monuments).

Modern embalming—and this is true as well for styrofoam, saran wrap, depleted uranium, and other contemporary hazardous wastes—was invented through a union of science and war. Although embalming enjoyed some limited popularity in Europe during the Crusades to send home bodies of slain Crusaders, it wasn't until the American Civil War that embalming became something more than a marginal practice. The large number of soldiers killed in action, often dying far from home, drove morticians to find new ways to send bodies back to families for burial. At the time, there were no practical means of refrigeration to cool bodies. Enter Dr. Thomas Holmes, the "father of modern embalming."

A New York surgeon by trade, Dr. Holmes was familiar with medical preservatives and embalming methods that were, at the time, used mostly for anatomy specimens and medical cadavers. Dr. Holmes was concerned that the arsenic-, mercury-, and zinc-based preservatives of the era were hazardous to the health of medical students performing dissections. He had good reason to be concerned, of course. Such preservatives were essentially broad spectrum biocides, designed to kill living beings, with microbes of decay as their intended targets. Nonetheless, when war broke out and the Union Army engaged Dr. Holmes to deploy battlefield embalming stations, he used an arsenic-based fluid on the bodies. Poisonous embalming compounds, after all, are effective *because* they're poisonous.

Dr. Holmes's innovation was, in his particular technique, called arterial embalming. He took the bodies of soldiers, drained their blood, and pumped an arsenic-based fluid into the blood vessels so arsenic would permeate the entire body. Nineteenth-century embalming fluid recipes varied, and were often trade secrets, but some patented fluids required injecting as much as twelve pounds of arsenic into a single body.[179] After the war, Dr. Holmes returned to civilian life as an embalmer, and, curiously, ordered that his body not be embalmed after his own death.

Arterial embalming has become a common technique, along with other means of introducing embalming fluid into a body, such as by injecting fluid into the abdominal cavity, hollow organs, and underneath the skin. However, the fluid now used is different. Arsenic-based embalming fluids were banned in the early 1900s, because—as Dr. Holmes recognized—they were hazardous to embalming practitioners. Arsenic contamination of cemetery grounds was not a concern at the time, even though—assuming the lowest expected dosages of arsenic—a small-town cemetery likely accumulated hundreds of pounds of arsenic during the roughly three decades that arsenic use was commonplace.[180] If we assume higher dosages, the groundwater under those cemeteries could be contaminated with several tons of deliberately buried arsenic.

Currently, embalming fluid is a mixture of formaldehyde, methanol, ethanol, and other solvents.[181] Other ingredients include dyes, perfumes, anticoagulants, and disinfectants. Close to three gallons are pumped into the body as the blood is drained out.[182] At the same time, the body is washed and disinfected. Cosmetics are applied, and the body is groomed to give it a more "natural" appearance.

It's estimated that in the US alone, about 5.3 million gallons of embalming fluid are buried every year, enough to fill eight Olympic-sized swimming pools.[183] That's not good. Although less overtly toxic than arsenic, formaldehyde is a recognized toxin and carcinogen. Its use in Europe has been banned. Embalming aside, formaldehyde is a common indoor air pollutant. Formaldehyde resins are used in many construction materials, which, once installed, can slowly outgas formaldehyde. Methanol, a common antifreeze ingredient, is also toxic.[184] Some embalming fluids even contain chloroform, a recognized organochloride carcinogen, which has a very long half-life in groundwater.[185] Other hazards include lead chromate, toluene, methylene chloride, trichloroethylene, hexane, glutaraldehyde, and phenol, a compound used for lethal injections in Nazi concentration camps.[186] It's the same problem that Dr. Holmes had—effective embalming agents are generally harmful to living creatures.

Of course, embalming fluids are only one part of burial waste. In North America, bodies are commonly placed in wooden caskets or coffins, and then these are placed in metal or concrete "burial vaults" underground. These accoutrements require resources to be extracted, too. It's estimated that every year in the United States, more than one hundred thousand tons of steel, 30 million board feet of temperate and tropical hardwoods, and 1.6 million tons of reinforced concrete are buried in cemeteries.[187]

As common misconceptions to the contrary, measures like embalming and burial vaults don't actually stop the body from decaying. The body stays preserved for the funeral, yes, and the *way* that the body decays changes. But it still decays. A body in a sealed buried vault has no access to air, so it can't compost aerobically. Instead it putrefies, gradually changing into a semi-liquid residue doped with toxic preserving agents.

At pretty much every funeral and burial I've been to, a priest or preacher has stood over the casket and intoned, without irony, "ashes to ashes, dust to dust," as though people in the modern world commonly recognize and appreciate the cycle of human life and death. I guess "ashes to formaldehyde, dust to toxic sludge" doesn't have the same ring.

Over the past few days, as I've been preparing to write this section, a story has been running through my head over and over. It's something a friend told me as a child. I don't know if it's based on any factual truth, but it has come back to me decades later. A hardened and violent criminal, my friend told me, was once captured and put in prison. In prison, he fought with other prisoners and made no friends. But one day while out in the exercise yard, the prisoner found a baby bird who had fallen from the nest. He snuck the bird back into his cell, fed it, and cared for it until it could fly. The bird stayed with him, often perching on his shoulder and keeping him company. But one day the prison warden found out about the bird. Prisoners weren't allowed to have pets, so the warden ordered the prisoner to set the bird free. When the warden sent the guards to take the bird away from the prisoner, the prisoner grabbed the bird from his shoulder and crushed it to death. If he couldn't have the bird, he declared, no one could.

The prisoner's attitude is echoed in the attitude of Westerners toward their own bodies after death. We all borrow our body's nutrients from the living world, and eventually we all die and must return them. We eat the bodies of other creatures to live, and the very least we can do is return our own bodies with a modicum of grace when we can no longer use them. But in the dominant culture, this is not done. We can't stop ourselves from being eaten, eventually. But like sore losers, like the prisoner, in an approach falling somewhere between extreme selfishness and psychopathy—and of course this approach describes this entire culture—we poison those who would help return our bodies and nutrients to living cycles: if we can't have our bodies, no one can.

Not all cultures haves such an attitude toward death and burial. Traditional Muslim, Jewish, and Ba'hai beliefs forbid embalming. And one of the most interesting, even beautiful, burial practices is the traditional Tibetan practice of *jhator*, also called sky burial. In Tibetan, the name literally means "giving alms to the birds." In *jhator*, the body is ritually cut into small pieces and placed on a mountaintop. The bones may be smashed into small, edible pieces, and barley flour, tea, and yak butter may be added to the mix. The mix is then exposed to the elements and eaten by animals, especially birds of prey.

The act, as the name suggests, is explicitly considered a gift. The majority of Tibetans are Buddhists who believe in reincarnation. "When the body dies, the spirit leaves, so there is no need to keep the body," explains one monk. "The birds, they think they are just eating. Actually they are removing the body and completing part of life's cycle."[188]

Some Western observers and occupying Chinese bureaucrats have called *jhator* "barbaric," "brutal," "primitive," even "insane." But what is really insane? To participate in the community of life? To gift your body as a gift to the land? Or to poison that body in an attempt to keep something that's no longer even useful to you?

/ / /

There is at least some good news. Embalming is much less common outside of North America. And even inside North America, there's a significant movement for green or "natural" burial that deliberately minimizes the ecological impact.[189] Cremation is often promoted as a greener alternative, and certainly it produces less groundwater pollution. But it also produces air pollution and uses the energy equivalent of forty-seven gallons of gasoline.[190]

I wish I could say that green burials solve the toxicity problem. But even without being pumped full of formaldehyde, our bodies are no longer the precious gifts to scavengers they once were. Oh, we still get eaten, but now the meal is laced with poisons. A lifetime of exposure to low levels of hazardous chemicals and pollutants means that a person's body has accumulated a significant load of persistent organic pollutants and heavy metals.

We can generally divide chemicals in the body into two groups, water soluble and fat soluble. Water-soluble compounds are less likely to accumulate, because they leave the body whenever we excrete water, especially when we pee. Not so for fat-soluble chemicals. Entering our bodies when we're exposed to them in the environment, or when we eat foods containing fat, fat-soluble toxins accumulate in our bodily tissues and have few routes of escape. We don't normally excrete large amounts of fat, with one notable exception: breast milk.

In her book *Silent Snow*, Marla Cone relates a story about an epidemiologist named Dr. Eric Dewailly who tested breast milk from indigenous women living in the Arctic, thinking that they might be useful as control samples for other tests, samples without detectable levels of pollution. But after sending the samples to the lab, Dewailly received a phone call. There was a problem with the breast milk. The samples—every single one of them—must have been accidentally contaminated in transit. They were swarming with polychlorinated biphenyl (PCBs), with chemical concentrations literally off the charts. But the lab did more testing, and confirmed the results. Far from being pollution free, the Arctic women actually had far higher PCB levels in their breast milk than women living in cities much further south.[191]

That was in 1987. Subsequent research has confirmed that the women in that test weren't exceptional. In fact, research on indigenous women across the Arctic in Canada, Greenland, and even Siberia, have breast milk high in not just PCBs, but various other persistent organic pollutants as well as heavy metals like lead and mercury.[192]

It may seem paradoxical that women living so far from industry would have this problem. But despite being far from large sources of industrial pollution, the Arctic is the destination of many pollutants. Persistent organic pollutants (POPs) are often volatile—they evaporate easily and can be transported by wind or ocean currents. Upon reaching the Arctic, POPs accumulate for a variety of reasons. In cold weather, POPs that evaporated in warmer climates may settle out and collect. Slow growing cold-weather plants may have a long time to accumulate toxins before being eaten by herbivores such as caribou, who will then pass on a larger load to predators, a process known as biomagnification. Diet is also an important factor. Many traditional food animals of northern indigenous peoples often have large amounts of fat or blubber to protect them from the cold. Many of our most important vitamins are fat soluble, making energy-dense fat an especially valuable food in colder climates. In fact, it's common for Inuit women to *increase* their consumption of traditional foods when pregnant or breastfeeding. But when POPs come into the picture, a valuable and nourishing food also becomes a carrier of biomagnified poisons.

The problem is so bad, writes Marla Cone, that the fat and breast milk of indigenous northern women could technically be considered hazardous waste.

Of course, it's not just people in northern climates who have pollutants in their own bodies. It's people everywhere, of every age, all over the world.

There's a term for that accumulated pollutant contamination: it's called the chemical "body burden." Chemicals that build up in our bodies and stay over the long term are the "persistent body burden." Different chemicals have different rates of elimination. Arsenic, despite being overtly toxic, is mostly eliminated from the body in only a few days. Other chemicals are more insidious. Organochloride compounds, like dioxin and many pesticides, may take decades to filter their way out. And that, of course, is assuming you're no longer exposed to them. The PCBs I've consumed in my lifetime are gradually leaving my body in minute amounts, but I'm taking in more of them every day. Even though I organically grow most of my own food, I can't stop tiny amounts of PCBs from accumulating on grassy pastures, being eaten by cows, and making their way into the glass of raw milk I'm drinking at this moment. Don't get me wrong: the benefits I'll get from nutritious milk far outweigh the risk from the PCBs in this particular glass. The problem is that it's not even something we should have to evaluate.

Further, the chemical body burden doesn't start accumulating only when we start drinking milk: it starts almost as soon as we're conceived. In 2004, the Washington-based Environmental Working Group spearheaded a study of umbilical cord blood from newborn babies.[193] What they found was disturbing: although the placenta can protect new born babies from certain infections and harmful substances in the mother's bloodstream, it cannot stop persistent organic pollutants and heavy metals. Instead, these flow through the umbilical cord to the growing fetus. The blood samples revealed "pesticides, consumer product ingredients, and wastes from burning coal, gasoline, and garbage."[194] The tests detected 287 different chemicals, including 180 different carcinogens, 217 neurotoxins, and 208 chemicals known to cause birth defects or impair normal development.

Studies of adults have also shown a wide variety of industrial chemicals. In fact, the number of chemicals you can find seems to have more to do with how many you can afford to test for, rather than how many are actually present, since even testing for a few chemicals in one individual can cost thousands of dollars. It's safe to say, however, that something like 700 chemical contaminants can be found in the body of the average person.[195] These chemicals would be the more common and persistent types. Industry makes use of something like 80,000 different known chemicals, but most of these are used in such small quantities that they're not likely to be detectable in most humans.[196] Of course, that shouldn't necessarily make us sigh in relief. Only a small number of those chemicals have been tested for human health impacts. An even more miniscule number have been tested for their impacts on nonhuman species and biomes. Even for those that have been studied, we rarely know enough to predict long-term effects with any accuracy. And furthermore, virtually no testing has been done on interactions between these tens of thousands of chemicals. What new compounds are formed when they're free to react with each other in the environment? We don't know, and almost no study is being done on the matter.

When I die, I want to give my body back to the land. For all that the land (and my body) has given me, that's the least I can do. I won't be embalmed, or entombed in a metal and steel vault. I'll do my best to return my body as fully and gracefully as possible. But despite all my efforts, my flesh will contain a laundry list of industrial toxins, some of which were banned before I was even born. And there's a good chance that by the time I die my flesh will contain carcinogenic compounds that haven't even been *invented* yet.

And all that makes me very, very angry.

BODIES

In our own beginnings, we are formed out of the body's interior landscape. For a short while, our mothers' bodies are the boundaries and personal geography which are all that we know of the world. . . . Once we no longer live beneath our mother's heart, it is the earth with which we form the same dependent relationship, relying . . . on its cycles and elements, helpless without its protective embrace. —LOUISE ERDRICH

THE MOST INTIMATE, FUNDAMENTAL GIFT we can and must and do give each other is our bodies. Our bodies are the most ancient, most vital of all gifts.

This gift is far older than language, far older and deeper than words. It is as old as life itself, if not older. It is older than birth, perhaps older than death, as old as metaphor, possibly even as old as dreams.

This giving, and taking, of bodies, of flesh on flesh, flesh on stone, stone on root, root in stone, stomach on stone, skin, soil, is within and before all life, if there is a before all life. It is older than sex, older than that joining and unjoining then joining to become a third: different, new. It is older than this creativity. It is older than the bee, the pollen, the nectar, the pistil. It is older in this way than the wind. It is older than trees, older than ferns, older than algae. It is older than animals, older than mushrooms, older even than mycelia.

This gift is old. Old, like waves falling onto rocks, and rocks falling into waves. It is as old as mud, old as clouds, or even older. It is old, like water, like air, like everything.

Bodies shared, through sex, through touching, through breathing, through absorbing, eating, being eaten, becoming one or becoming another: all this is the gift of life, in all physical truth.

Bodies sustain us. They are us. They support us. We consume them. They become us, are us, as we become them, now, and later.

This gift, this support, this becoming, is where we come from, and where, whether we wish or not, we go.

✶ ✶ ✶

There is no waste in nature. One person's shit is another person's food. And one person's body is another person's food. We give, and we take, and we give. This is how life has always been. This is life.

✶ ✶ ✶

Reciprocity is the key to survival. Reciprocity is the essence of life. It is life. It is what we *do*. It is what we *all* do.

We are told, more or less incessantly, that survival is based on being the meanest, strongest, most selfish, best able to exploit.

But those who say that are wrong. They have forgotten—or do not care to remember—that nature loves a community.[197] This is true on every scale, from the largest to the most personal. It's simply true that nature loves a community more than nature loves you or me. Nature loves a community more than nature loves a community-destroying culture. Nature loves a community more than nature loves industrial civilization.

I'm sorry to report that it is not true that all of evolution has taken place to bring you or me into being. It is not true that all of evolution has taken place so that humans will exist. It is not true that all of evolution has taken place so that for a short time a relatively few (fiscally) rich humans can look at computers, watch televisions, and buy (and throw away) cell phones. It is not true that all of evolution has taken place so that humans can create industrial civilization. It is not true that all of evolution has taken place so that industrial civilization can deform humans to fit the needs of industrial civilization. It is not true that all of evolution has taken place so that humans can destroy life on this planet in the service of industrial civilization.

I'm sorry to have to be the one to deliver that news.

This culture is extraordinary, but not so much for the reasons so many people like to pretend: its high technology; its gadgets; its relentless expansion; its vast military capabilities; its art, literature, music, science, philosophy (such as it is). Instead, it is extraordinary in that it does not give back to the land, the water, the air, the nonhumans, the vast majority of humans.

This culture is even more extraordinary in that many of its members seem to think they can continue to not give back, and survive.

Or maybe they *can* survive, all the while keeping their jetskis and RVs, their gold and brass rings, their interstate highways and disposable diapers, their aircraft carriers and superdomes. They can keep this culture until they die.

Personally.

Only if they are already very old.

The planet is collapsing. Now. This culture is causing this collapse.

I'll say it again, since not enough people seem to be listening: this culture is killing the planet.

This culture is killing the planet.

Don't listen to me. Listen to the planet.

But a refusal to listen is part of the problem. We're taught (some explicitly, all implicitly) to become masterful at refusing to listen, and then to become just as masterful at refusing to acknowledge—to others, but most especially to ourselves—that not only are we not listening but that there is even anything to hear in the first place.

Many members of this culture—evidently an overwhelming majority, given the relatively small number of people actually doing *anything* to stop the destruction—simply don't care. Many of them were taught (once again, some explicitly, all implicitly) that if they ignore (in fact, foreclose) all possibility of relationship, and if they don't mind harming those around them (in fact making life impossible for many of those around them, and if they don't mind pretending not to notice the harm they cause (and then pretending they're not pretending not to notice, and then pretending they're not pretending they're not pretending not to notice, and so on *ad omnicidium*), and if it doesn't bother them that they're destroying the land and air and water that those who come after them will need to survive, then they can take advantage of the short-term competitive advantage that not giving back gives them, and thus they can more effectively dominate,

enslave, exploit, or simply kill all those who *do* give back (and who therefore must be inferior), and who have the misfortune to come into contact
with them. Then when they reach to every part of the earth—meaning that
more or less *everyone* has the misfortune of coming into contact with
them[198]—as this narcissistic nonreciprocal culture now does, they will kill
the planet that (or rather, who) supports them. But many of these individual narcissists will die before then. And so they can say, and mean, that
statement most famously said by King Louis XV's mistress Madame de
Pompadour: when the king's ministers complained that her extravagance
(and the extraordinarily expensive wars her advice helped cause) was going
to lead to their own destruction, she laughed them off with the phrase,
"*Après nous le deluge,*" literally translated as, "After us, the deluge," more
loosely translated as, "When we are dead the deluge may come for aught
I care."[199] And she was right. She died in 1764, some fifteen years before
the deluge of the French Revolution and nineteen years before the Reign
of Terror, with its deluge of blood—both royal and otherwise—flowing
from the guillotine.

/ / /

Après nous le deluge.

I guarantee this statement will be far more accurate and deadly for those
who say it now than for those who said it before. This time, as the entire
world collapses, it is not just the French but everyone who pays. Everyone.

PART II

MORALITY

WE ALL HAVE TO PAY OUR WAY. But so many members of this culture don't pay their way. They may or may not understand—and in many ways it doesn't matter whether or not they understand—that this debt *must* and *will* be repaid, but they don't give a shit—literally—who pays, so long as it isn't them.

I've been thinking a lot lately about what seems to be a fundamental guiding principle behind many—I don't think I'd be off-base to say the overwhelming majority—of this culture's moral, ethical, legal, economic, political, military, technological, and sexual decision-making processes, which is: if you're not caught, it doesn't count. (Or, if you're rich or otherwise politically powerful, you can get caught, but so long as you're not fined more than your marginal profits, it doesn't count.) In other words, the morality is almost entirely external.

Many of the indigenous have commented on the contrast between this culture's "morality," and the morality of human cultures. I think of Black Hawk, who said, "The white men despise the Indians, and drive them from their homes. But the Indians are not deceitful. The white men speak bad of the Indian, and look at him spitefully. But the Indian does not tell lies; Indians do not steal. An Indian who is as bad as the white men could not live in our nation; he would be put to death, and eat up by the wolves. The white men are bad schoolmasters; they carry false looks, and deal in false actions; they smile in the face of the poor Indian to cheat him; they shake them by the hand to gain their confidence, to make them drunk, to deceive them, to ruin our wives. We told them to let us alone, and keep away from us; but they followed on, and beset our paths, and they coiled among us, like the snake. They poisoned us by their touch. . . . The white men do not scalp the head; but they do worse—they poison the heart; it is not pure with them."[200]

The Powhatan-Renape-Lenape man Jack Forbes brings the issue of morality to the present. He writes, "The life of Native American peoples revolves around the concept of the sacredness, beauty, power, and related-ness of all forms of existence. In short, the 'ethics' or moral values of Native people are part and parcel of their cosmology or total world view. Most Native languages have no word for 'religion' and it may be true that a word for religion is never needed until a people no longer have 'religion.' As Ohiyesa (Charles Eastman) said, 'Every act of his [the Indian's] life is, in a very real sense, a religious act.'"

Forbes continues, "'Religion,' is, in reality, 'living.' Our 'religion' is not what we profess, or what we say, or what we proclaim; our 'religion' is what we do, what we desire, what we seek, what we dream about, what we fan-tasize, what we think—all of these things—twenty-four hours a day. One's religion, then is one's life, not merely the ideal life but the life as it is actu-ally lived. . . . Religion is not prayer, it is not a church, it is not 'theistic,' is it not 'atheistic,' it has little to do with what white people call 'religion.' It is our every act. If we tromp on a bug, that is our religion. If we experi-ment on living animals, that is our religion; if we cheat at cards, that is our religion; if we dream of being famous, that is our religion; if we gossip maliciously, that is our religion; if we are rude and aggressive, that is our religion. All that we do, and are, is our religion."[201]

Forbes is describing an internalized morality that guides one's behavior. One does a certain thing because it is right and good and moral, and one does not do some other thing because it is wrong or bad or immoral. These actions (or inactions) are not guided by a fear of punishment, or a concern for what others might do if they found out, or by calculations of precisely how much one can get away with.

I contrast that with this culture, where people will so often do whatever they can—and I mean whatever they can—to increase their power or the amount of money they have. I think of the words of Red Cloud: "They made us many promises, more than I can remember. But they only kept but one. They promised to take our land and they took it."[202] I think of the actions of businesspeople and their hirelings, who lie and cheat and steal to make a buck, and whom we *expect* to lie and cheat and steal to make a buck. I think of the actions of politicians, who lie and cheat and steal to support their corporate sponsors (so those who run these corporations can make a buck), and whose corruption is so brazen and ubiquitous that most of us have ceased to regard their corruption as anything other than what it is: an utterly normalized state of affairs.

I think about a developer operating close to my home who my neighborhood has been fighting for the last two years.

I've been hesitating to write about the developer because I hate him so much. I've wondered if I'm too close to that situation to write about him effectively.

While of course I do not believe in the desirability, utility, necessity, or even possibility of being "objective," I also know that, at least in my own case, there often exists an optimum level of anger, joy, outrage, compassion, and so on, as well as an optimum closeness to or distance from a subject, that generally leads to my best writing, that is, my most precise, lyrical, compelling, persuasive, vital, honest, moving, emotionally accurate, real writing.

If I don't feel strongly enough about some subject, or if I have too much distance, the writing is often flat, sometimes dead, nearly always forced, and always difficult and painful to write. If, on the other hand, I'm too pissed off about the subject, or too giddy about it, or too much in love with it, or too sorrowful about it, and most especially if I haven't had time to metabolize these emotions and perhaps more importantly the events that led to them—if I haven't had time to take them into my body and to allow my body to determine which parts of these issues, these emotions, these experiences it wants to make part of its very cells, and which parts it wants to let pass through to be excreted and left behind—then the writing sometimes ends up not communicating quite as well as I would like.

In order for my writing to be the most effective, you have to be able to trust that when I write that I hate this culture, I really do hate this culture; that I am not projecting unexamined, unconsumed, unmetabolized fears. For you as a reader to be able to trust that, I as a writer must be able to trust it first. And for me to trust it, I must have examined it, examined it again, thrown out everything I thought I knew and even everything I knew I knew, and examined it again, then again, and again.

In this case you can trust it: I do hate this culture.

Which brings us back to the developer, and to my hesitation to write about him. I hate him. No, I *really* hate him. I hate him enough that it's got me wondering if I can write about him without simply ranting, much as I want to fill page after page with expletives (and you know they're some serious expletives if even *I* don't just say the swear words but demurely pass over them) and with fantasies of him and his equally destructive buddies meeting spectacularly symbolic ends.[203]

But I *have* to write about them because their actions so perfectly exemplify so many processes by which the planet is being dismembered, and I think I've been able to metabolize my hatred and anger toward these people sufficiently that I'm pretty sure I'll be able to describe these pricks without letting my anger get the best of me, and without using unnecessary obscenities to describe these motherfuckering assholes. I think I can do that, don't you?

And besides, I'm not really writing about them to write about them. I'm writing about them to illustrate a point.

And that's the point.

The developer's corporate name is Red Cloud. I'd never have the guts to make up such an ironic name. The owner's name is Dale Smith. Smith hired a forester (shouldn't they really be called deforesters, since that's what they do?), Jim Sawyer, and a biologist, Ed Schultz.

Together, these three manifest precisely what I wrote near the beginning of *A Language Older Than Words*: "In order to maintain our way of living, we must tell lies to each other, and especially to ourselves. It is not necessary that the lies be particularly believable, but merely that they be erected as barriers to truth. These barriers to truth are necessary because without them many deplorable acts would become impossibilities. Truth must at all costs be avoided. When we do allow self-evident truths to percolate past our defenses and into our consciousness, they are treated like so many handgrenades rolling across the dance floor of an improbably macabre party. We try to stay out of harm's way, afraid they will go off, shatter our delusions, and leave us exposed to what we have done to the world and to ourselves, exposed as the hollow people we have become. And so we avoid these truths, these self-evident truths, and continue the dance of world destruction."

Even though I've written many books about the dysfunction and destructiveness of this culture, and even though as a longtime activist I've heard more than my share of lies told to buttress and expand people's fortunes (or in the case of bureaucrats, who use those lies to keep bureaucracies functioning smoothly, in other words, to keep their jobs), part of me is still, each time, surprised at the brazenness with which these lies are so often told, and the readiness with which they are so often believed, or at least "believed" well enough for the lies to do their job.

The developer and his two hirelings have lied throughout this process with a stunning consistency. They have lied about things that matter, and they have lied about things that do not. They have lied about things that are verifiably inaccurate, and they have lied in ways that are more nebulous or harder to verify. They have lied with the eagerness of fresh faced amateurs, and they have lied with the dogged consistency of pros. They have lied with straight faces that remind me not so much of conmen or even attorneys as of sociopaths. Their proclivity makes me think they are wasting their talent in this small town, and really should take their lies to the next level, to a national stage.

Frankly, anywhere but here.

In other words, they lie about as much as a good portion of the members of this culture, about as much as the overwhelming majority of the ruling class, about as much as many members of a culture with only an external morality, with no functioning internal morality.

In many ways the story starts with three slender salamanders who live on land Dale Smith wants to "develop"—in other words, destroy. I have seen these salamanders. Two are males and one is a female. She is gravid. I have seen the eggs through the skin of her belly.

Slender salamanders are common here, and I see them often. They're tiny—two inches long—and they look like worms with stumpy legs you can barely see. They live under decaying logs and in duff, eating creatures who decompose the forest—who turn the dead into usable food for soil, then for plants, then for herbivores, predators, and then once again for decomposers (and then for salamanders, who in turn will be eaten by predators or decomposers, who will all in turn be eaten by the soil).

Or maybe it starts with coho salmon, who have been here far longer than time—not so long as trees, not so long as redwoods, cedar, cascara, alder, spruce, willows—but long enough, and who for as long as they have been here have given their bodies, given that most ancient gift, to trees, to soil, to those others who were here before them. The salmon have also given their bodies to those who came after, given that most ancient gift to their own children for generation after generation, for literally uncounted and uncountable generations. The salmon who today are only barely hanging on, trying to survive this wretched and exploitative culture, trying to survive until this culture collapses so they can once again simply live, as they did before. Coho salmon, whose babies I often see in Elk Creek, black marks on their sides, wide eyes, and pale pink anal fins.

Or maybe it starts with Pacific or river lampreys, who, too, have been here far longer than time, and who, too, give their bodies to the forest, to the soil, to the land; lampreys who, too, stay here in their infancy, then swim to the sea before coming back years later to make their way upstream and begin a new generation. Pacific or river lampreys, whose babies I see in Elk Creek, tiny black worms of fish who live beneath the bed in the sand and soil, receiving the gifts of the bodies of others, then growing so that one day they can pass on the gift of their own bodies to yet others, to the descendants of those whose bodies earlier they ate.

Or maybe it starts with sand, with sediment from logging begun not long after the arrival of this culture, by people probably not unlike Smith, not unlike Sawyer, not unlike Schultz, who cut the ancient redwoods, redwoods 2000 years old and 300 feet high. These people did not cut the trees as part of that ancient exchange of gifts where body would help body, one now and another later. They cut them for money. The people scalped the forests here in the same manner and for essentially the same reasons as many of these same sorts of people (and probably many of these same people) who have scalped their way through the forests of this continent, scalped their way through the nonhuman inhabitants of this continent, scalped their way through the human inhabitants of this continent. Because of the scalpers, water that once fell as rain and ran into twisting roots of trees to be carried up trunks and into limbs and leaves and needles to be transpired to rain down again and start this cycle anew (for these gifts must and always do go in cycles, in circles and spirals and large and small loops of literally unthinkable complexity, cycles and circles and spirals and loops which must never be broken, which are never broken, except by these people like Smith, like Sawyer, like Schultz, like these other scalpers, like so many members of this culture, who are unweaving the entire tapestry of life by the simple act of taking and not giving back), now lands on bare soil and runs off into streams, carrying with it the soil the forest had built—through bodies, through litter, through shit—for thousands and tens of thousands of years. Soil runs from these scalped zones and clogs streams, fills spaces between stones, spaces where small fish swim, spaces where insects and worms and mollusks live, spaces where the eggs of coho breathe. This soil covers these spaces, kills the spaces, kills the small fish and insects and worms and mollusks and eggs, kills the streams.

Or maybe the story starts with a small dam on Elk Creek. I found this dam one day years ago when I was walking the stream. I've walked Elk Creek many times. I know this stretch better than any other living human, which is sad, because I do not know it all that well. I know it well enough to have seen signs of aplodontia, the oldest living rodent, and to have seen baby coho and lamprey, to have seen the signs of bobcats, mountain lions, and bears. I know it well enough to have fallen in and gotten soaked many times, to have absorbed the water into my body, and for the stream to have washed my skin, to have carried away with it what it would. I know it well enough to have seen the dark tannin tea of one branch of the creek mix with the clear water of another, and to have tasted both, to have taken both into my body, as one day I hope the creek will take my body into it. But I do not know Elk Creek as well as I know the films of Alfred Hitchcock, the songs of UFO or Beethoven, the tricks of *Half-Life 2*, the books of Dostoyevsky or Stephen King, the peccadilloes of celebrities.

One day walking the stream I came across the remains of an old wooden dam, now reduced to a couple of rotting wood posts and a few splintered crossbeams. I later learned it had been used to move logs. The people would (will, actually, because it is still done in some places) dam a stream and create a pond, fill this pond with logs, blow the dam, and let the flash flood carry the logs (and sediment, and any creature unfortunate enough to be in the way of this commercial activity) down to a mill (did you ever wonder why so many sawmills are next to streams or rivers?). This of course scours and then suffocates the stream or river, killing everyone and everything in its path, but when profits are to be had, the health of the stream (or the river, or the world, or anyone or anything standing in the path of profits) be damned.

Or maybe the story starts with another kind of dam. A few hundred yards downstream from my home a berm rises about three feet above the bank's normal height. If you climb this berm, then climb a little up a dead tree, to maybe twice your height, so you've risen above the huckleberries, salal, salmonberries, and sedges, you can see an area, perhaps thirty by sixty feet, swimming with huge skunk cabbages. They are growing, as skunk cabbages do, in muck. In the dry season it is remotely possible to push through this, if you don't mind exhausting yourself and getting more or less nowhere. In the wet season, you may as well stay home. At the far side of the muck, flowing into it, is a stream.

This stream—unnamed by us—is supposed to flow into Elk Creek. It used to do exactly that. It doesn't do that anymore, because of the berm. But the berm shouldn't be there. The berm is a result of logging: logging debris packed tight with logging-induced sediment during logging-induced flooding. Someday I want to take out this berm, and let the stream channel its way through the muck and skunk cabbages, let the stream not be forced to go underground to once again merge with Elk Creek. I want to open up the stream so the coho and lamprey can find their way back up that tributary, so coho, lamprey, and stream can be together again as they were before the logging. The tributary flows perhaps a mile. That would be another mile of habitat for these imperiled fish.[204]

Or maybe the story starts with the unnamed stream itself, which is not a trickle, but is about six feet wide and eight inches deep during the dry season. This stream flows through the land that Dale Smith wants to destroy, but because the stream's mere existence would stand in the way of Smith's profits, Smith, Sawyer, and Schultz all simply declared that it does not exist.

But I'm getting ahead of the story.

I think the story begins, as so many horrible stories do, with asphalt.

My neighbors to the west live on a private dead-end road. The road is narrow and gently winds between tall trees. Until two years ago it was unpaved. As often happens to unpaved roads in wet climates, their road, to use an odd and oxymoronic phrase, filled with potholes. The neighbors got together and decided to pave. But before they did, they wanted to make a hundred percent sure of something they already knew, which is that the property beyond the dead end did not have legal access down their private road. The last thing they wanted was a developer showing up the day they finished paving and suddenly claiming he had access. They contacted a title company, which reassured them that they were correct: the property beyond the end of the road did not in fact have access. So they paved their road.

I'm sure you can predict what happened next. The day—the very day— they completed the paving, a developer showed up saying he had just bought the property beyond the end of the road, and, you guessed it, claimed he had access. One of the neighbors told him to get off of her driveway, and when he refused she called the sheriff. The developer returned a few days later with a note from the title company telling the neighbor that the title company was sorry, but it had made a mistake and the truth is that the developer *did* have access.

I think we can make a stab at the real story: the title company didn't simply make a mistake, but rather the developer talked his buddies at the title company into interpreting the title his way. When the neighbors tried to talk the title company back into seeing it the way everyone (including the previous landowners and the county tax assessor) had seen it before, the title company was, to no one's particular surprise, intransigent. As is so often true within this culture, there is one set of rules and interpretations for those on the inside—in this case the developer—and another for the rest of us.

I went into the real estate office where the property had been listed to see if I could ferret out more information about the developer and his intentions. A small, elderly agent—I hate to say this but he reminded me of Gollum—greeted me, took my questions. When I mentioned the land, he started to grin—leer would actually be more accurate—and to shake. His demeanor reminded me of nothing so much as an addict looking at a fix, or maybe a pedophile looking at a child. He was actually rubbing his hands together like a bad actor playing a character from a Charles Dickens novel. I could tell he was getting off as he told me how wonderful it would be to divide up this land and put in luxury or investment homes, and how much money the developer stood to make.

I left the real estate office shaking, but for an entirely different reason. I was thinking about legacies, and what we leave behind. This man was definitely on the downward slope of his life. What will he and people like this developer leave behind? Wrecked forests and ravaged streams, wild communities destroyed. And of course money.

What sort of legacy is that?

I poked around and discovered that the developer has a history of avoiding environmental and planning laws and regulations, and basically doing whatever he can to maximize his profits, with essentially no concern for (especially nonhuman but also certainly human) communities. In other words, he is a typical developer. He put in one subdivision without bothering to get the necessary septic permitting, and those who bought homes there found themselves in a nightmare of sewage and contaminated water supplies. Elsewhere he built a house in a basin, and sold it during the dry season: it presumably would have been much harder to sell during the wet season, when, well, what do you think happens when you combine a house, a basin, recently denuded coastal temperate rainforest, and five inches of rain in a couple of hours?

But hey, he made money off the deal, so it must be okay, right? That's the only variable that seems to matter in this equation.

Smith hired his friends Sawyer and Schultz. Sawyer's job was to come up with a plan to deforest, and Schultz's job was to use his certification as a biologist to lend credibility to the lie that whatever Smith and Sawyer did wouldn't harm the land.

Many people who know and care about forests hate foresters, for the reason listed before: they really are *de*foresters, serving large or small timber interests, perceiving trees as dollars on the stump, and causing great harm to forests. There are probably no more than a couple of dozen foresters in the world who, if faced with juries consisting of trees and other forest-dependent plants, animals, fungi, and bacteria (or those who care about them), wouldn't cut them down faster than you can say, "This action will cause no significant harm to the forest." Despite extremely rare exceptions like Orville Camp and others who really do practice restoration forestry, I'm guessing forests and those who love them would probably choose to clearcut the whole bunch. Given all this, it's saying a lot that Jim Sawyer is quite possibly the most hated, corrupt, deceitful, unscrupulous (de)forester in California. His contempt for anything that stands in the way of deforestation (and making money)—living beings, laws, decency, and the desires or needs of those (humans and nonhumans) in the community he is about to destroy—is extreme enough that the state Board of Forestry[205] suspended his license, citing that he had "committed acts of deceit, misrepresentation, material misstatement of fact, incompetence, and/or gross negligence in his practice as a Registered Professional Forester. . . . [He failed] to flag a Class II watercourse despite his certification to the contrary on the submitted exemption form, and [he failed] to discover, disclose, and protect the Great Blue Heron Rookery." Also cited was his failure to adequately address watercourses, wet areas on three Timber Harvesting Plans, and his failure to completely disclose all past projects related to cumulative impacts analysis. Even this language isn't strong enough, however. Sawyer knew great blue heron rookery was on the land; he had been told it was there and he had seen it was there (and you can't miss a heron rookery because of the sight and fishy smell of heron poop), and he knew it was illegal to cut it down, and he did so anyway. When law enforcement officials confronted him about his actions, he, of course, denied any knowledge of the rookery.

Once again, I'm sure you can see why, if he were to face a jury of his victims, there would be a lot of trees willing to volunteer their sturdy branches for the rope.

He's also the head of the local Republican Party, if that means anything.

Ed Schultz is just as bad. He is what many call a "biostitute": a biologist who lies to serve the financial interests of those who pay him. *Biostitute* is not a term I generally use, in part because I don't think it's apt, and in part because it demeans prostitutes. This means I had to come up with a new word. First I tried *biopimp* since that's pretty accurate to what they do. *Bio* means life, and they turn life—the sacred—into commodities and force individuals—nonhumans, in this case—into servitude for money. They pimp life. So *biopimp* works, but it's not quite specific enough. They also pimp their knowledge, the *ology* part of biology, so maybe they should be called *pimpologist*. But that wouldn't be specific enough, either, because pimpologists would include geologists who pimp the natural world to mining corporations and climatologists who work for oil companies. Then the answer came to me: *biopimpologist*. It covers both of those bases, and as a bonus it sounds silly, which is what this whole bloody business would be if it weren't so deadly.

I mentioned Schultz's name to another biologist, who responded, "I can't understand how he sleeps at night. He causes so much damage. Why would he become a biologist if he didn't love animals? And if he loves animals, how can he hurt them so much with his lies?" Silence, then, "I also can't understand how he gets away with it. I've talked to so many biologists who hate this guy's guts. We all know he's lying, and he even sometimes gets called on it, but the projects he supports still always seem to go through: no matter how many times we show his statements to be lies, he somehow never loses credibility with decision makers." I asked a local official about Schultz, and he responded a bit more crudely: "That motherfucker lies every time he opens his mouth. And he gets himself appointed to every possible committee, where he does incalculable harm. If there were any way I could get that liar off those committees or at least lessen his harm, I would do it."

Now, pretend for a moment you are Ed Schultz. I hope that is not too painful, and I hope that you have access to a shower afterwards, to clean off the slime. Pretend at one point you did care about animals, about plants, about the wild. That's why you went into biology. Pretend that when you finished your degree and entered "the real world"—the "world" that is more important to most people in this culture than the *real* real world— you found that your best chance to make some decent money was to work for a resource-extraction corporation, doing surveys for wildlife in areas where this corporation was going to log or mine. Pretend that you soon found yourself being subtly and not-so-subtly rewarded for not finding plants or animals who would impede resource-extraction, and subtly and not-so-subtly penalized for finding them. It might be a shared look between you and an older biologist when you see an endangered salamander, a look that somehow lets you know that neither of you are to report this sighting. It might be that you see who gets promoted and who does not. It might be that you see who gets "let go" and who does not. You find yourself in a social setting where resource-extraction is rewarded, and the failure to extract resources is not. So it's no wonder that you do what is rewarded.

Sometimes you think about the love you used to have for wild creatures. You still love them, of course, and would do whatever is appropriate to protect them, but more and more you're growing to realize that environmental regulations are far too restrictive, and that those damn selfish environmentalists have already locked up too much wilderness and that *something* has to be left for the men and women who work for these corporations: they've got to make a living; they've got to support their families. More and more you realize that our entire way of life is based on resource extraction, and would collapse without it; as the bumper sticker says, "If it's not grown, it has to be mined." How do those fucking environmentalists expect for people to live? More and more you grow disgusted at the stupidity of those so-called environmentalists who wouldn't know a redwood from a cascara. Don't they realize that trees grow like weeds? Hell, don't they realize that trees *are* weeds? What's the difference between a redwood and a dandelion, except that you can sell a redwood for a hell of a lot more money. And don't they realize that if left to itself, a forest will just grow overmature and decadent? Don't they realize that a managed forest is a healthy forest? And more and more you come to understand the wisdom of survival of the fittest: if some creature can't survive a little logging here and there, then there really is something wrong with that creature. It's sad, but true. You cannot stop progress. Besides, extinction is natural: all creatures eventually go extinct. So those environmentalists who *say* they want to protect these creatures really do hate nature: they hate extinction, which is natural, which means they *must* hate nature. Sometimes you can't believe how stupid some people are.

Pretend that is your life.

Now, pretend that you decide to follow the American dream. You want to run your own business. You want to help your community. You see a need for a biologist whom people can hire to survey properties before putting in subdivisions, or before logging, or before putting in big-box stores. You start your own business. It doesn't take you long to realize that you're far more likely to receive referrals for more work when you tell the landowner what the landowner wants to hear. The landowner wants for there to be no wetlands on his property? Fine, there are no wetlands. The landowner wants for there to be no endangered species? Fine, there are no endangered species.

Sometimes you look back on your younger self, and you're amazed and somewhat embarrassed at how naïve you were.

Besides, your son needs a new car, and your daughter is getting married this fall. And you have that second mortgage you have to pay off. Oh, and there's the big screen high-definition television, and the hot tub, and that trip to Cancun next January.

Pretend that is your life.

Or maybe we've got it all wrong. Maybe you—as Schultz—never cared about wildlife at all. Oh, sure, you convinced yourself you did, and you actually believed you did (and still believe you do) but your love of wildlife is the same as the love of those foresters who say they love the forest as they destroy it, the same as the love of those pornographers who say they love women as they exploit them, the same as the love of so many in this culture who do not know how to love, who do not know the difference between love and exploitation, and who have been able to talk themselves into believing that a desire to control and exploit *is what love is*. So pretend you can tell yourself that you love the wild as you systematically destroy it.

Or maybe we've still got it all wrong. Maybe we're making this way more complicated than it really is. Maybe you—as Schultz—never even pretended to love wildlife. Maybe you chose biology as a career path, nothing more and nothing less. You like walking around outside, and math was never your strong suit, so biology seemed a better idea than accounting. And maybe the reason you lie about the presence of wetlands or endangered species is not because you're jaded, not because you had your love worn away, and not even because you believe the false sense of love that has been handed to you by this culture, but rather because you don't give a shit about anything but money. Maybe you have no functioning internal morality. Maybe you're just greedy, and that's really all there is to it.

Or maybe it's all even simpler than this. Maybe, and this is something that those of us who care about life on this planet *must* learn, and soon: Schultz's motivations and history don't matter nearly so much as his actions. His motivations ultimately don't matter any more than, say, Albert Speer's, Adolf Eichmann's, J. Robert Oppenheimer's, or those of any other technician of atrocities.

So let's talk about Schultz's actions. Pretend your business is growing quickly enough that you're able to hire assistants. In fact you have no choice. One day a new fellow moves to town. You're swamped—Ha! This type of wetlands you like!—and so you hire him to do spotted owl surveys for a medium-size timber company that wants to log some overmature red-wood (what those environmental obstructionists call "old growth"). This new guy's credentials look good, so you send him out by himself. He comes back a day later, tells you he called the owls, and got several calls back. He's pretty excited: his first job in this town and he found an iconic endangered species. You say, "Are you sure?"

"Positive."

"Sometimes barred owls can mimic other owls."

"I know the difference. I've done lots of surveys elsewhere."

"And you found spotted owls in these surveys?"

"Absolutely."

You think, *Why didn't you tell me this before, you little prick?* You say, "Could you please go and check again. My understanding is that there are no spotted owls there."

"I heard them."

"I *said*, 'My understanding is that there are no spotted owls there.' Go check again. Do we understand each other?"

Pretend the new hire repeats the survey. Pretend he comes back two days later. Pretend he tells you that his survey was correct, that there are in fact spotted owls out there. Pretend this will jeopardize the entire timber harvest. Pretend you cannot afford this right now. Not right now.

What happens next?

Pretend you've faced this situation before, and you know what works. Pretend that some people aren't smart enough to read your subtle instruc-tions, and for them you have to be a little more obvious. Pretend you get a little angry. Pretend you get in this newbie's face. Pretend you tell him he must have heard something else. Pretend he doesn't back down. Pretend you have to protect your financial interests. Pretend you fire him. Pretend you file a report saying that there are no spotted owls present.

Pretend further that you convince the United States Fish and Wildlife Service—which used to at least do a mediocre job of protecting endangered species, but now is more or less entirely captured, and so didn't really take that much convincing—that since you've searched so extensively for northern spotted owls and never found any here, no more local surveys need to be done.

Pretend this makes you proud.

Pretend this makes you money.

Pretend this is what you do.

Pretend this is who you are.

Pretend this is your life.

It should come as no surprise that when Smith sent Sawyer and Schultz to this property he wanted to "develop," they found no streams, they found no wetlands, they found no "wet spots." They found nothing that would impede Smith's planned "development."

Smith went to the county planning commission to get a permit to put in roads, clear homesites, and conduct perc tests for septic tanks. To conduct a perc test, he would dig a fairly large hole, then put water in it and see how long it takes for this water to percolate into the soil. He also said he wanted to widen Dundas Road. When one of the neighbors complained that his new, improved road would run through her living room, down the middle of her sofa, he responded, "I can't help it if your house is on my right-of-way."

We in the neighborhood turned out *en masse* to this planning commission meeting. We provided photos of the nonexistent stream, the nonexistent skunk cabbages, and other nonexistent indicator plant species of nonexistent wetlands.

Faced with indisputable photographic evidence of their lies, Smith, Sawyer, and Schultz did what nearly all other developers, (de)foresters, and biopimpologists would do: they kept lying. I was there, and in the time since have played the scene again and again in my mind, but I still don't understand not only the brazenness—*What are you going to believe, me or your own eyes?*—and the utter lack of both pride and shame, but even the sentences themselves. I mean, I certainly understand each of the words individually: *There. Is. No. Stream. There.* I even understand them if you put them in a sentence: *There is no stream there.* But to have that sentence said in response to being confronted with a photograph of the stream makes my brain hurt. The least they could have done is say, "Oh, shit. I screwed up." Or even, "You caught me!" But they didn't.

Of course.

That was not the only incomprehensible lie. They also said that they weren't going to put in a development (which would have carried with it requirements that were at least nominally more stringent). Instead they were merely going to put in roads and do perc tests. Oh, and they would clear house sites. And then they would put in houses. But that doesn't mean they're putting in a development.

Even at this remove, all of that spinning makes me dizzy.

I guess such statements from members of a culture that is killing the planet that is their—our—only home, and consider making money more important than the life of the planet, shouldn't surprise me as much as it does. But each time, it surprises, appalls, horrifies, saddens, and infuriates me anew.

Faced with photographic proof of the developer's lies, the planning commission voted 3–2 to deny his permits.

And yes, that means that for two members of the planning commission, physical reality, honesty, legality, communal desire, communal well-being, the natural world, and common sense were all less important—individually and collectively—than guaranteeing this man a profit. The dissenting members of the commission said as much: *He bought the land, which means he can do what he wants with it. And besides, he's not actually putting in a development: he's merely putting in roads, perc tests, house sites, and. . . .*

Thank goodness that for us to prevail, only three of the five commissioners needed to be capable of anything even remotely resembling cognition.

Smith appealed to the board of supervisors. This time he said that while, yes, the photographs do show a stream, that this stream only sort of kind of cuts across the tiniest bit of one corner of the property. He must have thought that story stood a better chance of convincing the supervisors than the truth, which is that the stream runs directly through the property. Of course expediency is always in this rubric infinitely more important than physical reality. Further, his argument went, the photos don't do the stream justice: the stream is small, so small. Since Smith couldn't argue that our photographs of skunk cabbages were misleading—skunk cabbages are huge—he instead attacked the "so-called biologist" we had hired as not having nearly as strong a reputation as Schultz (reputation for what, Smith did not say). Smith's allies also said that the land was so degraded that it couldn't possibly degrade it more if he put in roads, conducted perc tests, cleared house sites, put in houses. Besides, he wasn't actually putting in a development. . . .

He lost. Or so we thought at the time. Evidently so did Ed Schultz, because on the way out of the supervisors' chambers, he approached me, made his hand into the shape of a pistol, and verbally threatened me. Later that day he called another biologist we'd cited and yelled at him, threatening his job.

Having twice lost, the developer did the standard next move in the developers' playbook: he found a way to circumvent the law. He got a buddy in the planning department to declare that the property was laced with pre-existing roads, and that Smith didn't actually need to *put in* roads, which the planning commission and the board of supervisors had both explicitly disallowed. Rather, Smith and his buddy declared that any work he would do would merely be *maintaining existing roads*. And of course them declaring it makes it true, does it not? Never mind that these "roads" were old long-abandoned logging roads (one with a huge old-growth log lying smack across it), skid trails, illegal off-road vehicle (ORV) trails, and even completely overgrown game trails. And as long he was in there maintaining these roads, it would be a waste—almost a sin—not to dig some teeny, tiny holes for perc tests. His planning department friend—an employee, by the way, and not an elected official—agreed that this was only reasonable.

I asked one of the supervisors why they were allowing this employee to unilaterally overturn their ruling, something I later learned he does routinely, and, as I learned as well, with impunity. I also asked, "What's the point of requiring permits from the planning commission or board of supervisors if somebody in planning can just wave his hand and make this activity 'legal'?"

The supervisor responded that she wished she could stop him, but didn't know how. Evidently she and the other supervisors had never heard the sentences, "Your ass is fired for insubordination, for assuming power you don't have, and for violating state laws and county regulations. You have fifteen minutes to clean out your office. We'll be watching you. Don't steal any pens on the way out." I spoke with supervisors from two other counties who said they would have fired him the moment they'd heard of his actions.

But I'm not sure that's true, because similar stuff happens all the time, with no meaningful repercussions to the perpetrators.

And that's really why I'm telling this story. The point is not and has never been Smith, Sawyer, and Schultz, no matter how sleazy they may be. Unless you are one of those slender salamanders I mentioned, or one of their neighbors, these specific events—and the lies which led to them—will probably never affect you personally. But if we change the details while maintaining the central theme—that of people who shamelessly lie to increase their fiscal wealth or power at great cost to communities, and who get away with it time after time—this story could just as easily not be about those three but instead about George W. Bush, Dick Cheney, and their sock puppet general of the day, whether it's Powell or Petraeus. Or it could be about former Secretary of State George Schultz, who said with a straight face that the most recent invasion of Iraq had nothing to do with oil. Or it could be about chemical and mining corporations and the toxification of the total environment. Or it could be about corporations, governments, and global warming. Or it could be about Big Timber, the Forest Service (or Bureau of Land Management), and the Fish and Wildlife Service conspiring to deforest. Or it could be British Columbia extirpating spotted owls by some biopimpologizing—if I may continue with this word fabrication—that would make Ed Schultz proud.[206] Or it could be the mayor of Fortuna, California, former president, director, and manager of the corrupt timber corporation (are there other types?) Pacific Lumber, who since "leaving" Pacific Lumber and entering "public service" in this former timber town ravaged by PL, has received more than $800,000 from PL, with one requirement of his receipt of this money being that he never acted in any way that "materially and adversely affects the best interests of Pacific or any other member of the Affiliated Group or any person with whom Pacific or any such member has a substantial economic relationship."[207] Or it could be about the developer in Los Angeles whose biopimpologist told the planning commission there that, miraculously, endangered California red-legged frogs were just above where the developer wanted to put in 2000 houses, and just below where he wanted to put in 2000 houses, but I'll be jiggered 'cuz there ain't no frogs right where he wanted to put in 2000 houses. A member of the planning commission asked, "How stupid do you think we are?" The answer to this question ultimately became clear when the developer put in 2000 houses, frogs, laws, planning commission, and everything else be damned.

I sent what I've written so far to several friends. One wrote back, "I'm sorry this whole developer bullshit is so terrible, and I'm sorry I didn't comment to you sooner. It's probably because it hits me on such a personal level. Everything I've read sounds painfully familiar because my old neighborhood underwent the same processes: I often wonder if developers take a class called 'Destroying Communities, Destroying Lives, 101.' [Yes, by the way, they do. It's called high school and college. It's called advertising. It's called the entire reward system of this culture, where nature- and community-destroying activities are rewarded, and opposing those activities is demonized. It's called the legal system. It's called corporations. It's called regulatory agencies. It's called the government. It's called this entire culture. It is both a miracle and a testament to the resiliency of our natural heritage as living beings bound to landbases that any of us at all end up as anything other than as wretched, greedy, dishonest, and destructive as these developers.] We eventually settled and they have yet to do even one thing we agreed upon. They sold the contract to a new developer who won't even acknowledge the agreements. Then they had the audacity to name the road to the development Conservation Drive. I don't know one person (myself included) who has dealt with any sort of moral or sane developer. They are insatiable and I guess we've got to match that insatiability with an intensity of our own."

We have to match the insatiability of all those whose legacy is the murder of the planet with an intensity of our own. And there are a lot of them. An entire culture's worth. And they lie. And then they lie. And then they lie. And they keep pushing and pushing, until you get tired and say, "Okay, take it. Just leave me in peace." But they never leave you in peace because they want everything, and they will keep lying, and they will keep pushing, until there is nothing left. And then if you stop them and then stop them again and then stop them again, they will keep pushing and lying until they find some tiny opening they can push through to take everything they want. They want everything. And they will take it. Like Red Cloud said, "They made us many promises, more than I can remember. But they only kept but one. They promised to take our land and they took it."[208] The same is true today in the neighborhood I share with slender salamanders, coho salmon, lampreys, redwoods, cascara, golden chinquapin. The same is true in your neighborhood. At this point the same is true everywhere. It's what this culture is about.

It's what makes this culture proud.

It's what makes this culture money.

It's what this culture does.

It's what this culture is.

It's why this culture kills life.

Sawyer, the deforester, filed a Timber Harvest Plan. Of course he lied. All throughout the THP he lied. He said studies were performed that were not, and he lied about the results of the studies that were performed. He lied about Smith's intentions, saying that Smith was no longer going to put in a development (because, as was true earlier at the planning commission, THPs leading to development have more stringent requirements) while Smith and his allies were busy telling seemingly everyone except the California Department of Forestry (CDF) that his plans had never changed. Sawyer lied about us, saying, for example, that we had spelled out threats to him with spruce cones and sticks. We were never quite sure whom he was accusing: perhaps it was the retired environmental health specialist and his wife out walking their dachshunds, or the fundamentalist Christian who doesn't even swear, or her retired prison guard husband. He also, amusingly enough, said we were a "contentions [sic] group," and repeated the attacks on our biologist, saying, "During the [public hearings] process the neighbors complained about everything bring in there [sic] own 'experts.'" Yes, you read that correctly, and no, literacy is not a requirement to be head of the local Republican Party, and yes, I forgot to mention that Sawyer got himself installed as head of the local Grand Jury, and no, literacy is not a requirement for that position, either.

Yes, we are all deeply screwed.

At first CDF rejected his THP. Sawyer told CDF that the average age of the trees is about seventy years. He said this not because it's true, but because CDF won't allow deforesters to clearcut trees who average less than sixty years. CDF found that the average was forty-eight years. Even the most incompetent deforester should be able to tell that difference, which means that even Sawyer should have been able to tell that difference, which means that Sawyer took a shot at getting to clearcut, and it didn't work out for him. No harm done (to him), though, because CDF imposes no penalties for lying (or incompetence). CDF just tells the deforester how to revise. The deforester then resubmits the THP, and if the deforester is smart enough to follow CDF's lead, the THP gets accepted. Sawyer was evidently not quite smart enough to get it in one try, but given enough chances even the poorest student can eventually figure out what to do. A couple of tries later CDF said the THP was good to go.

To give an idea of *how* captured an agency CDF is, the forester essentially writes CDF's final response to public concerns. Yes, you read that correctly as well. In Sawyer's case, they cleaned up his grammar and spelling, but kept and endorsed his lies.

This isn't unusual.

This is how the system works.

This is how the world is murdered.

So nice and legal. So fully permitted.

It's got to stop.

We've got to bring accountability to these murderers.

Each and every one of us.

To each and every one of them.

I've been thinking again lately about the Declaration of Independence, and about how it says, "That whenever any Form of Government becomes destructive of these ends [Life, Liberty, and the pursuit of Happiness], it is the Right of the People to alter or to abolish it. . . ." Even if we leave off Liberty and the pursuit of Happiness for right now, we only need mention that this culture is killing the planet. It is not very possible for this culture—and I'm thinking specifically of the corporations and the governments that serve them—to be more destructive of Life than it already is. If we agree that this Form of Government is destructive of Life, then what does that suggest we should do?

Throughout all of this, Sawyer spent a lot of time hanging out near some of the neighbors' homes. It is perhaps significant that he hung out near the homes of women, most especially single women or women whose husbands were at work. The women complained. He ignored—or perhaps enjoyed—their complaints. The men in the neighborhood erected a gate, locked it. Sawyer cut off this lock and added one of his own so that the people who lived behind the gate couldn't get out. Sheriffs were called. They refused to remove the lock or to cite Sawyer, but they did call him and tell him to open the gate. When he arrived, he told everyone there that this was not his lock, that he had not done this, that we had done this to make him look bad. After incredulous stares by sheriffs and neighbors alike, he opened the lock that was not his, that he had not put there. He then said he had no recollection of putting that lock there, and added that sometimes when he gets mad he loses control and does things that later he does not remember.

The women tried to get personal restraining orders against Sawyer.

In response, Sawyer made sure to let the women know—in writing—that he carries a gun.

We all got the message.

Both of the judges in our county recused themselves because they're Sawyer's buddies. One woman expressed her fears to the district attorney. The district attorney also refused to get involved, stating that he, too, is a longtime close personal friend of Sawyer's, and that Sawyer is a very fine person, a good citizen.

Of course he's a good citizen of this culture. He's precisely what we would expect.

I feel the need to point out the obvious: if I took to hanging out near Sawyer's home when he was not there, peering in his windows at his wife, and if I told her that I sometimes lose control and do things that later I do not remember, and if I sent her a letter informing her that I carry a gun, and if I went up to Schultz in the board of supervisors' chambers and formed my hand into the shape of a pistol and threatened him, I would now be in jail.

Sawyer is not in jail. Nor is Schultz.

We sued to stop the logging and development (because the THP was so bad, not because of Sawyer's stalking). We went to court, with a visiting judge. It was extremely clear that our case was better than the other side's. It was also clear that our attorney— who literally *wrote the book* on the California Forest Practice Act,[209] and who has argued forest defense cases before the California Supreme Court—was infinitely better than theirs, whose research skills can best be summed up by the fact that he didn't even bother to Google the lead plaintiff, me.[210] Their response to our lawsuit cited essentially no case law, and was riddled with inconsistencies (such as stating that they're not going to develop, and at the same time suggesting to the court that our lawsuit is nothing more than an attempt to prevent them from developing), and of course the same old falsehoods, with some new ones thrown in.

So given all this, we won in court, right?

Well, to believe so—as at one point I believed would happen—would be to manifest an unforgivable naïveté and an embarrassing ignorance of how the judicial system works. It would be to believe that the judicial system is primarily about justice, or that it's primarily about laws, or that it's primarily about making decisions based on a preponderance of evidence, or that it's primarily about debating positions, or that it's primarily about physical reality, or that it's primarily about common sense, or that it's primarily about a search for truth. It is none of these. It is primarily about maintaining the current social order. It is primarily about property. It is primarily about profit.

But you already knew that, didn't you?

Once we were in the courtroom I was reminded of what I already knew: in this system profit trumps law, trumps science, trumps logic, trumps truthfulness, trumps the natural world, and so on.

The judge's first question to our attorney was: "If we presume that he is not going to develop, how does that alter your case?"

That was the moment I knew we'd lost. Smith had pushed, and lied, and pushed, until people—first the county employee, then CDF, then the judge—were eager to believe his lies.

The judge's attitude—and I have to say that he was a very personable fellow who was not *entirely* unsympathetic to what we were saying, although he was completely ignorant of environmental law, and it did not seem to concern him in the slightest when the other side would reference nonexistent laws and our attorney would respond by citing (and producing hard copies of) precise case law showing how the other side's comments were both inaccurate and misleading—perfectly manifests this culture's attitudes toward the natural world. He looked softly toward where I sat next to our attorney, and said, "You have an interest in what happens on this land." He then turned and looked just as softly at the developer and his attorney, and said, "And we can *all* see that you *certainly* have an interest in what happens on this land." That's it, right there. We can accept that people who live next to a piece of land, who love that piece of land, who love the salamanders and salal and redwoods and huckleberries and skunk cabbage and red-legged frogs who live on that land, have an interest. But someone who wishes to destroy this land to make a buck, someone who

does not even live here, someone who has no interest in this land whatsoever except insofar as its destruction makes him money, we can all see that
he *certainly* has an interest.

There you have it.

We settled the case, and so far as settling, we did well, for now. The
developer will cut—kill, murder for money—the trees he has marked for
death on half of the property, and will not be able to develop that for a year.
The other half was also saved temporarily. If he tries to develop after that,
we can (and will) of course still fight him.

There's a sense in which we won, but even when we win these victories, the trees and salamanders and birds still lose.

And even when we win I still feel like crying.

TAKING IT PERSONALLY

The great wilds of our country, once held to be boundless and inexhaustible, are being rapidly invaded and overrun in every direction, and everything destructible in them is being destroyed. How far destruction may go it is not easy to guess. Every landscape, low and high, seems doomed to be trampled and hurried. Even the sky is not safe from scath— blurred and blackened whole summers together with the smoke of fires that devour the woods. —JOHN MUIR

IT IS LONG BEFORE DAWN, the morning after the trial, I awaken from a dream. In this dream I'm on the edge of a cliff, and think I might fall off. Then I climb away from the cliff, and see that there is a building on the top of the mountain. I go inside, where a bunch of people are about to play a poker tournament. The person who runs the apartment says it will cost twenty dollars to enter. I get out my wallet and see that I have a few hundred dollars. Everybody else sees that too. I give him twenty. I see a table with chocolates. I eat a piece, and sit down to play poker. I have a hard time figuring out their chip system: they have lots of weird chips that don't make any sense to me. They're terrible players, and are playing really slow. We play three hands. I blink a couple of times, then say, "Can we play faster? This is going to take forever."

They all look at me, and one says, "You're out of the tournament. You lost all your chips on the last hand."

"I didn't even play the last hand."

"Yes, you did. You played a terrible hand, and we took all your chips."

"I don't play terrible hands. I only play hands where I've got a good shot at winning."

"You lost."

I pick up a telephone, and call some sort of hotline. I tell them what happened. They say this place is known for that. The people who run this place had drugged my chocolate, then when I was unconscious they'd taken my chips. I look in my wallet. It's empty. Everyone in the room is looking at me intently, and I know that if I confront them they'll kill me.

I shake my head and walk out of the room.

This is how I feel about the legal system.

/ / /

It just doesn't fucking stop. We settled Friday afternoon. Saturday morning before nine Sawyer was trespassing on one neighbor's property, outside her (please note the feminine personal pronoun) door inside her fence, preparing to widen the road (explicitly excluded from the agreement). She confronted him, told him to leave. He refused. When our attorney complained in writing, Sawyer responded that he was within his rights, and strongly (and falsely) implied that the neighbor is an alcoholic.

Because part of the agreement is that they only work on weekdays, and another part is that they not widen the road, it took them less than eighteen hours to violate the settlement.

/ / /

Of course none of this is unique. This is how this culture works. This happens all over.

/ / /

I don't know how much more of this I can take.

/ / /

I don't know how much more of this the world can take.

/ / /

I'm talking to my sister about my work.

She asks how I keep from getting angry.

I tell her I don't think anger is a bad thing.

She asks how I keep anger and sorrow from consuming me.

I tell her my standard answer, which is that I feel it, and I aim it, and I don't take it personally.

This is all true.

But this particular assault is killing me, more than much bigger atrocities elsewhere, more than much bigger atrocities I've dealt with, more than when the system has destroyed equally beautiful and much larger wild places I've worked to defend. And it's because this time I'm taking it personally.

/ / /

This is all precisely what anyone who opposes this culture deals with every day.

/ / /

Maybe part of the problem is that too often we *don't* take it personally. Maybe I should have taken it personally when my grandfather died of cancer. Maybe I should have taken it personally when a developer destroyed a prairie dog village near where I grew up. Maybe I should have taken it personally when the Forest Service destroyed tens of thousands of acres of old growth I was working to protect. Maybe I should take it personally that Dick Cheney stole water from the Klamath River to give to farmers because he wanted their votes, and the Klamath River salmon suffered a monumental fish kill. Maybe I should take it personally that islands of plastic trash suffocate oceans. Maybe I should take global warming personally, not only because it will inundate this land I love but because it is killing the planet. Maybe I should take the toxification of the total environment personally.

If I did that, if I took all of this personally, how would that affect my actions?

If you took this personally, how would it affect your actions?

Maybe it would hurt more, but maybe we would do more about it.

Maybe we would fight back.

/ / /

It's a week later. I return to the land. It is well after working hours. I am trespassing. The attorney has told me that if I am caught trespassing, it will harm our case in future trials. We are both aware that the other side is allowed to violate laws with impunity. If we are caught violating laws once, our credibility is lost forever.

But right now that's not as important to me as the fact that someone needs to ascertain whether Sawyer and his crew are violating the terms of the agreement. I know CDF won't do it. They won't even return our telephone calls.

I'm not sure what I will do when I discover violations. Sawyer's lies haven't cost him before. Why should they now?

But I need to know.

So I walk in. The cliché I'm supposed to invoke is that what's left of the forest looks like a battlefield. But it doesn't, because I only see casualties from one side. It looks more like a massacre. I walk over and among the dead and dying bodies of trees. I say only two things, and I say them over and over: "Fuck," and "I'm sorry. I'm so sorry."

I walk to where I saw the slender salamanders. The area, like so many others here, like so many others all over the planet, is devastated. I cannot see how the salamanders could have been anything other than crushed.

I notice, of course, that the trees have been cut in clusters, creating clearings. I notice as well that these clearings are conveniently centered around the perc test holes. But I'm sure that must be a coincidence, since time and again Smith and his allies have told us they aren't going to put in a development. And we have no reason to disbelieve them, do we?

Of course I see THP violations left and right. To choose just one egregious yet not atypical example, they've allowed—or caused, actually, since were they not there this would not have happened—mounds of dirt to fall into the stream. The THP stated explicitly that they would not allow—cause—even a speck.

But none of this will make CDF or the courts or the county government or Fish and Wildlife stop the slaughter, for the simple reason that an essential function of CDF and the courts and the county government and Fish and Wildlife is precisely to facilitate, rationalize, and protect these destructive activities.

I walk among the downed trees. I look at their still-green needles. "Fuck," I say. "I'm so sorry."

It's getting dark. I walk home. I hate the people who did this, and I hate the system that serves them and those like them. And I am ashamed of myself and those like me for not standing strong enough, for not holding those who would destroy accountable, for not stopping them from destroying the places we love, and for not following the Declaration of Independence and abolishing these institutions that are so destructive of Life.

I am ashamed that we are not stopping this culture from killing the planet.

I am tired of saying, "I am so sorry."

To say we are sorry is not good enough.

It is time to stop these destructive activities.

MORALITY REVISITED

What is morality in any given time or place? It is what the majority then and there happen to like, and immorality is what they dislike.
—ALFRED NORTH WHITEHEAD

WHEN I SAID THAT MEMBERS of this culture don't have a functioning internal morality, I was wrong. Everyone has an internal morality. I do. You do. Smith does. Sawyer does. Martin Luther King Jr. did. Richard Nixon did. Josef Stalin did. Abraham Lincoln did. Ted Bundy did. We all do.

Earlier in this book I mentioned R. D. Laing's *The Politics of Experience.* The fundamental point of that book seems to me to be that people act according to the way they experience the world, and if we can understand their experience we can understand their behavior, no matter how non sensical it may seem to us on the outside. So within his skewed perspective, Josef Stalin's actions made sense. The same can be said for Dick Cheney; for Ted Bundy; for Herb Mullen, the serial killer who murdered people because voices told him this was the only way to prevent California from sliding into the ocean in a massive earthquake; for Richard Trenton Chase, the serial killer who blended his victims' blood and organs, then drank this mixture because he heard voices telling him this was the only way to keep his own blood from turning to powder. From the outside the actions of these people are clearly insane (and immoral), but if you're the one hearing those voices, the actions can begin to make sense to you.

We can say the same for a culture that values gross national product over the life of the planet. From the outside it is clearly insane and immoral, but from the inside there is a sort of internal consistency: it can be made to make a kind of sense.

The psychologist Lundy Bancroft also helped me understand that everyone, no matter how abusive or depraved, has an internal morality that guides actions. Bancroft works extensively with perpetrators of domestic violence. He wrote, "A critical insight seeped into me from working with my first few dozen clients. *An abuser almost never does anything that he himself considers morally unacceptable.* He may hide what he does because he thinks *other* people would disagree with it, but he feels justified inside. I can't remember a client who ever said to me: 'There's no way I can defend what I did. It was just totally wrong.' He invariably has a reason that he considers good enough. In short, *an abuser's core problem is that he has a distorted sense of right and wrong.*"[211]

So from an abuser's perspective, his wife refusing sex may be wrong, and him beating her because of this may not be wrong.

From a slaveowner's perspective, a slave refusing to work may be immoral, or wrong, and beating the slave because of this may not be wrong. If it's necessary to teach other slaves, or potential slaves, a lesson, hanging one or two from the nearest railroad trestle may in this perspective not be immoral, but an entirely appropriate response to the threat that nonworking or uppity slaves present to the current righteous social order.

From the perspective of the standard European or later American of the last several centuries, Indians refusing to give up the land they call home (but don't properly use) is immoral, wrong, and selfish. Killing these Indians and taking their land is not only right, but is divinely ordained. It is, of course, one's Manifest Destiny.

One's perceived sense of entitlement can certainly influence the perceived morality of one's actions. Let's say you're white, you're rich, and you live in South Carolina in 1857. Society tells you it's acceptable for whites to own blacks, that blacks are inferior—not really fully human—and are meant for servitude. God has ordained it such, given you dominion over them. You perceive blacks as resources. Your entire economic system is predicated on this perception, this servitude. And besides, this perception, this servitude, makes you money. You perceive—honestly perceive, as honestly as Herb Mullen perceived that voices were telling him to kill in order to prevent an earthquake—not only that God approves of what you're doing (with this approval sometimes manifested by you making lots of money) but also that slavery is moral, and that attempts at slave liberation are immoral. You do have an internal morality. The fact that others may find your actions immoral does not mean you don't have a functioning internal morality.

Now, let's say you're white, you're rich, and you live in Crescent City, California, in 2007. Society tells you that it is acceptable for humans to own land, and to own trees, and to profit from the flesh of murdered trees. Society tells you that trees are not real beings—they're not subjects, but objects—and that their value is derived from how much money you can make off of them. God has ordained it such by giving you dominion over all nonhumans. You perceive nonhumans—in this case trees—as resources. Your entire economic system is predicated on this perception, this use. And besides, this perception, this use, makes you money. You perceive—honestly perceive, as honestly as the slave owner, as honestly as Richard Trenton Chase perceived that voices were telling him to kill in order to keep his blood from turning to powder—that this perception of others as resources, this slaughter of others for money, is moral, and that attempts at stopping the economic system that is based on this perception, this slaughter, are immoral (that is, when you don't perceive them as incomprehensible and silly). You do have an internal morality. The fact that others may find your actions immoral does not mean you don't have a functioning internal morality.

The bottom line is that it's inaccurate and not helpful to say that some people have no functioning morality. Every action and inaction reveals your morality. Every action and inaction *is* your morality. Let's revisit Jack Forbes's quote, this time changing the word *religion* to *morality*: "'Morality,' is, in reality, 'living.' Our 'morality' is not what we profess, or what we say, or what we proclaim; our 'morality' is what we do, what we desire, what we seek, what we dream about, what we fantasize, what we think—all of these things—twenty-four hours a day. One's morality, then is one's life, not merely the ideal life but the life as it is actually lived." He continues, "'Morality' is not prayer, it is not a church, it is not 'theistic,' is it not 'atheistic,' it has little to do with what white people call 'morality.' It is our every act. If we tromp on a bug, that is our morality. If we experiment on living animals, that is our morality; if we cheat at cards, that is our morality; if we dream of being famous, that is our morality; if we gossip maliciously, that is our morality; if we are rude and aggressive, that is our morality. All that we do, and are, is our morality."

Profits are this culture's morality.

"Development"—read, the destruction of living natural communities and their replacement by artificial structures—is this culture's morality.

Highways, automobiles, and long-haul trucks are this culture's morality.

Oil tankers are this culture's morality.

Oil spills are this culture's morality.

Wars to gain and maintain access to resources (including oil) are this culture's morality. In other words, wars of conquest are this culture's morality.

Theft of resources is this culture's morality.

Perceiving others as resources is this culture's morality.

Colonialism is this culture's morality.

Exploitation is this culture's morality.

The conquest of nature is this culture's morality.

The destruction of the natural world is this culture's morality.

The destruction of local cultures is this culture's morality.

Eco-groovy headquarters for transnational corporations that make money from sweatshops are this culture's morality.

Slave labor is this culture's morality.

Global warming is this culture's morality.

The extinction of passenger pigeons, great auks, baiji, and so many more is this culture's morality.

The looming extinction of polar bears, penguins, great apes, spotted owls, and so many more is this culture's morality.

Great rafts of plastic choking the life out of the oceans are this culture's morality.

Yellow Boy is this culture's morality.

Toxified streams are this culture's morality.

Toxified bodies are this culture's morality.

The abrogation of our most fundamental exchange, that of body for body, is this culture's morality.

The murder of the planet is this culture's morality.

All that said, I think it's more accurate to say that each person will have two fairly distinct moralities (actually each person will have a near-infinitude of distinct moralities, but let's keep it simple for now): internal and external. My internal morality suggests things I will or won't do because

I consider them (or to me, they *are*) moral or immoral. For example, I would consider it immoral to steal from an independent bookstore. I would not consider it immoral to steal from Wal-Mart. Or to remove the morally charged language from these statements: I would consider it immoral to walk out of an independent bookstore without paying for the book, and I would not consider it immoral to walk out of WalMart without paying for the objects. I would not normally consider it immoral to throw toilet paper or other decomposable materials onto the ground. I would normally consider it immoral to dump crank case oil onto the ground.

My external morality would be things I would do because I want others to see me do them, as well as things I would not do because I would not want to get caught. To continue the above examples, I don't shoplift at independent bookstores because it would be wrong to do so. I don't shoplift at WalMart because I don't want to get caught.

More examples. I don't remove large dams not because I would find it immoral, but because I don't want to get caught. I don't sexually abuse women because I would find it immoral and repugnant. I no longer drop toilet paper on the ground because I don't want people (at least those people who've not read this book) to think I'm strange.

Smith has an internal morality, and an external morality. So does Sawyer. So did Hitler. So does George W. Bush. So did Ted Bundy. Even someone as Machiavellian and destructive (which in my mind adds up to the single word *evil*) as Dick Cheney has an internal morality: although he has caused the deaths of hundreds of thousands of human beings and countless nonhumans, I'm guessing that even he would find it immoral to shoot his own friend in the face. Oh, sorry, bad example.

You get my point.

But the problem[212] is that the internal (and for that matter external) moralities of so many in this culture, especially those who run it, are really fucked up.

A consequent problem is that so often even those who commit the most heinous acts can justify, or at least rationalize, these acts. Even those who commit the most heinous acts can feel good about these acts—or at the very least can commit them with the self-righteous sorrow of one who is simply doing what must be done, and the suffering (of others, of course, not oneself) caused by these acts is simply a price that must be paid. They can consider themselves moral, righteous, and good. And from their perspective, they *are* moral, righteous, and good.

This makes them all that much harder to stop.

LEGACY

Whatever I have said about my deeds and words in this trial, I let it stand and wish to reaffirm it. Even if I should see the fire lit, the faggots blazing, and the hangman ready to bring the burning, and even if I were in the pyre, I could not say anything different. —JOAN OF ARC

WHAT WILL BE YOUR LEGACY? What will you leave behind? Will it be regions deforested by your actions and inactions? Will it be the wreckages of forest communities you destroyed or allowed to be destroyed? Will it be lots of money? Will it be forests you did not work to save? Will it be the remnants of forests you worked to save (as I worked to save this one) but did not succeed in saving (as I did not succeed with this one)? Or will it be forests you saved? How many families of slender salaman ders will continue to live because of who you were and what you did? How many coho salmon? How many green sturgeon? How many warblers, whippoorwills, chickadees, phoebes, bobwhites? How much deeper will be the soil because of you? How many tons of greenhouse gases will have not been emitted because you lived, and because you acted? I'm not asking how few you caused to be emitted—I'm not letting you, or me, off the hook that easily. I'm asking how many you caused to be not emitted because of your actions. There is all the difference in the world.

Will your legacy be a world who is healthier, stronger, more resilient, more diverse, than had you never lived? If not, then the world would have been better off without you. If not, the world would have been better off had you never been born. This is neither an accusation nor a threat, but a simple—even tautological—statement of fact.

Further, will your legacy be a world that is healthier, stronger, more resilient, more diverse, than it was before you were born? What will it take for that to be your legacy?

If you would like for this to be your legacy—and this improvement, this strengthening of the world, is the legacy of the overwhelming majority of humans and nonhumans who were or are not members of this culture (and if you disagree, then simply ask: how did this world get to be as diverse as it was, and for now still is, unless the inhabitants of this world were making it so by their existence and actions?)—how will you achieve that? What will you do to leave behind healthy soil, healthy forests, healthy oceans? What will you do to stop those who would destroy, and what will you do to help this planet—this planet who is our only home—flourish? What will you leave behind? What will be your legacy?

/ / /

Would the world be better off had you never been born?

I don't mean the culture. I mean the world. The real, physical world. Would it be more resilient, stronger, more diverse?

Is the world better off because you were born?

I don't mean the culture. I mean the world. The real, physical world. Would it be more resilient, stronger, more diverse?

I cannot think of a deeper, more solid—more *real*—foundation for any morality than this question, put both ways. What is more real than the real world? What is deeper, more solid—more *real*—than life on earth?

For a morality to be conscionable, forgivable, livable, *necessary*, it must be based on the health of the world, because without a healthy world no one has any morality whatsoever, because there is no life whatsoever. In all physical truth.

In *Endgame* I described clean water—and more broadly the fact that I am an animal, and that without clean water, clean air, livable habitat, I will die—as the basis for a livable morality. I wrote: "If the foundation for my morality consists not of commandments from a God whose home is not primarily of this Earth and whose adherents have committed uncountable atrocities, nor of laws created by those in political power to serve those in political power, nor even the perceived wisdom—the common law—of a culture that has led us to ecological apocalypse, but if instead the foundation consists of the knowledge that I am an animal who requires habitat—including but not limited to clean water, clean air, nontoxic food—what does my consequent morality suggest about the rightness or wrongness of, say, pesticide production? If I understand that as human animals we require healthy landbases for not only physical but emotional health, how will I perceive the morality of mass extinction? How does the understanding that humans and salmon thrived here together in Tu'nes [the Tolowa name for where I live] for at least 12,000 years affect my perception of the morality of the existence of dams, deforestation, or anything else that destroys this long-term symbiosis by destroying salmon?"

I liked that moral foundation at the time, but now I see it as only a partial foundation—necessary but grossly insufficient—because while it does indirectly express concern for one's larger community (in that clean water can benefit not only that individual but others, including fish, amphibians, and so on), it is still fundamentally self-centered. It would be entirely possible for people to trash landbases, ruin rivers—ruin the planet—yet in the meantime maintain fantastically expensive water treatment factories that provide somewhat clean water to those humans who can pay for the services of water treatment (or bottled water) corporations. Of course this would only be workable in the short run—and even then only more or less workable, and even then only workable at all for those on whom the externalities are not forced—but in that short run those who own these fantastically expensive water treatment factories; those who operate them; those who consume their product; and those in the government who oversee (and in fact order, and pay for through taxpayer subsidies) the toxification and destruction of natural water supplies (also known as lakes, rivers, and aquifers) can feel as though they're performing a moral act by providing somewhat clean water to the humans who can pay for that product at the expense of the larger community.

But these questions of whether you make the world stronger, more diverse, healthier—and these questions hold for *any* relationship: if your partner isn't stronger, healthier, more resilient because of your relationship, that person would be better off without you (and conversely, if your partner doesn't help you to be stronger, more resilient, healthier, then you should, in the immortal words of Little Charlie and the Nightcats, "Dump that Chump"; further, if your culture doesn't make you (and the world) stronger, more diverse, healthier, then get rid of it, too)—point the way past self-centeredness. These questions point us toward a morality that works for the common good. The *real* common good. In other words, they point us toward a *real* morality.

Would the world—the real physical world—be a better place had you never been born?

Is the world—the real, physical world—a better place because you were born?

Those really are the most important questions.

What are your answers?

/ / /

Why do you answer as you do?

/ / /

What, if anything, is the relationship between your answers to these questions and your internal morality?

What, if anything, is the relationship between your answers to these questions and your external morality?

/ / /

Why do you answer as you do?

/ / /

Once again, is the world a better place because you were born?

I don't think that putting plants on Ford truck factories, designing eco-groovy Nike headquarters, and designing "sustainable airports" is sufficient to make the answer *yes*. Rhetoric of "sustainability" aside, I don't think these actions make the world a better place: a truck factory—native plants or no—is one of the last things we need. A transnational athletic shoe company—especially one using sweatshop labor—is another. An airport is a third.

Turning the mirror back on myself, I don't think writing books railing against this culture's destructiveness is good enough. I don't believe that the world is a better place because I was born.

Sure, there are many thousands of acres of forest still standing in part because I helped protect them through my activism (although they probably would have been saved anyway—I only helped—and in any case unless this culture is stopped it will ultimately destroy these forests anyway). And sure, right now I am helping my mom to protect (and serve) this land we both live on, this land which would surely have been cut and "developed" by someone as sleazy as Sawyer and Smith were we not protecting it. And sure, I helped catalyze the neighbors to partially stop or at least slow Smith in this case. And sure, even if the land is eventually destroyed, there will have been at least a few generations of salamanders and butterflies and frogs and slugs who will have been able to remain in their homes. And sure, people tell me that my work has inspired them in their own activism, which means a bit more of the wild saved (for now).

But living within this culture necessitates at best a complex calculus of harm and healing. How many acres have been cut to make my books (as well as the books I read)? And using recycled paper (on the occasions publishers even choose to do that) doesn't entirely nullify the damage caused by the industrial fabrication of my books (or we could say the industrial mass reproduction of my ideas). Collecting recycled paper takes energy, as does repulping it, as does transporting it, printing on it, binding it, transporting the books, and so on.

People who live simply within this culture somewhat reduce the harm side of the equation, but just because *they* live simply doesn't mean they should get too self-righteous or excited. The industrial economy is inherently destructive (it takes from the land, not only failing to give the land what it needs, but even worse, poisoning it), and if they participate in the economy at all, they cause great harm.

It doesn't help any of us to pretend that there is any participation in an industrial system that doesn't harm the planet. For it does. How much ecological harm occurs just because I eat? I don't feel *too* bad about the cow I just purchased from a local rancher: I live in a reasonably wet area, and the cow was pasture-raised, which means that the cow wasn't too unhappy, watering the cow didn't draw down the aquifer too much, and the cow contributed his shit to the soil. Of course the pasture where the cow was raised used to be a redwood forest, it *should* be a redwood forest, and it *needs* to be a redwood forest, yet the cow's grazing helps prevent the forest from returning, so once again I shouldn't feel too self-righteous. I shouldn't feel self-righteous at all about the peas and corn I had last night, both of which (whom) came from the grocery store, which means they came from factory farms, which means the plants had miserable lives and their growing[213] toxified the soil and drew down the aquifer. And it doesn't much matter whether they came from Iowa or the San Joachin Valley: in either case they're from beautiful and fecund regions devastated by agriculture.

/ / /

This is a point I've hammered in all of my books, but it's a point that needs to be repeated. Not every human culture has damaged its landbase; not every human culture has been unsustainable. In fact, most, until they were conquered, *were* sustainable: they didn't damage their landbases. Not every culture has twisted even simple, necessary, beneficial, and beautiful acts such as eating into harmful activities. To turn acts like eating or shitting into unsustainable acts is extraordinarily stupid, for reasons I hope are obvious. If at this point they still aren't obvious we're in even worse shape than it would otherwise seem. But hell, this culture *is* killing the planet, which is literally the most stupid action possible—even imaginable—so I'll be explicit: if even the most fundamental and necessary daily activities harm the landbase—we're not talking about luxuries here—there is no chance at all for sustainability under the current system.

But you knew that already, didn't you?

/ / /

Coho salmon make the world a better, richer, more diverse, stronger, healthier place by their existence and actions, by their eating and being eaten.

Red-backed voles make the world a better, richer, more diverse, stronger, healthier place by their existence and actions, by their eating and being eaten.

Northern spotted owls make the world a better, richer, more diverse, stronger, healthier place by their existence and actions, by their eating and being eaten.

Torrent salamanders make the world a better, richer, more diverse, stronger, healthier place by their existence and actions, by their eating and being eaten.

The Tolowa Indians, on whose land I now live, made the world a better, richer, more diverse, stronger, healthier place by their existence and actions, by their eating and being eaten.

I need to as well.

/ / /

I just got a note from one of the neighbors. She finally got a hold of someone at CDF. Her note read: "The CDF agent said he'd inspected Smith's site more than any other; I told him that was good as it was under so much scrutiny by the neighborhood. He said he'd cited Red Cloud for failing to file paperwork on the creek easement in a timely manner, but that was the only violation. [There was, of course, no mention of the soil in the stream. Why does that not surprise me?] He also said the road looked pretty muddy. I asked about a water tanker for cleaning the vehicles to prevent the possible transmission of Port Orford cedar root rot, and he said he wasn't aware of a need for that. He looked it up in the plan, found it, and thanked me for pointing it out. But the next thing he said is that it didn't matter, because it had been dry. It took me a moment to even understand what he was saying, because of what he'd said earlier about the road being muddy. But before I could say anything, he said that it also didn't matter because unless he actually sees a violation as it occurs he can't cite them. So even though the road is a mess, and even though we've asked him to watch the site, and even though we've witnessed them violating the law, there is nothing, he says, that he can do."

Don't you wish the law treated us all the same? There is only one CDF inspector for the entire county, and if he doesn't see a violation with his own eyes, it's treated as though it didn't happen. So let's pretend that analogously there was only one cop in all of this county, and let's pretend the rules were the same: you can't be arrested unless this lone cop happens to actually see you commit the crime. How would your actions be different? Would single mothers who were poor take their hungry children to corporate grocery stores so they could eat their fill? If they needed a little cash, would they go to corporate banks and withdraw the banks' money? After all, if the lone cop never actually saw them do it—remember, we had pictures of Smith's violations—they wouldn't have to worry about where next month's rent comes from.

But of course it doesn't work that way. Cops, like courts, like the whole judicial system, like entire governmental systems, do not have fairness as a fundamental guiding principle. We all know this. These are all set up primarily to protect the profits of the already-powerful.

If someone were to destroy Sawyer's equipment, I would be arrested—after all, didn't he already accuse us of spelling out threats?—whether or not the cop saw me do it, and even if it wasn't me who did it anyway. Sawyer, Smith, et al. violate law after law, and what happens? They make money.

That's how the system works.

"They made us many promises, more than I can remember. But they only kept but one. They promised to take our land and they took it."[214]

/ / /

Two of the three days before my neighbor talked to the CDF agent—two of the three days he said the road was dry (although somehow still muddy)—it rained. Hard.

Still he saw no violations.

/ / /

Everyone I talk to locally says this CDF employee is the most diligent they have ever encountered.

/ / /

What does this say about this culture's morality?

THE REAL WORLD

Reality is not protected or defended by laws, proclamations, ukases, cannons and armadas. Reality is that which is sprouting all the time out of death and disintegration. —HENRY MILLER

How many legs does a dog have if you call the tail a leg? Four. Calling a tail a leg doesn't make it a leg. —ABRAHAM LINCOLN

There is an objective reality out there, but we view it through the spectacles of our beliefs, attitudes, and values. —DAVID G. MYERS

Belief in God? An afterlife? I believe in rock: this apodictic rock beneath my feet. —EDWARD ABBEY

WHAT SHOULD, OR WOULD—or do—any of us do, living in this culture that is alienated from and destroying the earth, if—or when—we realize that this world would be better off had we never been born, or having been born, if we were to die?

For now, at least, I see several options that many people take.

The first option, taken by nearly everyone within this culture, is to do everything we can in increasingly frantic, desperate attempts to keep this realization at the unconscious and not conscious level. Thus jetskis and off-road vehicles, thus Disneyland, Walt Disney World, Magic Mountain, and Six Flags over Everywhere. Thus scuba diving and whitewater rafting. Thus the existence of hundreds upon hundreds of television channels, with movies and movies and movies and *Deal or No Deal* and *Dancing with the Stars* and basketball game after basketball game after football game after football game after baseball game after baseball game. More and more. Faster and faster. Thus the internet, with its ever-increasing ways—spectacular ways—to kill time. Thus *Doom 1, 2,* and *3*. Thus *Half-Life 1, Half-Life 2,* and *Half-Life* episodes *1* and *2*. Thus Second Life, MySpace, and YouTube. Thus the tidal wave of pornography, sports, and financial news, all with their simulacrum of diversity, all with titillation, all with excitement, all promising to transport us somewhere, somehow. Thus the obsessions with Britney Spears, Paris Hilton, Tom Cruise, Brad Pitt. Anyone but those in front of us. Thus the abuse of marijuana, cocaine, methamphetamines. Thus so many other addictions, like the stock market, the economy, politics. Thus the frantic-happy, frantic-smiling faces—all of them just alike—on the evening distraction—I mean, spectacle—I mean, news. Thus toys and more toys and more toys. Thus the obsession with playtime by adults who work at jobs they hate. Thus diversions to divert us from the diversions that divert us from the diversions that divert us from the myriad realizations we must never have if we are to maintain this way of living and to maintain our role in the ongoing destruction of all that is real. And beneath these myriad realizations are more diversions, and more. There is phony meaningless optimism and phony meaningless hope and phony meaningless actions like putting plants on truck factories, all keeping us from staring into the abyss of destructiveness that is right now staring straight at us. And all of these phony meaningless diversions divert us from the understanding that our failure to stare at this abyss will not stop it from swallowing us, as well as everyone and everything else. Beneath these diversions there are phony fears of despair, phony fears of hate, phony fears of rage, phony fears of sorrow, phony fears of love and loves: real loves, fierce loves of self and others that cause us to at all costs—and I mean all costs—defend our beloved. And beneath all these fears? A dreadful fear of responsibility, a fear that if we get to this point, if we survive the annihilation of the self that is so meticulously, so violently, so repetitively, so mercilessly, so relentlessly, so abusively, so obviously forced upon each of us in order to allow us to continue to breathe, to work, to labor, to produce, then we will need to take responsibility for our actions and for the wonderful and beautiful and stunningly extravagant gift of our life that this planet has given to us. Indeed, we will need to *act*, and to act in such a way that the world is better off because of our actions, because of our life, because we were born. And as with sustainability itself, what was at one point as easy as eating, shitting, living, and dying, is now more and more difficult.

We fear death. And not just the death that all experience, but another that scares us far more than the real death that comes at the end of our phony lives. This other death that we fear even more comes before the real death—sometimes long before—if it comes at all. This is the death of our socially constructed self. Once that self dies, then who will we be? We cannot face the possibility of actually living, of actually becoming who we really are and who we would be had we not been so violently deformed by this culture. We cannot face the possibility of being alive, of living, so we turn, to return to the beginning of this discussion, to jetskis and off-road vehicles, to Disneyland, Walt Disney World, Magic Mountain, and Six Flags over Everywhere. Most of us would prefer our real, physical selves die, and indeed the world die, rather than face the realization that, given our socialization, the world would be better off without all of us who allow our socially created selves to continue to breathe, to work, to labor, to produce—and that, of course, is the real point.

That is the most popular option for members of this culture.

And it is precisely the same inversion of what is real that we talked about earlier, with the interview by Andy the obnoxious radio host. Do you remember his inversion of what is real and what is not real, where dying oceans and dioxin in every mother's breast milk are not the real world, where the real world is industrial capitalism? It's easy enough to laugh at his stupidity and insanity—if insanity can be reasonably defined as being disconnected from the real, physical, material world, then I'm not sure how much more insane one can get than to suggest that such tangible physicalities as oceans and breasts are not the real world—but how hard will we laugh when we realize his stupidity and insanity are mirrored by nearly everyone within this culture, including nearly all of those environmentalists who even purport to care about the natural world?

Think about the prominent "solutions" suggested to help curb the worst of global warming. What do they have in common? I'm talking about *every* major "solution," from those proposed by Al Gore (compact fluorescents, inflating tires, reducing packaging, and so on); to James Lovelock (nuclear energy); to Newt Gingrich (giving polluters tax credits to lean them toward voluntarily reducing their carbon emissions); to the various ideas proposed and promoted by scientists, such as the idea of dumping tons of iron, or alternatively, tons of agricultural waste—how conveeeenient!—into the ocean in the hope that this will cause algae to flourish, absorbing CO_2 into the algae's bodies and, by the way, doing god knows how much damage to the already-being-murdered oceans; or that of injecting sulfur particles high into the atmosphere to reflect sunlight back into space; or a further refinement of this idea, put forward officially by the United States government, to put giant mirrors in outer space to reduce the sunlight that arrives here; or (and I can hardly believe I'm not making up these obscenely and insanely stupid ideas, but each one has come from a "respected" source and received a lot of mainstream media attention) an idea pushed by NASA scientists to move the Earth farther from the Sun.[215] I never thought I would see solutions presented that would make me pine for the relative sanity of plants on Ford truck factories.

What all of these "solutions" share—and of course the same is true for the "solutions" presented by people like William McDonough, Paul Hawken (who wrote *Natural* [sic] *Capitalism* and *The Ecology* [sic] *of Commerce*[216]), Al Gore, and nearly all of the so-called environmental intelligentsia (or "Bioneers" as some call them)—is that they all suffer the same stupid and insane reversal of what is real that Andy did. They all take industrial capitalism as a given, as that which *must* be saved, as that which must be maintained at all costs (including the murder of the planet, the murder of all that is real), as the independent variable, as primary; and they take the real, physical world—filled with real physical beings who live, die, make the world more diverse—as secondary, as a dependent variable, as something (never someone, of course) that (never who) must conform to industrial capitalism or die. Even someone as smart and dedicated as Peter Montague, who runs the indispensable *Rachel's Newsletter*, can say, about an insane plan to "solve" global warming by burying carbon underground (which of course is where it was before some genius pumped it up and burned it), "What's at stake: After trillions of tons of carbon dioxide have been buried in the deep earth, if even a tiny proportion of it leaks back out into the atmosphere, the planet could heat rapidly and civilization as we know it could be disrupted."[217] No, Peter, it's not civilization we should worry about. Disrupting civilization is a good thing for the planet, which means it's a good thing. Far more problematic than the possibility that "civilization as we know it could be disrupted" is the very real possibility that the planet (both as we know it and as we have never bothered to learn about it) could die. Another example: in a speech in which he called for "urgent action to fight global warming," and in which he called global warming "an emergency," UN Secretary-General Ban Ki-moon gave the reason he wants urgent action to combat this emergency: "We must be actively engaged in confronting the global challenge of climate change, which is a serious threat to development everywhere."[218] Never mind it being a serious threat to the planet. He's worried about "development," which is in this case code language for industrialization.

This is the same perspective of those who do not hide the fact that they are grotesquely antienvironmental. Just recently, Bjorn Lomborg, the latest in a long line of writers who are paid well to deny or understate the damage this culture causes to the natural world[219] finally acknowledged that global warming is happening, and that it is caused by industrial civilization. But his next move was mind-numbingly predictable: he immediately shifted to the fallback position of saying that nothing can (or should) be done about it, stating that it is "somewhat silly" to think that this culture can change.

And it's not silly to harm or destroy the planet you live on?

As always, it is this culture which is primary, permanent, immutable; and the real world that is secondary, and that (rather than who) must bend to this culture's will.

Trying to force sustainability onto a functionally unsustainable culture causes severe cognitive dissonance, and makes people suggest absurd solutions. No solution can be too absurd so long as it fulfills its primary purpose of keeping us from seeing that the culture can never be sustainable, and that to attempt to sustain this culture is to harm the world. The real, physical world.

Any solution that springs from the (most often entirely unconscious) belief that the culture is more important than the world (or that the culture is real and the real world exists only as a backdrop and a source of raw materials) will not solve the problem.

One more example. I posted this section of the book to a global warming listserv, and one of the activists there replied (and as with McDonough, I've put my responses to some particulars in endnotes), "If you feel civilization is the problem, then you need to have a realistic and practical solution. If your idea of a solution is to have 5 billion people[220] behave as lemmings and jump in the ocean[221] you could see why most people would respond with a 'you first.'"[222]

I find his response interesting because it reveals the precise inversion of reality we've been talking about. When I suggested that civilization needs to go, he immediately equated that with human suicide. I wasn't talking about humans being exterminated. I was talking about ending civilization. He and I don't speak the same language. The same is true for many members of this culture. We do not even agree on the answers to the most basic questions: What is real? What is primary? What are we trying to save? Hell, not only do we not agree on the answers, we don't even agree on what these questions mean. So I've started making a translation dictionary. Here are my first four entries: when I say *world* many people in this culture hear *industrial capitalism*; when I say *the end of the world* they hear *the end of industrial capitalism*; when I say *civilization* they hear *human existence*; and when I say *the end of civilization* they hear *the end of human existence*. But those are not at all the same. Worse, as we've laid out, within this culture the *world* is consistently less important than *industrial capitalism*, *the end of the world* is less to be feared than *the end of industrial capitalism*, *civilization* is more important than *human existence*, and *the end of civilization* is more to be feared than *the end of human existence*.

It's insane. Literally. I'm sorry to have to be the one to break this news, but the planet is more important than this fucking culture.

And *of course* these thinkers care more about this culture than the planet: that's how this culture has taught us to feel (or more accurately to not feel) and to think (or more accurately, to not think). This culture could not have gotten to this point of planetary crisis without inculcating most of its members into this perspective.

And this perspective is really fucking stupid.

And it's really fucking insane.

And it really doesn't work. Industrial capitalism can never be sustainable. It has always destroyed the land upon which it depends for raw materials, and it always will. Until there is no land (or water, or air) for it to exploit. Or until, and this is obviously the far better option, there is no industrial capitalism.

Industrial capitalism is a social construct. Civilization is a social construct.

It's embarrassing to have to write this, but you can't have a social construct—any social construct—without a real world.

The real world is the independent variable. Our social constructs—any social constructs—must be dependent variables. Our social constructs—any social constructs—must conform to the real world. Our social constructs must make the real world a better place, a more diverse, more resilient place, a healthier place. If they don't, they will destroy the world—the real physical world. And if we allow social constructs to destroy the real, physical world, well, then once again the world would be better off had we never been born.

We must destroy that which destroys the real physical world.

How do you stop or at least curb global warming? Easy. Stop pumping carbon dioxide, methane, and so on into the atmosphere. How do you do that? Easy. Stop burning oil, natural gas, coal, and so on. How do you do that? Easy. Stop industrial capitalism.

When most people in this culture ask, "How can we stop global warming?" that's not really what they're asking. They're asking, "How can we stop global warming, without significantly changing this lifestyle [or deathstyle, as some call it] that is causing global warming in the first place?"

The answer is that you can't.

It's a stupid, absurd, and insane question.

To ask how we can stop global warming while still allowing that which structurally, necessarily causes global warming—industrial civilization—to continue in its functioning is like asking how we can stop mass deaths at Auschwitz while allowing it to continue as a death camp. Destroying the world is what this culture *does*. It's what it has done from the beginning.

How can we stop global warming?

You know the answer to that.

Any solution that does not take into account—or rather, count as primary—polar bears, walruses, whippoorwills, bobwhites, chickadees,

salmon, and so on, and the land and air and water that support them all—
is no solution, because it doesn't count the real world as primary, and social
constructs as secondary. Any such solution is in the most real sense neither
realistic nor practical. Any solution that does not place the well-being of
nonhumans—and indeed the natural world, which is the real world, which
is the stable, healthy world (or was, before this culture began to systematically dismantle it)—at the center of its moral, practical, and "realistic"
considerations is neither moral, practical, nor realistic. Nor will it solve
global warming or any other ecological problem.

Ask yourself: what is real?

Industrial capitalism—Andy's real world, William McDonough's real
world, Paul Hawken's real world, Al Gore's real world, the real world of so
many—is killing the planet.

Do we want a living real world, or do we want a social structure that is
killing the real world? Do we want a living real world, or do we want a dead
real world, with a former social structure forgotten by everyone, because
there is no one left alive to remember?

You choose.

/ / /

Reality has a habit of intruding and revealing one's fantasies for what they
are: fantasies.

For example, I can fantasize all I want that I can fly, but if I jump off a
cliff, the real, physical world will backbreakingly intrude on my fantasy.

You could respond, "But you *can* fly. Thanks to high technology anybody who can afford an airplane ticket can fly."

There's a sense in which that would be right, and a larger sense in
which it wouldn't. The answer doesn't acknowledge the problems of high
technology, and more broadly, this culture from which high technology
emerges. First, it doesn't really fulfill these fantasies, but distorts and
makes toxic mimics of them. As Dennis Gabor wrote in *Inventing the
Future*, "Science has never quite given man what he desired, not even
applied science. Man dreamt of wings; science gave him an easy chair
which flies through the air. Man wanted to see things invisible and afar . . .
he got television and can look inside a studio."[223]

Second, these "fulfilled" fantasies inevitably serve rich and powerful humans better than they serve anyone else: in other words, the primary function of these applied technologies is inevitably to further centralize power (how does "flying" help subsistence farmers in India?).

Third, these "fulfilled" fantasies are inevitably used to cause a net transfer of money and power from individuals to big corporations. In my fantasies of flight I've never had to use a credit card to buy a ticket from a large corporation, nor have I had to remove my shoes and belt and walk through a metal detector under the eyes of an agent of the state, nor have I had to worry about being put on a "no-fly list" assembled by those in power.

Fourth, these "fulfilled" fantasies don't make our lives easier or better. Carl Jung weighed in on this one, and I quote it at length because it's so on point: "Our souls as well as our bodies are composed of individual elements which were all already present in the ranks of our ancestors. The 'newness' in the individual psyche is an endlessly varied combination of age-old components. Body and soul therefore have an intensely historical character and find no proper place in what is new, in things that have just come into being. That is to say, our ancestral components are only partly at home in such things. We are very far from having finished the middle ages, and classical antiquity, and primitivity, as our modern psyches pretend. Nevertheless, we have plunged down a cataract of progress which sweeps on into the future with ever wilder violence the farther it takes us from our roots. Once the past has been breached, it is usually annihilated, and then there is no stopping the forward motion. But it is precisely the loss of connection with the past, our uprootedness, which has given rise to the 'discontents' of civilization and to such a flurry and haste that we live more in the future and its chimerical promises of a golden age than in the present, with which our whole evolutionary background has not yet caught up. We rush impetuously into novelty, driven by a mounting sense of insufficiency, dissatisfaction, and restlessness. We no longer live on what we have, but on promises, no longer in the light of the present day, but in the darkness of the future, which, we expect, will at last bring the proper sunrise. We refuse to recognize that everything better is purchased at the price of something worse; that, for example, the hope of greater freedom is canceled out by increased enslavement to the state, not to speak of the terrible perils to which the most brilliant discoveries of science expose us. The less we understand of what our fathers and forefathers sought, the less we understand ourselves, and thus we help with all our might to rob the individual of his roots and his guiding instincts, so that he becomes a particle in the mass, ruled only by what Nietzsche called the spirit of gravity.

"Reforms by advance, that is, by new methods or gadgets, are of course impressive at first, but in the long run they are dubious and in any case dearly paid for. They by no means increase the contentment or happiness of people on the whole. Mostly, they are deceptive sweetenings of existence, like speedier communications which unpleasantly accelerate the tempo of life and leave us with less time than ever before. *Omnis festinatio ex parte diaboli est*—all haste is of the devil, as the old masters used to say.

"Reforms by retrogression, on the other hand, are as a rule less expensive and in addition more lasting, for they return to the simpler, tried and tested ways of the past and make the sparsest use of newspapers, radio, television, and all supposedly time-saving innovations."[224]

Fifth, these technological "fulfillments" of fantasies cause massive ecological harm. At this point I hope we don't have to talk too much about the harm caused by airplanes through the mining of the materials to make the planes; through the manufacture of the planes; through the infrastructures necessary to build and maintain these planes; through the extraction of oil and the refining of jet fuel, as well as the maintenance of the infrastructures necessary to extract, refine, and deliver this fuel, and of course through the harm caused by the military forces necessary to guarantee access to this fuel and its antecedents; through the effects of the emissions, which are by no means limited to global warming; through the massive use of airpower to bomb, or at the very least surveil, those who oppose the will of those in power, in other words, to increase the reach of those in power; through additional ecological effects caused by the ability of the rich to move goods (and themselves) very quickly; and so on.

Sixth, when I've fantasized about flying, I've been able to control my movement, and if I get in danger, it's because I chose to fly too fast, or through too-thick trees, or too close to the ground. But once I board an airplane, I have given over control of my life to those who build and maintain the airplane, to the pilots, to pilots of other planes, and so on. In short, with this and other technological "fulfillments" of fantasies, we give up control of our very lives to distant financial and corporate entities, and to their employees. With each new piece of high technology, we trust our lives more and more to those who own those technologies. I hope that at this late stage at least most of us can see how this applies to the larger issue of this culture killing the planet. Further, I hope that at this late stage, at least most of us can see that this is a very bad idea.

And seventh, no matter how high the technology involved, eventually the real physical world will, in one way or another, put a halt to our fantasy. Global warming is a response in part to the perverted fulfillment of this fantasy of flying.

We can perform this same exercise for so many other fantasies, and their "fulfillment" through technological means. People fantasize about leaving a mark on the world, and we end up with a swirling mass of plastic the size of Africa in the middle of the ocean. People fantasize about being able to bend the world to their will, and the world dies at their command.

The point is that in order to maintain these lies—that we are really flying, that we can exploit a landbase (or planet) and live on it, and so on—we must keep pushing away physical reality, and we must keep telling ourselves these lies again and again. The maintenance of these lies is incredibly expensive psychologically, emotionally, intellectually, physically, financially, morally, ecologically, and so on.

TAKING IT PERSONALLY,
VOLUME II

At this moment in history, we are all caught in the hell of frenetic passivity.
—R. D. LAING

NOW, BACK TO THE QUESTION, which was: What should, or would—or do—any of us do, living in this culture that is alienated from and destroying the earth, if (or when) we realize that this world would be better off had we never been born, or having been born, if we were to die?

So the first option is to distract ourselves from this realization, try to keep it unconscious.

The next option—allied to that first one—is to attack anyone who reminds us that there is the possibility of another sense of self, who reminds us that it is possible to live another way. That's one reason—in addition to wanting to steal their land—that members of this culture eradicate all indigenous cultures. It is also one reason they destroy or attempt to domesticate all wild creatures. All that is wild—uncontrollable—must be destroyed so it will never remind us that we were not always slaves.

These attacks can also be much less direct. Just this week the *Sun* magazine excerpted part of my book *Thought to Exist in the Wild: Awakening From the Nightmare of Zoos*. They received many positive letters about it, and two negative ones. Neither negative note responded to any of my content. Instead, one writer said he didn't like my tone and style of writing, and the other said it was "inconsistent" of me to write about zoo animals when I share my home with dogs. If you can't find ways to discount the message itself, attack the messenger.

When the first two options don't work—when neither distraction nor destruction suffice to keep the real world from intruding on our fantasy that we are separate from and above the natural world—the next option is to pretend that the damage isn't really damage. This is done both by those who are overtly antienvironmental, and also by many of those who otherwise seem, or at the very least pretend, to care about the natural world. An example of the former: in 1972, Chevron offered the Cofan people of Amazonian Ecuador candy and cheese for drilling rights to their land. The Cofan threw away the cheese, but consumed the candy, which led to Chevron claiming the right to drill. Chevron looted the oil and left behind massive oil spills and stinking pits of oil-field residue, all of which have contaminated rivers, aquifers, soils, plants, fungi, animals—both human and nonhuman—and so on. Human health problems include an epidemic of cancer, open pustules on those who contact the water, and bloody vomiting leading to death. The Cofan sued Chevron and its subsidiary, Texaco. A Texaco attorney responded, and his response unintentionally reveals much about the contamination of the entire planet by this culture: "And it's the only case of cancer in the world? How many cases of children with cancer do you have in the States, in Europe, in Quito? If there is somebody with cancer there, [the Cofan parents] must prove [the deaths were] caused by crude or by [the] petroleum industry. And, second, they have to prove that it is OUR crude, which is absolutely impossible." He also said, "Scientifically, nobody has proved that crude causes cancer."[225]

Well, then, I guess it's all settled.

But maybe not. If fifty different people poison you with the same poison at the same time, could we prove which person killed you? Using this attorney's logic—which of course is the logic of capitalists everywhere— if we could not prove which person's poison killed you, then all poisoners go free. Unfortunately it doesn't go the other way: if you and forty-nine others shot each of these poisoners in the head before they were able to poison you, I fear police, district attorneys, judges, and then later guards and executioners wouldn't care which of you fired the fatal shots: you'd all share responsibility, and would be executed by the state (by multiple people pushing buttons to start the lethal injection machine—multiple people so nobody knows who is the actual killer).

The illogic hurts my head. And the rest of me.

Or how's this: Salon.com has an advice column called *Since You*

Asked . . . People write in their concerns or questions, which are answered by someone named Cary Tennis. Tennis has been called "the Walt Whitman of advice columnists," who "responds to these problems at a deep level" and who "explores the big questions of life thoughtfully and maybe provides a little peace." I give this preamble so you know the guy has some cultural credibility, and is not, at least from the culture's perspective, a total wingnut. One question was: "How can I look my daughter, who is the light of my life, in the eye, smile and continue to follow through with the same things I always do, knowing the truth of the terrible world she will soon live within, which will try to crush her future?"

It's certainly a reasonable question.

The "thoughtful" response by this "Walt Whitman of advice columnists" was: "You are troubled by a vision of planetary death, which stems from fear of personal death. This fear also makes you ache for your daughter, whom you envision being left alone in a dying world. It is not a technical problem. It is a spiritual problem. You seem to find it hard to accept that. Perhaps you can accept this, though: You need a vacation."[226]

How typical, how manipulative, how nasty, how narcissistic, how useless (to the real world, but extremely useful to the processes of planetary murder): the world is being murdered, and this advice columnist denies the man's fear and empathy, and suggests a vacation.

How's this: "My daughter is being murdered. It's tearing me apart. What should I do?"

"Your daughter's murder bothers you because it reminds you of your own mortality. You need a vacation."

It's simply true that right now, living in and with a world being killed, we all have knives in our chests. The world is being murdered, and so are we. This man is saying that there is a knife in his chest, and one in his daughter's chest, too, and the response of this insane—I mean this literally—columnist is to suggest a vacation.

Those who at least purport to care about the natural world are sneakier—though no less headache-inducing—in their dismissal of the damage. Just two days ago I was talking to a group of students, and at one point I said that this culture is killing the planet (Oh, okay, you got me: I said it at *many* points). An activist about my age said, "But this culture can't kill the planet. Algae or something will be left, and then in millions of years evolution will move in another direction. The Earth won't die. It will just change."

I asked if any of the students there happened to have a knife, and if so, could I borrow it. The woman sitting next to me began to dig one out of her pack. I asked if it was sharp. She said yes, then gave it to me. I stood, opened it, walked to the activist, and asked, "Can I have your hand?"

He said no.

I said, "I'm not going to kill you. I'm just going to cut off your little finger at the base of the fingernail. Then I'll move up a knuckle. Then another. Then I'll start on another finger. Then I'll let you think about it for a while. Then I'll cut off another. And another. I'll move up your arms, and then I'll start on your feet, and move up your legs. You're not going to die. At least not for a while. You're just going to change."

He got the point.

The very next night I got an email from a thirty-something man from New York who had seen one of my talks on YouTube, and who said much the same thing. I cannot believe how many people claim this perspective. Nor can I believe how many variants there are. One variant is that the damage done to the planet right now isn't a big deal, because someday the sun will expand and kill the planet anyway. I always respond, "Someday you're going to die, so how about if I torture and kill you right now?" Usually they understand. Another variant is, "The earth is very resilient, and to think that the culture can kill the planet is to manifest the same arrogance as those who are actually killing it." Not only does this perspective ignore the very real possibility that industrial civilization-induced global warming could turn this planet into one like Venus, but more immediately it ignores the salmon, sturgeon, delta smelt, spotted owl, marbled murrelets, and so many others already being driven extinct. Further, it incorrectly, insanely, and absurdly conflates a fear and hatred of this culture's arrogance and destructiveness, with the arrogance and destructiveness itself.

Because I get so many emails expressing variants of this perspective— and because I see this perspective so often in print—I include a portion of this recent email, with a portion of my admittedly tired-of-this-bullshit response:

He wrote: "You've said that this culture will quite possibly kill the planet. How? This culture may be insane enough to kill all life on the planet or change our genetic structuring beyond all imagination, but I don't give humanity's insanity power over the planet.[227] This planet could end up like Venus or any other planet with no detectible life left on it. . . . Or we could be allowing the evolution of super bacterium to thrive. But I'm confident the planet will remain, whether in a steady state with little life on it or in a state of recovery or exhaustion with no hope for reconstituting itself remotely close to what it is now. The planet will be as good as dead with no biota but it will physically remain. Humans will suffer and perhaps even go extinct real soon but the planet is set. Various life forms have come and gone on this planet for billions of years. So I'm not sure it's about who needs 'saving' anymore."

I responded: "Okay, so someone breaks into your home, and begins to torture your son: are you going to stand back and say, 'Various life forms have come and gone for billions of years. So I'm not sure it's about who needs "saving" anymore.' No, you save your son. If we see an injustice we work to stop it. End of story. I've seen this sort of 'analysis' so many times and it's all rationalization for inaction. I'm not interested in inaction. I love the salmon. I love the sturgeon. I love the redwoods. I am fighting for them. This is my life. I am giving my life for these others. It's what we all should do.

"Let's try this again. *You're* tied up and being tortured. I have the opportunity to save you, and instead I choose to say, 'Various life forms have come and gone for billions of years. So I'm not sure it's about who needs "saving" anymore.' Your response would rightly be, 'Fuck you. You're no friend of mine.' Friends act to defend their friends."

He continued, "I think the planet has the capability to withstand much. At this point I'm concerned with what is being done to humans more than the destruction of the planet."

Of course he would be more concerned about what is being done to humans than the destruction of the planet. This value structure is central to this culture. It is a requirement of this culture. It *is* this culture. Those who wish to be fully vested members of this culture must accept, live by, and proselytize this value system. In fact for most people in this culture, humans are not only the sole beings worthy of concern, humans are the sole beings who even really *exist*.

And as we have seen, it's even worse than this, as for most people in this culture, not even humans are as worthy of concern as industrial civilization itself. As Frederick Winslow Taylor, founder of scientific management, put it, "In the past the man has been first; in the future the System must be first."[228] It's bad enough that the vast majority of people in this culture are so narcissistic they think that "the man" must be first—more important than everyone else, including woman, land, air, water, frog, tree, barnacle, life itself—but it's even worse that people are stupid enough to put "the System" above even "the man," and therefore above everyone else, including all life on the planet. At least the man who wrote this note did not do that.

And of course valuing "the System" over human life is a toxic mimic of a sustainable value system in which humans are seen as dependent upon and therefore by definition less valuable than the larger living community of which they are one member. Also, of course, an increase in perceiving that "the System" must come first is a direct and inevitable consequence of an increase in "the System" being forcefully inserted between us and these larger living communities, which leads (and this is both necessary and intentional) to an increase in forced human dependence upon "the System." In other words, those whose food comes directly from the land and whose water comes from, for example, a river, and who therefore are not dependent upon "the System" for food, clothing, shelter, and waste disposal, in other words, for their very lives, will rightly perceive the notion that "the System" should come first as insane, intrusive, and irrelevant; while those whose food comes from the grocery store, whose water comes from the tap, and who sit on a toilet which magically whirls away their shit, and who therefore are dependent upon "the System" for food, clothing, shelter, and waste disposal, in other words, for their very lives, will then be put in a position of almost undoubtedly perceiving that "the System" should come first. Never mind that "the System" is exploiting them, and never mind further that "the System" is dependent upon the larger living communities it is murdering.

He continued, "I have faith in the planet's capabilities and its ability to care for itself far better than we have been able to let it be. Why can't the Earth be listened to, really felt, as if feeling a pulse, and letting her speak."

He went on to say what this listening means to her: "Listening means just sitting there and hearing and feeling and sympathizing perhaps."

If before my response was sharp, now it became downright rude. I wrote, "NO! NO! NO! Goddamnit NO! Your son is tied up and being tortured, and he is begging and screaming for your help, and your response is going to be 'just sitting there and hearing and feeling and sympathizing perhaps'? That's really fucked up. The appropriate response is to fight like hell, to kill the motherfuckers harming your son, to untie him, and to get him the hell out of there. Listening means fighting like hell to protect those you love."

He sent another note, which reads in its entirety: "I want to ask you not to use my son in your analogies. What you said has to be cancelled from the universe so no harm comes his way. It's existentially bad for him (and it's taunting me) and I ask that you bring that back in and eliminate it. I don't want any such thoughts being associated to him (or me) in any way. Especially since I speak and envision the complete opposite for him and his testimony to date has nothing remotely close to that experience. I don't want it about him hypothetically, imagined, or otherwise, now or in the future, so I surround ourselves with better fortune, foresight, vision, imagination. Not that I'm in denial of what is out there I'm just practicing spiritual power and taking really good care."

What strikes me about this email exchange is that the man described the killing of all "biota" on the planet in terms that could be described, all things considered, as bland, blithe, blasé, academic, emotionless, and abstract, yet when I, instead of talking about the destruction of something "out there," made it about his son, *made it personal*, he strongly objected. As I read his reply to my note, I kept thinking, *this is why we must personalize it*. We must feel the murder of the planet as deeply as, if not more than, we do the murder of our families. And that's the point, isn't it? Most of us don't love the land, don't love the salmon, redwoods, prairie dogs, rivers, rocks, Joshua trees, tortoises, whippoorwills, chickadees, hammerhead sharks, and so on. Hell, for most of us these others don't exist, unless there's a way we can make money off of them. If we loved these others, we would stop this culture in a heartbeat.

We keep the real world from intruding on our fantasy that we are independent from it, that we don't need to love it, by convincing ourselves that the damage isn't really damage, either because nobody can prove that the damage we caused was really caused by us; or because the earth is someday going to die anyway and so killing it now doesn't matter; or because the earth is so strong it can deal with whatever damage industrial civilization (never called industrial civilization, by the way, but rather "humans," since "humans" and "industrial civilization" are and must be the very same) does to it; or maybe because the earth doesn't really exist but is the movement of God's eyebrows; or maybe because Jesus is going to come and take all the Christians to a better place (and if that's the case I wish He'd hurry the fuck up and get them the Hell out of here before they destroy everything that's left); and so on, with each frantic rationalization more absurd than the one before.

MAGICAL THINKING

I was thinking as small children think, as if my thoughts or wishes had the power to reverse the narrative, change the outcome. In my case this disordered thinking had been covert, noticed I think by no one else, hidden even from me, but it had been, in retrospect, both urgent and constant.

——JOAN DIDION

ANOTHER WAY TO REJECT THE REAL WORLD, practiced almost universally in this culture, is to resort to magical thinking.

Magical thinking can be defined at base as the conviction that thinking is equivalent to doing. For example, some believe that anyone can become a writer (actor/painter/athlete), if he or she just dreams it vividly enough. I could fantasize all I want about how great it would be if I were a writer, but if I don't actually write (and write and write) it ain't gonna happen. Similarly, I've heard too many environmentalists rhapsodize about singing back the salmon; we can all try to sing back the salmon, but if the dams still stand the salmon don't stand a chance. In both cases, fantasizing turns out to be grossly insufficient.

Because my experience with magical thinking has been more, er, practical than academic—I do it lots, but have never been precisely sure what it *is*—I looked up some definitions. I found that, to quote one of them, magical thinking "includes such ideas as the law of contagion, correlation equaling causation, the power of symbols and the ability of the mind to affect the physical world."

Law of contagion? That didn't help me understand. So I looked that up and found, "Objects or beings in physical contact with each other continue to interact after separation. Everyone you have ever touched has a magical link with you, though it is probably pretty weak unless the contact was

219

intense and/or prolonged or repeated frequently. Magical power is conta-
gious. Naturally, having a part of someone's body (nails, hair, spit, etc.)
gives the best contagion link."[229] Ah, so now I get it: this is why I need a bit
of someone's hair or fingernail whenever I try to put a hex on him (and
also explains the high black market value for Dick Cheney's nail clippings).

Next: correlation equaling causation. My nieces had larger vocabularies
at the age of twelve than they did at the age of one. They also had more cavities. Thus, larger vocabularies cause more cavities. Or here's another: have
you ever noticed that the more firefighters there are at a fire, the larger the
fire is and the more damage it causes? Ray Bradbury was right: the damn
firefighters sure do cause a lot of damage. And here's yet another: last week
I won two online poker tournaments in the same day, and so the next day
I kept wearing what I now presumed to be my lucky socks. I won two that
day, and then the day after, and the day after that. I'm still wearing my same
lucky socks, and I must admit they stink just a little, but don't laugh: I'm
still winning. And besides, you don't *even* want to get me started on you
and your lucky underpants.

The next part of the definition is the power of symbols to affect change
in the real, physical world. To continue the example from above, that
explains the popularity of Dick Cheney voodoo dolls. Since magical
thinking is generally considered to be delusional, you can go ahead and
get rid of that doll. On second thought, however, I don't see how it can possibly hurt to stick pins in it; or burn it; or deprive it of sleep; or force it to
hold uncomfortable positions for long periods. Give it a try. You never
know.

The last aspect of magical thinking is the ability of the mind to affect
the physical world. This is the power of positive thinking, or visualization.
I know for a fact that visualization doesn't work. Do you know how I know?
Because Dick Cheney's head hasn't exploded into a million pieces. If a billion human people and countless nonhuman people wishing him dead
isn't going to do it, then wishing just doesn't work.

Engaging in magical thinking is not the same as being delusional: the
former is a subset of the latter. All magical thinking is delusional, but not
all delusions are magical thinking. So, to think that the government serves
the interests of living beings over the interests of corporations is delusional, but it is not magical thinking. To write a letter to Dick Cheney

politely requesting he do something to stop global warming—or even that he do something that benefits humans over corporations—and to expect this to accomplish something in the real, physical world, is magical thinking.

Perhaps an example from my own life will help. My sole hickey in junior high or high school was the result of magical thinking. Knowing that my romantic experiences in those years consisted almost exclusively of what is technically known as "wishful thinking" (sometimes called "a rich fantasy life," which meant, and you know I had this all worked out, that if anyone would have ever *asked* about my relationship life, I could have winked and said, "fantastic," and I wouldn't *precisely* have been lying) might give some clue as to why this would be the case, but it would be a false clue. Instead my hickey involved the Denver Nuggets professional basketball team. I was in ninth grade, and was alone in our family's living room, drinking water and listening to the radio. My (then) beloved Nuggets were in a tight playoff game. Late in the third quarter they went cold, didn't score a few times down the court. If they kept this up I knew they'd lose. I had to do something to help. But what? It seemed clear the best thing I could do—really the only thing I could do—was to send positive energy. I did. They still didn't score. What more could I do? Maybe it would help, I thought, if I manifested this energy through some tangible action, some powerful manifestation of my intent and desire. But what could that be? I took another sip, and the perfect action came to me: I noticed that the glass fit snugly with one edge under my chin and the other over my lower lip, and realized that if I sucked on the glass it would stick to my face (and I could even defy gravity: when I leaned forward and moved away my hand, the glass still stuck!), so as a symbol of my solidarity with the Nuggets I vowed I would keep the glass on my face until they scored again. Evidently the basketball gods didn't find my offering sufficient, or more likely they were laughing their asses off, calling all the other gods over to watch and saying, "We were going to have David Thompson hit a running one-hander the next time down the court, but let's stretch this out and see what happens." The Nuggets didn't score for several minutes, and even when they did I had no idea what I had done (to myself, not the Nuggets). I didn't learn until I went into the bathroom and saw myself and my discolored chin in the mirror. At first I was mystified—although I'd seen hickeys on some people's necks at school, I had no real idea what they were or how precisely one gained them—and then I was horrified. What would everyone think? Overnight, however, I ruminated on the physics and biology of hickeys—of blood coming to the surface of the skin through suction, creating a slight, painless bruise—and a light slowly flickered, then turned on bright inside my head (kind of how it happens when you turn on one of those earth-saving compact fluorescent bulbs) and then suddenly, as is the nature of epiphanies, I understood in perfect (theoretical) detail how one could gain a non-self-administered hickey. Further—and here's the part that's delusion instead of magical thinking—I saw, in one blinding flash of insight, that perhaps this could be turned to my advantage, reputation-wise, and disperse or even dispel the common belief at school that I was a nerd. Never mind that I'd never seen a hickey in precisely this location: in my fevered fifteen-year-old mind that was no problem at all: it might just mean that the girl who'd given me this glorious gift was especially daring and avant-garde, with a love so strong she's willing to risk public scorn (unfortunately, in this case, scorn aimed at me, but we'll choose to ignore that inconvenient truth) by throwing cultural convention to the wind and expressing her affection in whatever spontaneous manner (and in whatever spontaneous place, as in my chin) she might find appropriate. And no, nobody at school bought that either. Evidently the notion of someone giving me a hickey was as far from the minds of everyone at school as it was from physical reality: only two people seemed to notice; both looked at me concerned, with one asking if I'd had an allergic reaction to food and the other saying that was an unfortunate place to get a bug bite.

Oh, to be fifteen again, and to think with the mind of a fifteen-year-old. Of course these days I'm much more mature, and would never do anything so silly and delusional.

But now you'll have to excuse me while I go write a letter to my representative. It's not as bad as it sounds: I have tremendous faith that it's going to make a difference, mainly because I'm still wearing my lucky socks.

/ / /

In the 1940s, behavioral psychologist B. F. Skinner performed an experiment in which he starved a bunch of pigeons, then put them in a cage that dispensed a small amount of food at regular intervals. The pigeons started to demonstrate odd behavior, repeating the particular motions they happened to be doing when the food was dispensed. Skinner wrote: "One bird was conditioned to turn counter-clockwise about the cage, making two or three turns between reinforcements. Another repeatedly thrust its head into one of the upper corners of the cage. A third developed a 'tossing' response, as if placing its head beneath an invisible bar and lifting it repeatedly." And so on. The birds were, it seemed, confusing correlation for causation: *I was bobbing my head when the food appeared, which means my bobbing caused the food to appear, which means if I bob my head again more food will appear.*[230]

Clearly, in this perspective, the birds were manifesting magical thinking (I suppose we should be thankful that these scientists deigned to acknowledge that nonhumans can think at all, even if it is magical thinking)—or, to use Skinner's term, superstition—and many analysts have extrapolated this supposed genesis of superstition from birds to "primitive" humans—them and their rain dances and all of that—and from there to Christians—them and their prayers and all of that—with the strong implication that the only ones free of this sort of magical thinking are the scientists themselves.

Oh really. Shall we talk about giant mirrors in outer space? How about changing the earth's orbit? How about inventing the internal combustion engine and believing there would be no consequences? How about the commonly held notion that technology will provide some magic bullet that will solve problems caused by, you guessed it, technology? How about the notion that you can live on a planet that you are destroying?

Do you see a relationship between this experiment and everything we're talking about here? Pigeons don't exhibit these weird, repetitive motions in nature, but only when they've been confined, isolated, deprived, and forced to depend on machines for survival. Since they don't understand how the machines work, and they don't know how to escape, they start to believe that their own unrelated thoughts and actions are somehow responsible for what the machines do.

Sound familiar?

It has to do with being powerless (or at the very least perceiving oneself as powerless), and with desperately attempting to find some way that one can influence physical reality from within this frame of powerlessness. For example, I could not quickly hitchhike to Denver, run into the arena, dash onto the court, grab the basketball, and knock down a twenty-five-foot jumper. So instead I sent good vibes, and then stuck a glass on my face. All trying to affect the world in a situation where I was ineffectual. Or here's another: I can play good poker, and I can get all my chips into the pot with my three-of-a-kind aces, and I can hope that some idiot calls me with his inside straight draw, but I am powerless to prevent the final card from being an eight and winning the pot for him (not that this has ever happened to me, and even if it had happened to me—recently, goddamnit—it wouldn't make me bitter or anything). So I wear my lucky socks to try to take control of a situation entirely outside of my control.

In those cases—and this was certainly true for the pigeons as well—the events *were* outside of my control. But even having at least some measure of control doesn't guarantee I won't fall into magical thinking. I've had flareups of Crohn's disease during which instead of going to the doctor or at least upping my meds I focused on positive thoughts. Not to say positive thoughts are bad, because of course they can and often do heal, but to think positively *instead* of acting effectively (presuming that going to the doctor or upping my meds are effective actions) is to engage in magical thinking, and to render myself powerless in situations where I actually am not. I could provide a long list of times when I have for this or that reason *chosen* (most often unconsciously or at most semi-consciously) to remain powerless and to think magically—which I guess could also be defined as the act of doing whatever it takes to help me gain and maintain the sensation of doing something effective while actually doing nothing, or, to return to a theme of this culture, doing whatever it takes to make sure that precious little me feels good about myself while I do nothing to help the external, real world—rather than to actually do something to rectify or improve the situation. I'm guessing that many of you could provide your own lists.

Are you a magical thinker? I know that too often I am.

With apologies to Jeff Foxworthy, let's play a little game.

If you put a bumper sticker on your hybrid Prius that reads *Visualize*

World Peace in the hope this will bring about world peace, you might just be a magical thinker.

If you buy a hybrid Prius in the hope this will slow global warming, you might just be a magical thinker.

If you think you personally not owning a car will significantly slow global warming, you might just be a magical thinker.

If you think buying compact fluorescent light bulbs will slow global warming, you might just be a magical thinker.

If you vote in the hope this will change anything—never forget Emma Goldman's line: "If voting changed anything, they'd make it illegal"—you might just be a magical thinker.

If you think the government has your best interests at heart, you might just be delusional—or maybe you're just incredibly stupid, incredibly rich, or both—and if you act upon this belief you might just be a magical thinker.

If you think those in power care about you or what you think, you might just be delusional (or stupid, or rich), and if you act upon this belief, you might just be a magical thinker.

If you think sending money to the Sierra Club, National Wildlife Federation, Audubon, World Wildlife Fund, the Environmental Defense Fund, or other big environmental corporations will significantly help the natural world, you might just be a magical thinker.

If you think Weyerhaeuser will stop deforesting because you ask nicely, you might just be a magical thinker.

If you think Monsanto will stop Monsantoing because you hold signs or sign petitions, you might just be a magical thinker.

If you think this culture will stop killing the planet without being forcefully stopped, you might just be delusional, and if you don't act to stop this culture, then you will be failing in your responsibility as a living being.

Although working within the system is extremely important—we need to do whatever we can to protect life on this planet—if you believe that working within the system is sufficient to stop this omnicidal culture from killing the planet, you might just be a magical thinker.

If you think recycling will stop this culture from killing the planet, you might just be a magical thinker.

If you think hanging banners (or writing books) will stop this culture from killing the planet, you might just be a magical thinker.

If you think giving land to state or federal governments will lead to that land's protection, you might just be a magical thinker.

If you think giving land to the Nature Conservancy or many other land trusts will lead to that land's protection, you might just be a magical thinker.

If you think convincing Congress to designate an area as wilderness will protect that area from oil and gas extraction, you might just be a magical thinker.

If you think pointing out to "developers" and to government representatives (who are, as you know, in actuality representing the "developers") that a piece of land is especially sacred to American Indians will stop the land from being "developed" (in other words, killed), you might just be a magical thinker.

If you think buying fair-trade products will save the earth, or will save indigenous humans (or nonhumans), you might just be a magical thinker.

If you think scads of cash and a nice house will protect you from the current collapse we are only now beginning to perceive, and that only dimly, you might just be a magical thinker.

If you think numbing yourself through alcohol, drugs, television, sex, socializing, computer games, sports fanaticism, political fanaticism, or other means will protect you from the current collapse, you might just be a magical thinker.

If you think those in power will scruple at torturing and/or killing anyone who significantly opposes them—and many others as well—you might just be delusional, and if you think your acquiescence to their plans will protect you, you might just be a magical thinker.

If you think not talking about the horrors of this culture will somehow protect you and your loved ones from these horrors, you might just be a magical thinker (How many of those you love have already died of cancer?).

If you think taking some sort of "enlightened" stance on the murder of the planet, where damage to the earth is not really damage because, for example, heaven is God's throne and the earth is merely his footstool (which I guess, now that I think about it, means that the earth is where God rests his feet, and heaven is where he rests his, well, we need take that Christian image no further), or because the earth is a place of suffering and Nirvana is where the real action takes place (or doesn't, depending on your definition of Nirvana), or because the sun will someday eat up the earth, or because the "planet turns itself over and some biota is lost but new life begins/grows/reconstitutes itself," or because the Earth is too powerful to be destroyed, and so on, will somehow protect you from the current collapse, you might just be a magical thinker.

If you don't care about the collapse or otherwise don't fight to protect the planet from this culture's rapacity because you're benefiting from the economic and social system that is killing the planet, you are beneath contempt.

If you don't care about the collapse or otherwise don't fight to protect the planet from this culture's rapacity because you're benefiting from the economic and social system that is killing the planet, the world would be better off had you never been born, or having been born, if you were now to die. This is not a threat, but a simple statement of fact, a syllogism so obvious it's almost tautological.

If you think using so-called alternative energies will stop this culture from killing the planet, you might just be a magical thinker.

If you think the proper course of action through the collapse is merely to protect yourself and your own human family, then you leave me shaking my head at your self-centeredness and lack of gratitude toward this planet that (or rather, who) gave and continues to give you and your family life.

If you think not acknowledging that war is being—and has long-since been—waged against the natural world will stop this culture from waging that war—"Oh, you shouldn't use that language because it's too violent and divisive"—then you might just be a magical thinker.

If you think you can fight evil with good thoughts (or, as Peace Pilgrim put it, "When there is no attack but instead good influences are brought to bear upon the situation, not only does the evil tend to fade away, but the evil-doer tends to be transformed"), you are most definitely delusional *and* a magical thinker. Worse, you are acting in direct support of the evil-doer—acting as an ally to the evil-doer—because you are telling the evil-doer's victims to do precisely what the evil-doer wants them to do: not resist, and to provide "good influences" to the evil-doer. That would have worked great with Hitler, would it not? And Ted Bundy? Forget imprisoning or killing him, just bring your good influence to him—love him enough—and he will be transformed! Hallelujah! Now, back to reality. This whole line of thinking that Peace Pilgrim was promoting is insupportable, codependent, emotionally unhealthy, ridiculous, and just plain inaccurate. It is typical of the absurd, inaccurate clichés so often thrown out by pacifists (and which should then be thrown out by the rest of us). It is, frankly, the worst advice one can give to one who is threatened by an evil-doer. It is a recipe for further victimization. It is precisely the advice that any abuser, any narcissist, any sociopath, would want potential victims to follow.

If you think you need not stop abusers or exploiters because "karma will get them in the end," then you might just be a magical thinker.

If you think visualizing salmon rushing up a stream—while assiduously avoiding blowing the fuck out of dams, halting industrial logging, halting industrial fishing, halting industrial agriculture, halting the murder of the oceans, halting global warming, and so on—will save salmon, you might just be a magical thinker.

If you think global warming, which is caused by the actions of the industrial economy, can be halted without halting the industrial economy, you might just be a magical thinker, and it's possible you're either a member of the eco-intelligentsia or a policy maker.

If you think global warming, which is caused by burning oil and gas, can be halted without halting the burning of oil and gas, you might just be a magical thinker.

If you think participating in carbon offset schemes, using "clean coal™" technologies, acting on "free-market environmentalism™," "natural capitalism™," or whatever other scams are put forward to give us the illusion of making change while perpetuating the same mindset and same system that are killing the planet, will somehow stop this culture from killing the planet, you might just be a magical thinker, and, as above, might even be a member of the eco-intelligentsia or a policy maker.

If you think the hundredth monkey story—where, once a certain numerical threshold of people have heard of some idea or learned some new ability, that idea or ability suddenly and magically spreads to the entire population—is anything other than new age bullshit, designed, once again (as so much in this culture is), to give us the illusion of making change while giving us an excuse *not* to make change, not to act decisively to protect the earth, will somehow stop this culture from killing the planet, you might just be a magical thinker.

The hundredth monkey story has been debunked thoroughly and repeatedly enough that only the most tenacious of the magical thinkers can still cling to it. So these days, those too terrified to effectively act have had to come up with another allegory (which they also claim is real), this one even more absurd than the hundredth monkey, yet with the advantage of being "cellular," which not only sounds more "scientific" and "fundamental" than a monkey story, but is also—even better!—completely beyond the ability of most people to verify with their own eyes.

The new story is of the transformation of caterpillars into butterflies through what they call "imaginal cells." This allegory was popularized by Elisabet Sahtouris, whose work, according to her website, "shows the relevance of biological systems [sic] to organizational design in business, government and global trade." She gives talks on such topics as, *Why biology is good for business; How organic models can meet corporate needs; How quality of life and profits can improve together;* and *The Internet: self-organizing system and key to human evolution.* You can see why both New Agers and corporate sponsors nuzzle up to her. It's the same old McDonough message of attempting to naturalize industrialism, except this time she doesn't even bother to include native grasses.

Sahtouris insists, as do so many New Agers and Green Businesspeople, that, "The Globalization of humanity is a natural, biological, evolutionary process."[231]

Once again, she's trying to naturalize—justify—this culture's destructive activities. Further, this statement of course repeats the racist and imperialist notion of cultural maturation leading inevitably to this culture's transformation into some greater state. Further still, it ignores ecological reality, and ignores what is most central to every sustainable indigenous culture: place. Two impolite words for "globalization" are *genocide* and *ecocide* in that globalization is by definition a singular society across the entire world (which means it has eliminated significant cultural differences), and is by definition not based on place (which means it will never be sustainable). By now we should all know this, and more of us *would* know this if there weren't always too many people like Sahtouris ever-too-ready to tell us these lies that too many of us believe. Capitalism—and more broadly industrial civilization—is based on globalization. Every traditional indigenous person I have ever spoken with at length has emphasized their deep (and often violent) opposition to glob-alization.

Now to the allegory. As Sahtouris says, "If you see the old system as a caterpillar crunching its way through the eco-system [sic], eating up to three hundred times its weight in a single day, bloating itself until it just can't function anymore, and then going to sleep with its skin hardening into a chrysalis. What happens in its body is that little imaginal disks (as they're called by biologists) begin to appear in the body of the caterpillar and its immune system attacks them. But they keep coming up stronger and they start to link with each other. As they connect, as they link with each other, they mature into fully-fledged cells and more and more of them aggregate until the immune system of the caterpillar just can't function any more. At that point the body of the caterpillar melts into a nutritive soup that can feed the butterfly.

"I love this metaphor because it shows us why, first of all, we who want to change the world [we've been through this too many times: the word should be *culture*, but she clearly conflates this culture with the whole world: how's that for narcissism?] are co-existing with the old system for a while and why there's no point in attacking the old system because you know the caterpillar is unsustainable so it's going to die [and of course take most or all of the world with it, but why should we let life on the planet get in the way of a stirring though absurd metaphor] What we have to focus on is 'can we build a viable butterfly?'"[232]

Ah, so we finally get to the point, which is the same fucking point these people make every fucking time: "There's no point in attacking the old system." Anything, *anything*, to justify not stopping this system. Anything, *anything*, to keep people from actually fighting back to defend those they love. Anything, *anything*, to justify cowardice.

There's really nothing new here: it's nothing more than a New Age retelling of the same old Christian rapture story: things suck now, but if you remain meek enough, if you don't fight back, if you're Christian enough, if you accept what this culture does to you and to the planet (which after all is natural since caterpillars are so voracious), then someday Jesus—or in this version the Great Butterfly—will magically appear and make things all better.

Bullshit.

Further, as hinted before, this metaphor of the caterpillar and the butterfly is extraordinarily racist, in that it states explicitly that this culture's consumption of the planet is natural. The metaphor's point is the notion that this culture's destructiveness is the necessary prelude to a transformation to some seemingly better state—this implies that traditional indigenous peoples are stuck in some primitive or immature caterpillar phase. I'm stunned that so few people can see the obvious ethnocentrism and racism in that progressive perspective. But perhaps I shouldn't be so stunned: people will do anything, say anything, to avoid looking at the fact that this culture is killing the planet, killing all who are wild and free, including indigenous humans.

And finally, this metaphor has no basis in physical reality. I asked Aric, who evidently remembers more from biology labs than I, because he responded, "The myth, though popularized by Sahtouris, originates from a book by Norie Huddle, called *Butterfly: A Tiny Tale of Great Transformation*, published in 1990 by Huddle Books (i.e., probably self-published). Huddle isn't an entomologist, or even a scientist, so I don't know where she got this stuff. Anyway, her metaphor was featured again in the *Institute of Noetic Science Magazine* in August 2000. Of course the Institute for Noetic Science is a new age outfit, dedicated to keeping people working on their enlightenment instead of actually doing anything to stop the murder of the planet.

"In reality, the driving tissues behind metamorphosis are 'imaginal discs'—though presumably individual cells in that disc could be called 'imaginal cells.' These imaginal discs are small structures already present in the caterpillar—they don't just arise. And the immune system doesn't attack them (at least, not in a healthy individual). The discs are essentially little clusters of stem cells. During metamorphosis, most of the body undergoes apoptosis—'programmed cell death'—and dissolves. The imaginal discs differentiate, absorb the liberated nutrients, and 'telescope' out into adult structures like legs, wings, antennae, eyes, and genitals.

So instead of the imaginal cells acting as agents of change in the body, the rest of the cells essentially commit suicide for the benefit of the imaginal tissues (and of course the new agers don't really want to delve too deeply into this aspect of their metaphor, although I'm sure polar bears, pygmy smelt, mountain gorillas, and more or less every other wild being would be delighted if members of this culture committed mass suicide for the benefit of the rest of the world: hmmmm, do you think we can talk the New Agers into promoting this new myth?). In fact, according to one study, the development of imaginal tissues is not suppressed by the immune system, but by something called 'juvenile hormone,' that prevents the caterpillar from developing into an adult prematurely. Without that hormone, imaginal tissues will develop even if the caterpillar is starving to death.

"And of course, imaginal cells have the same genome as the rest of the caterpillar. They don't have some 'butterfly genome,' because the butterfly and the caterpillar have the same genome, with different genes active.

"As a side note, the name, *imaginal*, doesn't come from *imaginary*. It comes from *imago*, which is a stage in the lepidopteran life cycle.

"It's stupid myth, of course, but why would we expect people who promote these myths to have even the most rudimentary understanding of biology, or for that matter, the most rudimentary understanding of real life?"

So, if you think that this butterfly metaphor is anything other than racist new age bullshit, designed, once again, to give us the illusion of making change while giving us an excuse *not* to make change, not to act decisively to protect the earth, will somehow stop this culture from killing the planet, you might just be a magical thinker.

How about three more?

First, "Baring Witness": "On November 12th, forty-five women in West Marin County, Northern California dared and bared all in protest against impending war. Lying down naked on a field in the rain, they formed the word PEACE with their bodies, spelling out their convictions for all to see.

"The photograph of their protest became the shot seen around the world, once it hit the news wires and the Internet. It has aroused passion and inspired women and men nationwide to take action, speak their minds and express their frustrations at not being heard by those in power. Many of these new activists have never taken part in a protest before. Some have never written an e-mail to anyone about a political issue. Such is the persuasive power of the vulnerability of the naked female body.

"That power is seduction and it may be the deciding factor in creating support for peace. . . . As simplistic as it sounds, the movement can make our rulers stop and listen, even if only for a second. That one second could be the difference between their pushing the button and listening to their hearts."

And more: "It is no accident either that women would choose to get naked for the sake of peace and justice. For Baring Witness is about using the greatest weapon women have, the power of the feminine, the power of our beauty and nakedness to awaken our male leaders and stop them in their tracks. [How insulting—and frankly pornographic—to suggest "the greatest weapon women have" is the "power" of their "beauty and nakedness" to cause the "seduction" of "our male leaders." What about women's intelligence? Courage? Fortitude? What about their ability to use their voices to convince? What about their ability to kick the shit out of oppressive motherfuckers? What about the guns they've been target-shooting at silhouettes of muggers, rapists, CEOs, politicians, and other oppressors? What about their gifts for painting, drawing, organizing, hacking, singing, bomb-building, lawyering, counseling, teaching, and so on? All of these are lesser weapons than their beauty? Help me understand how this is not the same old pornographic and patriarchal mindset that values women primarily for their bodies, and in fact also considers women's beauty a weapon to be used against men.] In this way Baring Witness is about heightening the awareness of human vulnerability.

"By risking with our nakedness—our charm and beauty and vulnerability—in service of peace we are exposing the flesh all humans share. We are casting off the old dominant paradigm of aggression and restoring the power of the feminine to its rightful place as the protector of life. [If we're going to picture a naked woman as an agent of change, what about a naked woman carrying an uzi? Or would that be too scary? Would that be too threatening to those in power (and more to the point, too threatening to members of the pseudoresistance)? Hell, if we're serious about bringing down systems of oppressive power, and not merely causing the "seduction" of "our male leaders" (and reinforcing pornographic ideals), let's just drop the nudity and keep the uzi.] It is time for women to deter the men in their lives from violent acts, as nurturers, as guardians of our families and as voices of reason."[233]

If you think taking pictures of naked women formed in the word *peace* will stop those in power from waging war (against humans or the planet), you might just be a magical thinker.

Evidently getting naked and taking pictures isn't quite cutting it so far as stopping US military aggression, so these same organizers have raised the stakes in their struggle against war. Their new campaign (which of course receives lots of press) is about having orgasms for peace. On winter solstice, December 22, as many people as possible are supposed to have orgasms simultaneously, and this will, according to the theory, somehow stop or slow war.

From the website: "Our minds influence Matter and Quantum Energy fields, so by concentrating our thoughts during and after The Big O on peace and partnership, the combination of high orgasmic energy combined with mindful intention for peace could reduce global levels of violence, hatred and fear."

Here's what they're aiming for: "To effect positive change in the energy field of the Earth through input of the largest possible instantaneous surge of human biological, mental and spiritual energy."

But wait! The orgasms are not supposed to be merely on that day, but specifically at 6:08 a.m. (Greenwich Mean Time), which "means that Iraq, Iran, Turkey, Palestine and Israel get some choice morning hours. A gift with the intention of peace from Global-O.org."

Iraqi men, put down your weapons and pick up your penises! Women, stroke your clitori! Whacking off will be far more effective in driving the invaders from your land than fighting back!

Yeah, right.

One last note from the website: this is the second year, because "Yes, after its overwhelming success in 2006, the Global O is back!"[234]

Overwhelming success? I'm guessing the measures of "overwhelming success" do not include: a) whether the US war machine has been stopped from murdering people all over the world; and b) whether the industrial machine has been stopped from murdering the planet. I'm guessing that instead the measures of success include: a) how good their orgasms for peace felt; b) how good they got to feel about having their orgasms for peace; c) how much press they got; and d) whether they got enough additional foundation funding to continue their orgasm activism.

And finally, for those who find even orgasms too much work, there is the following:

"The Green Circle presents . . . The 9 p.m. Earth Watch.

"When: Every Night—Ongoing—Virtual Location.

"Where: Whately, Massachusetts.

"Nightly Meditation to Heal the Earth.

"Event Details: Meditate with us on Healing the Earth.

"Envision a clean, verdant planet with plenty of Oxygen and not so much carbon dioxide . . . healthy animals and plants, lots of tall trees, thriving rainforest, clean water, refrozen polar ice caps, reduced emissions, cooler temperatures and a long future.

"Feel the love for this beautiful home of ours coursing through you.

"Tell everyone of positive spirit who[m] you know to join in, too. [Lord knows we wouldn't want anyone of negative spirit to crush our mellow, and Lord knows we wouldn't want any anger to mess with our vibe.] The more, the better—if the minds are joyful and optimistic. [Baring Witness had this same stricture, writing (with bold face in their original): "We have the choice of continuing on the present path of destruction and dominance by a powerful few or, *without blame*, stepping towards a new structure. . . ." What is it with these people that they are so afraid of anger, and that they don't even want us to assign blame to those who are blameworthy?] We can make this happen if enough of us join our vision and our wills!

"Don't get too distracted by wondering how the healing will be achieved. [By all means don't get "distracted" by silly little things like what we're supposed to *do* about it or how we're going to go about it: don't get too distracted by the fact that you've got two feet on the ground and you need oxygen to live; don't get too distracted by the fact that doing nothing won't accomplish anything: gotta protect the mellow!] Just picture the end result desired, and as with our other magic, the path towards it will automatically be generated. It's not too late, if we all take action. [Action? Is this the pathetic level to which our pseudoresistance has fallen, where now sitting on your ass and thinking positive thoughts is considered *action*?] Be the 100th Monkey! Help spread awareness of the devastating ecological disaster[s] that threaten our planet.

"So Mote It Be!

"Event Location: Your Personal Temple on the Astral Plane in Whately

"Event TIME Details: 9:00 pm - 10:00 pm Eastern Time

"Directions: Go to your favorite mediation place and sit or lie down. Try to be calm and focused. Enter trance state and rendezvous with us. Think of it as a psychic conference call. If you can make it, dial in. If you can't, no big deal. Some of us will always be there, every night during this hour. Hopefully, more and more of us as time passes."[235]

In one form or another, nearly all of our so-called resistance consists of magical thinking.

It's all about learned helplessness. The person perceives a lack of real control over an unpleasant environment. Except instead of being passive, the person engages in meaningless action under the false belief that the action actually affects the environment. It's a way of coping.

The world doesn't need for us to cope with our feelings of helplessness. The world needs for us to fight back.

In learned helplessness experiments, a minority of subjects manage not to become helpless. Some psychologists have attributed this to the ability of that minority to identify the source of the problem as outside of themselves. I think that's what has to happen with magical thinking—people have to realize that actions are not the same as thoughts, and that the problem is not primarily inside of them. Otherwise the pseudoactions taken will just be mechanisms to displace responsibility.

COMPLEXITY

The thinker makes a great mistake when he asks after cause and effect.
They both together make up the indivisible phenomenon.
—JOHANN WOLFGANG VON GOETHE

THAT'S ALL TRUE, but in avoiding the trap of magical thinking we shouldn't fall into the trap of linear, rationalistic, reductionist scientific thinking, which is just as delusional, only operating under a somewhat different set of delusions.

A fundamental delusion underlying magical thinking is megalomania, in that magical thinking presumes our minds are so powerful that our thoughts by themselves (or with actions incommensurate with the longed-for outcomes; for example, orgasms stopping military aggression) significantly influence the real, physical world, such that there is no need for us to commit ourselves to tangible, physical action based on cause and effect. For example, in magical thinking mode, I might believe that if I visualize a river running free this might somehow cause the dams on that river to disappear. Non-magical thinking would imply that if I want a dam removed, I need to follow some sort of chain of cause and effect, where I (and perhaps others) take the necessary actions to either go step-by-step through the legal processes of decommissioning a dam (and before that, learning how to effectively participate in these processes), procuring funding for removal, waiting too many years while antienvironmental politicians hold up dam removal in attempts to gain concessions from those they know love the river, finally watching the dam come down, then celebrating the river running free; or going through the step-by-step process of acquiring explosives (and before that, learning how to effectively work with dangerous materials), setting the charges, firing them, and then, once again, celebrating the river running free. Or maybe I can just have an orgasm and hope things turn out all right. Only a megalomaniac—and a powerless one at that—would think that one's thoughts and desires are so potent they substitute for effective action.

A fundamental delusion of science, on the other hand, is an equal megalomania, in that science presumes our minds are so powerful that we can fully understand what causes lead to what effects, even when cause-and-effect associations in the real world are quite often not only more complex than we think, they are more complex than we are capable of thinking.

To take a fairly simple example, what causes flare-ups (or even good and bad days) of Crohn's disease? As someone who has nearly died from this disease several times, who has lost several feet of my colon to this disease, and who more or less constantly monitors my condition, my pain levels, my stool frequency and runniness, the relationships between what I eat or what I do and how I feel, I still have a hard time determining cause and effect. What caused this flare-up? Was it because I got too tired? Was it because I was in a stressful situation? Was it because I ate this or didn't eat that, or came in contact with this or didn't come in contact with that? Was it because of the antibiotics I took for some other condition? Was it nothing I did or didn't do, but instead the disease acting on its own? So, I eat a certain food. Tomorrow I feel much better. Is it because I ate this food, or is it not? How do I establish causality when there are so many variables?

The same was true a few years ago when I had a terribly painful prostate infection. I had good and bad days. Were the good days due to drinking a daily glass of vinegar, as one remedy suggested, or was I trying to make correlation equal causation? Did I drink glassfuls of apple cider vinegar for nothing? Did orgasms help (at least the prostate has something to do with orgasms, so this does make more sense than orgasms for peace)? Why did prostate massage sometimes seem to help and sometimes seem to hurt?

I'm not saying we can't establish causality. I know that herbs prescribed by a great Chinese herbalist eventually got rid of most of the prostate infection, and so far as Crohn's, I know that eating meat makes me feel better, and eating fruits and vegetables makes me feel worse. I know that some medicines help (but sometimes at severe cost: a potential side effect of two of the drugs I take together is, as my doctor puts it, "an especially nasty form of lymphoma that causes, well, death"), but even that isn't a one-to-one correspondence: sometimes they help and sometimes they don't. Why is that?

In the real world, causality can be extremely difficult to assign, even when the complexity consists only of one body (which, or who, is of course made up of millions of different entities sharing that skin). When we increase the complexity to include much larger-scale communities of many different forms of beings, causality becomes that much more difficult to establish. What happens when you remove *armillaria ostoyae*, a so-called parasitic mushroom, from the forests of Cascadia? What happens when you remove chickadees? What is the relationship (or rather, what are the relationships) between chickadees and tree growth? What happens to a forest if you remove salmon? What about passenger pigeons? What are the relationships between the extirpation of passenger pigeons and the near-extirpation of American chestnuts? What are and will be the full effects of plastics on oceans? What are and will be the full effects of global warming on tidewater gobies, Dungeness crabs, pikas? What are and will be the combined effects of plastic, global warming, industrial agriculture, dams, and industrial fishing on smelt, salmon, brown pelicans?

How does science deal with this complexity? It systematically excises it from its equations (and of course equations in the first place excise free will from among those subjects to be equated: those subjects who have been perceptually turned into objects by those doing the equating). Years ago I talked about this with the philosopher Stanley Aronowitz, who said, "For some scientists everything outside the box—defined by the rules of scientific discourse—must be ignored. And sometimes they get very agitated when you call them on the game they're playing."

I responded, "And the game is . . ."

"Religion. Teleology. Control. The desire for prediction, and ultimately the desire to control the natural world, has become the foundation of their methodology of knowing truth.

"Think about it. What is a laboratory experiment? At the beginning one must select from the multiplicity of objects and relations that constitute the world a slice to study. How do you conduct a laboratory experiment? The first thing you do is factor out the world. You factor out emotion. You factor out ethics. You factor out nature, if you want to put it that way. You factor out the cosmos. You create a situation of strict abstraction."

In other words, you reduce cause and effect to just one cause and just one effect. That is the *point* of laboratory experiments.

He continued, "From that, we think we can extrapolate propositions which correspond to the world and its phenomenon. Or rather scientists think that. And these propositions do correspond to the world, so long as we ignore the actual physical world and its context."

I asked him, "What are the social implications of this?"

He said, "The point of science—and this may or may not be true of individual scientists—is to make the world subject to human domination. If they can abstract, and then they can predict on the basis of that abstraction, then they can try, at both the human and natural levels, to use that prediction in order to exert control.

"Let's use genetic engineering as an example. The ideology underlying its conceptualization is that we cannot and will not depend on nature to yield its own productivity, both in terms of its own development and human need. We're going to intervene, because the process of maturation has to be faster, because the output has to be more plentiful, because production has to be cheaper, because humans have to be more in control of the process."

I need to say this again, since science is at this point a fundamentalist religion within this culture, and any criticism of science leads quickly to people misinterpreting what was actually said, and leads also to the frenetic quivering of so many sphincters: I am not saying that it is impossible to determine with *some* degree of accuracy cause and effect (in some cases); nor am I saying that the tools of science are not useful for gaining some pieces of information; nor am I saying that we should not attempt to understand cause and effect. I am merely suggesting humility. I am saying that it is arrogant, narcissistic, megalomaniacal, to think that we can even *begin* to comprehend the vast multiplicity of subtle and not-so-subtle associations of cause and effect in complex natural communities. And it's even more arrogant than this to perpetrate mass changes on these complex communities—to destroy these communities—without regard for the harm those changes—that destruction—causes. And even more arrogant than this is the belief that just because you don't see—or can't comprehend even the *existence* of—cause and effect associations between some action and a possible reaction, they don't exist.

The fact that there are many cause and effect associations we don't understand doesn't mean, of course, that having an orgasm will likely have much effect on the US war machine, or that singing back the salmon can in any way substitute for dam removal. I'm merely saying that the fact that magical thinking is silly doesn't imply that magic itself is silly. Magic may very well consist in part of entering into relationships with powers—and entering into relationships with cause and effect associations—not normally discernible by scientific methods or by those whose minds are too constrained by those methods.

/ / /

For the last couple of weeks I've been thinking about a line by science fiction writer Arthur C. Clarke: "Any sufficiently advanced technology is indistinguishable from magic."

We're often told that when Europeans first invaded the Americas, various indigenous peoples perceived many European technologies or trappings—including big ships, domesticated horses, armor, guns, and so on—as magical or as sent from gods. Likewise today when we think about extant indigenous peoples encountering airplanes (or pop bottles falling from the sky), movie theaters, televisions, telephones, or other pieces of modern technology, we can sometimes believe—rightly or wrongly—that indigenous peoples continue to perceive these technologies as forms of magic, or as the work of gods (crazy or not). I'm not sure if this is always true—not only because I'm not indigenous and so don't know their experience, but also because I've seen videos of indigenous peoples firing arrows at helicopters, which suggests they saw these helicopters not so much as being magical as being intrusive—but it does seem to me that it would make a certain amount of sense, in that when we—any of us—see something new, our first impulse is often to attempt to categorize this new thing according to our current perceptual framework. Of course. Until and unless one's current perceptual framework is broken or otherwise abandoned (insofar as a perceptual framework can ever be fully abandoned), it will in great measure determine our interpretations of everything we see. So a person who believes in magic and who sees the divine in everything will quite possibly perceive new things within that framework, perceive them as magical, as manifestations of the divine. Likewise, a person who believes in capitalism will quite possibly perceive new things through the lens of how he or she can make money off of them. We see this latter all the time, as those who believe in capitalism—which sadly, is most people in this culture—will not only attempt to make money off of, for example, global warming (see the moneylust with which the capitalist press is describing the potential for profit as melting icecaps open a northwest passage as well as new oil fields), but will also in general project their greed and propensity to exploit onto the natural world. Or another example: nice people often perceive others as nice until their perceptual framework is smashed, and more hostile people often are quicker to perceive hostility. And another: if I believe nonhumans are sentient, I will, all other things being equal, be more likely to perceive some new action by a frog or tree or river as a sign of this other's sentience, and if I believe nonhumans aren't sentient, I will, once again all other things being equal, be more likely to perceive that same action as reinforcing my belief that these others don't think. And so on. It's pretty straightforward: believing is seeing, until something dramatic happens to shake my original belief.

There's another way, then, to view the original contacts between the civilized—those who rely on the technology of machines—and the indigenous, who are generally considered by nearly everyone within this culture to be technologically backward (after all, they never invented chainsaws), and whose cosmologies are considered by many of the civilized to be based on superstitions, that is, based not on sound scientific principles, but rather on magical thinking, on such nonsensical actions as rain dances or conversing with plants, nonhuman animals, spirits, ancestors, and so on. Many of their cosmologies are based on what to the scientific mind would not be considered principles involving direct cause and effect. In short, many of the civilized look down their noses at the indigenous, and can say, voices dripping with either scorn or condescension and pity, "They believe in magic."

I know that the word *magic* is used in this sense pejoratively, but what if we remove the implied insult and ask, what if these people are right? What if traditional indigenous people *do* believe in magic? And more to the point, what if Arthur C. Clarke's statement, that any sufficiently advanced technology is indistinguishable from magic, is also correct? What are the implications?

Years ago Jeannette Armstrong went to northern Russia to stay with traditional indigenous peoples there. The people were hungry because the caribou had not shown up. Then one day one of the people in the village who was skilled in these matters declared the caribou were in a valley some distance away. Hunters set off, found the caribou, killed some to eat, and brought back food and skins. Jeannette asked the man how he knew where the caribou were, and he responded, "How do you know where your hand is?"

I have read credible accounts of indigenous peoples conversing with rain clouds, rivers, mountains, trees. I have read credible accounts of one indigenous culture—now driven extinct by the dominant culture—in South America where the members of their communities routinely shared dreams, by which I don't mean they talked about their dreams on awakening, but rather that everyone in the community *dreamed the same dream*. How did they do this? And how did indigenous peoples discover that certain poisonous plants can be turned into powerful medicines through complex preparations (grinding, curing, boiling, skimming, and so on)? Was it trial and error, as some scientists suggest, or was it because, as the indigenous say about their own processes, the plant told the people what to do?

Or how about this: many traditional indigenous peoples were able to meet human needs while actually improving the health and biodiversity of the land where they lived (as do other wild beings, such as salmon, cedars, waxwings, grosbeaks, and so on).

Is that magic enough for you?

Jeannette told me another story of the indigenous. Early in the Zapatista uprising in Chiapas indigenous peoples from all over the world gathered for ceremonies. One ceremony took place in a school gymnasium. Jeannette noticed that at the start of the ceremony the lights hanging from the ceiling started to sway. She elbowed the person next to her, looked up. He followed her gaze, then nodded. As soon as the ceremony ended, the lights stopped swaying. Again she elbowed him, looked up. Again he followed her gaze. Again he nodded. Then he said, "And that is why they [the civilized] want to kill us."

Is that magic enough for you?

But maybe it's not magic at all. Maybe it's just a technology sufficiently advanced to seem like magic to *us*. Oh, I'm not talking about any technology so primitive as to require *gadgets*. I'm talking about technologies involving songs, dreams, the interpretation of dreams, awareness of one's surroundings, the ability to communicate with—which definitely includes listening to, believe it or not—one's surroundings. How did that man know where the caribou were? Was it a lucky guess? Was it because he smelled something? Was it because he knew the caribou so well that he knew where they would be? Was it because the caribou gave him a dream? Was it because he had been taught by his elders how to interpret a dream from the caribou? Was it because he had been taught how to smell the air, how to listen to the wind? Was it because, as he said, he and the caribou are part of the same body? How had he learned to perceive this, and to work with it?

Advanced technologies—whether or not they involve machines—are complex and demanding. They must be learned. They must be supported by a communal infrastructure. They don't just happen.

Think about it. You've probably seen an automobile. You may even have driven one. But could you make one from scratch? All by yourself? I know I couldn't. For me to build a car—this piece of advanced technology— requires complicated mining, energy, transportation, and economic infrastructures in order to deliver the materials to my home. It requires the knowledge of how to form metal parts into necessary shapes (as well as the ancillary technologies, equipment, and infrastructure to form them), and then the knowledge of how to fit them together. It requires the proper tools to fasten them, as well as the fasteners themselves. It requires significant commitment and sacrifice on my part.

Why should we not expect as much about these other technologies? Why, apart from ethnocentrism and arrogance, would we presume these other non-machine, "magical" technologies do not also require communal infrastructures? Why would we presume they wouldn't require knowledge that has been passed down and built upon for generations, analogous to how the dominant culture has been building upon its knowledge for generations (except that many indigenous cultures are far, far older, with at least the potential then for older lineages of understanding and technology)? Why would we presume that to work with these technologies would not require years of training, sacrifice, dedication? Why would we presume they would not require tremendous precision? Many ancient figurines and paintings, for example, do not show people in random poses, but rather are precise instructions for exact positions people can assume in order to induce trance states, with different positions shown by different figurines and paintings yielding radically different trance experiences.[236]

There's a sense in which our magical thinking can be seen as efforts—however feeble and untrained—to reconnect to these ancient technologies, and the near-ubiquity of this impulse for magical thinking can be seen as a marker of not just our propensity to live in denial (although that propensity is certainly overwhelming, at this point), but rather as a marker of the strength of our ancient, embodied connection to these "magical" forces and processes.

That's one sense. There's another sense, however, in which much magical thinking is not only lazy—how much less work is it to try to sing back the salmon than to remove a dam?—and not only is it a toxic mimic of real magic, it is grossly disrespectful of the complex technologies required to perform real magic. To say that people can merely "dial in" if they feel like it (and "if you can't, no big deal") is to say that anyone can operate these advanced and complex technologies, no commitment or knowledge (or effort) required. Would we say the same about constructing and flying an airplane? Would you want to ride in an airplane built and piloted by people with such minimal understanding and dedication? I wouldn't. So why should we expect these dabblers to be any more successful at the technologies of "magic" than we would their equivalents if they dabbled in aeronautics or aviation? And why should we trust them any more?

Magic happens. It's all around us. Sometimes it leaks into our lives. But these leaks reveal mainly the power of magic or the numinous, such that even those who know nothing about magic, who have no communal infrastructure to support it, and who have made no particular commitment to it (and in some cases are actively committed to worldviews that disavow it completely) can still perceive it, if only vaguely. To fully enter into relationship with these technologies requires a commitment as deep and broad and abiding—if not moreso—than the commitment required to fabricate machine-based technologies.

Of course the question arises: if indigenous peoples had (and some still have) such advanced technologies, how has civilization conquered nearly all of them (and will probably conquer the rest as soon as the civilized decide to steal the resources on these people's lands)? Doesn't that prove that the civilized have more advanced technologies? Isn't that the general rule, that those with more advanced technologies generally conquer those with more primitive technologies?

Well, they do and they don't. It depends on what you mean by technology, and it depends on what are the functions of your technologies. The question ignores the fact that different technological strains have different functions. A straightforward example should make this clear.

Let's say you and I are going to be locked in a room where we will fight to the death. We each get to bring one and only one piece of technology to this fight. Would you rather have a relatively primitive 1873 "Peacemaker" Colt Single Action Army revolver (in perfect working condition with plenty of ammunition)—or if necessary, the even more primitive sword or club— or would you rather have what is supposed to be currently the world's most technologically advanced laptop computer, the Macbook Pro (complete with a 2.6GHz Intel Core 2 Duo, next generation 802.11n wireless, gorgeous upgraded displays, wickedly fast NVIDIA graphics, and a beautiful 17-inch monitor)? If the laptop doesn't seem likely to do the job for you, you can instead choose an iPhone. They're pretty darn advanced. Or if you don't think computers will help you, you could instead choose a bag of remicade, which is a highly technologically advanced medicine made of a combination of mouse and human genetic materials. It works wonders on arthritis, Crohn's disease, and even ankylosing spondylitis. Or how about the most modern vacuum cleaner? A CD player? Microwave oven? No? I'm sure you can see that all of these technologies are far more advanced than a 135-year-old gun (and infinitely more advanced than a sword or club), so they should serve you better in your attempts to kill me, right?

I'll take the gun, please.

The point is clear: to compare a gun to a CD player is to compare apples to oranges. The same is true when we compare this culture's technologies to the technologies of many other cultures. Different technologies have different goals, and whether one person is able to kill another using some piece of technology is no indication of which person holds more technologically advanced tools. Nor is it an indication of which culture is more technologically advanced, more evolved, or, to get to the point, smarter. It might be an indicator of which has more technologically advanced means to kill, and it might be an indication of which culture has a greater propensity to kill, regardless of technology.

The fact that we even have to talk about this—the fact that a common belief is that one reason this culture has conquered most of the rest of the world is that this culture has more advanced technologies—says much about this culture's relationship to technology, and what is the primary thrust and purpose of this culture's technologies: it makes clear something we don't often like to talk about, which is that the *raison d'être* of so much of this culture's technology *is* conquest. This shouldn't really surprise us, of course, since this culture is based on conquest (as Stanley Diamond famously wrote, "Civilization originates in conquest abroad and repression at home"); it could not be what it is without conquest; it could not continue without conquest; and as we'll see over the next few years (because we live on a finite planet and there are fewer and fewer places remaining for it to conquer), it will collapse quickly without constant conquest and theft. So *of course* this culture's technologies will be primarily technologies of conquest, of domination, of control (as George Draffan and I made clear in *Welcome to the Machine*, the function of a machine is to convert raw materials to power: this is no less true for this entire machine culture than it is for a particular machine such as an automobile or forklift). This culture's economic system is based on conquest, domination, control, exploitation, theft, and slavery. This culture's governmental systems are based on conquest, domination, control, exploitation, theft, and slavery. This culture's religions are based on conquest, domination, control, exploitation, theft, and slavery. This culture's epistemology (these days, science) is based on conquest, domination, control, exploitation, theft, and slavery (and if you don't believe me, just ask Francis Bacon). This *culture* is based on conquest, domination, control, exploitation, theft, and slavery. So we'd be foolish to expect this culture's technologies to follow a different path.

But other technologies, other epistemologies, other religions, other forms of governance, other economic systems exist, some of which are based on principles other than conquest, domination, control, exploitation, theft, and slavery. Some are based on long term (as in thousands of years) mutually beneficial relationships. In fact these other technologies, epistemologies, religions, forms of governance, economic systems, and so on, are quite natural, and until this culture began to destroy them, lasted far longer than this culture has.

I have a friend who lived for a time in northern Pakistan. While she was living in Hunza Valley, she had conversations with a ninety-eight-year-old-shaman who told her (among other things) that what we call *magic* isn't magic at all, but in fact forces who have existed from the beginning of time and who can be harnessed, "but only if you listen and show respect." My friend told me that "He compared his belief in these forces to things like this culture's discovery of magnetic fields and electricity, saying if hundreds of years ago anyone had claimed that lights would turn on at the flip of a switch or that you could talk with someone thousands of miles away using a c-shaped utensil, people would think of it as 'magic.' The biggest difference, he said, between believers in science and believers in magic is that the former use the forces they harness to advance 'the human project' only, with no respect for the forces with whom they're interacting, but rather an attitude that these forces should be bent to their will and made to 'perform.' . . . He told me that eventually these forces will tire of us and 'bite us in the arse.' He laughed out loud as he said this last bit. Practitioners of magic, on the other hand, mainly interact with the forces they understand to exist with a respect and awe for the forces themselves and with the understanding that these forces may or may not choose to respond. Further, the use of this 'magic' was not only for the benefit of humans but all 'things which have spirit.'"

My friend dreamed of Hunza the night before we talked about magic. She told me, "I was standing outside my room looking at Rakaposhi (as I used to). I saw the peak change from white to grey and in my dream I was terrified. Usually when I dream of Hunza I wake up happy. But I realized that this dream was the peak letting me know its extreme distress. It needs our help. This culture is killing everything, and we need to stop it."

DESPAIR

Life begins on the other side of despair. —JEAN-PAUL SARTRE

ALMOST EVERY INDIGENOUS PERSON I've ever worked with has said that our first and most important act of resistance must be to decolonize our hearts and minds. A significant part of decolonizing our hearts and minds is separating ourselves from this culture, breaking our identification with it and identifying instead with life. I didn't build dams. They're not mine. I need to dismantle them. Likewise, I didn't invade Iraq. They're not my troops. And also likewise, I didn't create car culture. I drive a car, but *car driver* is not my identity. There's all the difference in the world— and I mean all the difference in the world—between those who drive cars, and those who identify with car culture. It's the difference between those who drive cars simply because it's expedient (and who eagerly anticipate and are actively working toward the end of car culture), and those who would (and will) find the end of car culture not only inexpedient (and undesirable) but a threat to their very selves, or more accurately, to their perceived selves. It's the difference between those who propose solutions to global warming that take the world as primary, and those who propose solutions to global warming that take industrial capitalism as primary. I am committed to ending car culture, and there are longer levers I can find than simply not driving a car.

Perhaps the identification I'm talking about is somewhat akin to what some Buddhists call attachment. I drive a car, but I'm not attached to driving this car, or to car culture. If car culture ends (or rather, when it ends, because it will end) I will look around, befuddled (and happy), and then I will shake my head to clear my thoughts, and after that I will get on with my life.

And that's where, if I understand all of this correctly, I may part ways with at least some Buddhists. I see a fundamental difference between being attached to (or in my words, identifying with) car culture and being attached to (or identifying with) life on this planet.

If those in power, or more broadly the culture in general, can get us to identify with them, "I am the disease as well as the cure," we will not be able to fully fight them, because we will perceive ourselves as fighting ourselves. Contrast the effectiveness of actions that emerge from saying, as we've all heard so many people say, "I am the problem as well as the solution," to those that emerge from, "I am the solution, so you evil motherfuckers better watch out, because I'm going to stop you using any means necessary, and I'm going to stop you now." Or better, saying nothing and simply getting the job done.

/ / /

When people realize that magical thinking ain't gonna cut it, they sometimes return to one of the previous phases, distracting themselves increasingly frantically, destroying with ever more vehemence those who would even incidentally remind them there are other ways to be, or becoming more and more smug (and more and more brittle, and more and more volcanic) in their insistence that we needn't worry about the damage. Sometimes they do all of these at the same time.

But some people move beyond these phases, and for them sometimes the next step is a bone-crushing despair. This despair, as I've described in several books, can be a necessary part of the process of decolonization. This despair is, at least in my own case, intimately tied to death. It began with death, it led me to and through death, and when I emerged from this death, death was still my constant companion, only now everything was different.

For me, the despair began as anguish over the murder of so many whom I loved. When I was a child, a "developer" converted dry fields flanked by cottonwood trees near where I lived into an upscale neighborhood. Anthills, prickly pears, and tall native grasses were bulldozed and replaced with pavement, sidewalks, and Kentucky bluegrass. Meadowlark songs were replaced by lawnmowers. Pop and rock blared from backyard stereo systems. My friends the bull snakes, tiger salamanders, and western toads disappeared. I couldn't blame them. Eventually, so did I.

Fast forward twenty years, to my late twenties. I was living now in Idaho, having fled Colorado because there were too many people. But here, too, people were building (a.k.a. destroying). I saw beautiful forests murdered by clearcutting. I saw vibrant streams lose fish, birds, in some cases even algae.

This drove me deeper into the sorrow and despair that had shadowed me all those years, sometimes more noticeably, sometimes in the background; a sorrow and despair about this culture's destruction of so much life, its conversion of the living to the dead. I had no outlet for this sorrow, this despair, this inchoate rage at these clear injustices and at the sheer stupidity of it all, so I turned this sorrow, despair, and rage inward.

I hated myself for participating in this unjust system into which I had been born (obviously without my consent), and for my failure to effectively oppose it (actually at that point, for my failure to oppose it at all). I was paralyzed not only by this self-hatred, but also by my growing perception of the magnitude, ubiquity, nonsensicality, and horrifying momentum of this culture's tidal wave of destructiveness.

And then I found action, and resistance. At first my steps were tiny, and timid: letters to the editor under a pseudonym because I was too scared to have a voice; protests where I stood silently because, once again, to have a voice was too frightening. But even these small steps encouraged me, made me happy. And they were really fun! Not only the actions themselves; and not only the camaraderie of standing shoulder to shoulder with others who were at least doing *something* to try to stop the horrors; and not only the solid and profound joy and ecstasy—and I mean ecstasy—of doing the right thing; but also the equally profound and ecstatic joy of, however timid and terrified one may feel on the inside, rising from one's knees, standing full upright, and saying, "No!"

For a while that joy of resistance helped stave off some of the sorrow. Or perhaps it's more accurate to say it counterbalanced the sorrow, since the sorrow was still there in full, only now I felt like I was doing something about the cause of that sorrow.

This only worked for a few years: in time, resistance alone was no longer sufficient. I needed *effective* resistance, and this was (and still is) in short supply. I needed to do far more than just "put up a good fight," win symbolic victories, win only defensive victories, or raise public awareness while the real world continued to be destroyed at ever-accelerating rates. No matter what victories we claimed, real forests continued to be murdered; real rivers were killed or enslaved; real plants, animals, and fungi were driven extinct.

And just as powerful was the hard-earned realization that the processes through which we were allowed to resist were rigged in favor of the destruction. Over the course of many books, and earlier in this book (especially the story of Smith and his accomplices, including the involved agencies), I've described how and why these processes are rigged, as well as some of the effects these rigged processes have on the real physical world, on human communities, and on our psyches.

One of the effects of defeat after defeat after symbolic victory (which meant a defeat that we called a victory because we were so tired of being defeated) was that the sorrow and despair once again grew.

The rage, grew, too, yet that, I'd long before realized, had grown all along. As a child that rage had been impotent. When I'd begun to resist, the rage had become joyous, ecstatic. And when I came to understand the ineffectuality of nearly all of our resistance and pseudoresistance, and more even than that when I grew to see the ubiquity of the lies we're told by those in power, the lies we tell each other, the lies we tell ourselves, and the lies we too readily believe, that rage turned more bitter. I could feel this rage growing, in my throat and dry mouth, in my heart and tight stomach, in my clenching and unclenching fists.

The sorrow, despair, and rage continued to grow year after year. So did my confusion: I could not for the life of me (or the planet) understand how so many more people could care about professional sports or the sex lives of celebrities or making money or accumulating power than the continued existence of life on this planet. Or how about this: governments subsidize commercial fishing fleets for more money than the fiscal value of their catch. Taxpayers pay to murder oceans. We'd all be better off if these commercial fishermen were paid to sit at home in their underpants watching *The Price is Right*. How does any of this make sense? My paralysis grew as well: if I and all the activists I knew were working as hard as we were and things were still getting worse; and if the processes we were allowed to participate in were rigged against us; and if every time we figured out how to use rules designed by and for those in power (commonly called "laws") to stop the destruction of the natural world, those in power simply changed the rules and did whatever the hell they wanted anyway, why bother doing anything? Why not just quit? Of course, discouraging dissent (and even moreso resistance) is one major reason those in power have set up public participation/oversight processes the way they have: if the processes were fair, and based on justice and a love of life, we would win every time; if the processes were entirely unavailable to us we would more easily see the system for the autocracy that it is; but if they can keep us participating in these rigged, sham processes, where, for example, we can point out lie after lie after lie to CDF (you can of course substitute the USFS, BLM, USFWS, or any other AoD [Acronym of Death] and you'll see analogous results), and CDF still approves timber harvest after timber harvest after timber harvest,[237] we may well continue to feel worse and worse until we give up, turn on the tube, and cheer for our favorite football team to win the big game or our favorite horse to win the big race or our favorite actress to win the big Oscar or our favorite politician to win the big election. As I realized this, my pain and anguish grew.

But these feelings couldn't grow forever. Something had to give. I broke. And that was a good thing. Of course I didn't know that at the time. I just knew it hurt. No matter how real, fundamental, and necessary this breaking was for my growth—and for the decolonization of my heart and mind, and for the survival of my soul; and for my future resistance—and no matter how harmful and deceitful (to self and other) the avoidance (conscious or unconscious) of these feelings may be, it is sometimes—often—hard to find anything redemptive about these feelings while still in the midst of them. At the time they just hurt like hell, and I wanted (and want, when they happen now) them to stop.

It can be pretty easy not to get past this phase of pain, sorrow, rage, and self-hatred, partly because it hurts so much to go through these feelings to the other side (and we have so very few models telling us how to get to the other side, or even that there are any other sides to get *to*). It can be so much easier to back away from this pain just a little. Of course the pain never does go away completely, no matter how much we may try to drown our understanding of the massive destructiveness of this culture (and our role in it) through chemical and non-chemical distractions. But the pain can ease off some, moving from an excruciating, acute crisis to a deep chronic ache.

Another reason that so many of us step away from this pain, this transition, is that we don't know what to do. I shared the above with a friend who is a longtime Greenpeace UK (mainstream environmental) activist. She responded, "I think I'm trapped in the 'pain, sorrow, and self hatred' phase. I always knew I wanted to devote my life to the planet but now that the direction I was heading in seems futile, I don't know what to do. I feel like I'm standing alone on a road with no idea which way to go. . . ."

She continued, "If I'm honest, there are two reasons I'm not yet fighting back: one is that I'm not yet sure of what to do: what would be effective? And two, and probably the main reason, is that I'm scared. And that fear is probably what keeps most people oppressed and stops them from doing anything. I care desperately about what's happening to the world, and feel rage and sorrow and despair every day, but at the same time it's not a recognizable *direct* instant threat to me. If someone suddenly came at me, or someone I loved, with an axe, I would just react, without thinking. And I think this is the difference: it's not that I care less; it's just that I have way too much time to think about the consequences of fighting back, and then that scares me. I also think most people among the general population don't want to take responsibility for anything: they've been taught from birth that those in power will make the right decisions for us, and I think many people truly still believe that. And even when you try and show these people, their denial is so deep that they don't believe it, or they want to keep their lifestyle in any case, and so they force themselves not to care."

So many activists have said similar things to me. Faced with the enormity and seeming intractability of this culture's destructiveness, what do we do?

The pain and confusion can certainly feel like good enough reasons not to persevere, but there's an even more deeply frightening and fundamental reason that holds many people back. It has to do with the forcefulness and completeness with which this culture has inculcated us to more highly value this culture than our own life, the lives of others, and life in general; to consider this culture more real than life; to identify more with this culture than with life; to be more attached to this culture than to life. Because

we've been so completely inculcated into identifying more with this culture than with life; and because we're inculcated into thinking this culture
is life; and because we're taught to believe this culture is *our* life; and
because the destructiveness of this culture is so overwhelming and has so
much momentum; and because we're told again and again that the harm
this culture causes is our fault because we drive, wear clothes, eat, shit,
consume ("You can save the earth by consuming less"); and because the
processes by which we're allowed—allowed by those in power, and more
importantly by ourselves—to even slightly rein in the destructive acts of
those in power are rigged in favor of, no big surprise, those in power; and
because these processes are sham processes, Potemkin processes, set up to
deceive and discourage us; and because we've been so thoroughly trained
to be submissive to authority, to be good little boys and girls, to never talk
back to our parents/teachers/bosses/leaders, to never get smart, to answer
the questions posed by our betters and to never reframe or reject those
questions (and certainly to never pose questions of our own), to only be
creative in ways we're trained to be creative and to not be creative in ways
we're trained (often without being explicitly told) to not be creative; and
because we've been so completely turned into imbeciles with high IQs,
like our parents/teachers/bosses/leaders, that it never occurs to most of
us that even though the processes in which we're allowed to participate are
rigged, we can create our own processes, stop the destruction our own way,
on our own terms, not theirs (picture a Gordian knot, picture a sword, picture a sword cutting the knot: now go find yourself a sword and start
hacking). Because we don't know what to do (won't someone please tell
us?); and because our defeats can be so very painful and discouraging that
it can so very easily become so very attractive to not feel those defeats by not
attempting to win in the first place; and because to stop this culture from
destroying the planet would be to stop that with which we have been
trained to most closely identify; and because to act against this culture can
feel like we are acting against ourselves, can feel like we are splitting into
a thousand pieces, can feel like we are disintegrating (in part because we
are; or rather, because our socially created selves are); and because we've
been taught to despise ourselves, other victims, and all those beneath us on
the hierarchy set up by this culture, and never to hate those above us on
this hierarchy; and because we don't want to go to prison, or be tortured

or killed; and because we know that those in power often imprison, torture, or kill those who oppose their (oftentimes psychotic) desires; and
because we do not want the full power of the state coming down on us and
those we love; and because we do not want people to disapprove of us
(never mind that it would only be others of the living dead of this culture;
while the real living, including nonhumans, including wild humans,
including others of the resistance, would love us all the more, would stand
up in respect for us); and because the whole bloody mess is so out of control, it can be so very attractive, once again, to simply try to control what
we can, and hope that God (or Buddha, or, far more to the point, our cul-
turally-formed consciences based on the value systems of this culture)
grant us the serenity to accept that which we (have been taught to believe
we) cannot change. And that which we "cannot change"—so we are told
again and again, in ways large and small, including, for example, every
mainstream proposal to "solve" global warming which takes this destructive culture as a given—is this culture. What we can change—"the only
thing we can change," we're told again and again—is ourselves. As with
the "solutions" to global warming that take this culture as primary and that
attempt to force the natural world to conform to it, we are once again made
to believe that this culture is immutable, and this time it is we personally
who must conform.

POWERLESSNESS

Personal change doesn't equal social change. —WARD CHURCHILL

UNLESS YOU'RE IDEOLOGICALLY BLINKERED, irredeemably selfish, or just plain stupid, it's pretty easy to recognize that every action involving the industrial economy is destructive. And because the continued existence of the industrial economy cannot be questioned, much less threatened, and because we must *always* be disallowed from realizing that the problem is the culture, not us (just as in any abusive situation all people must always be disallowed from realizing that the problems are caused by the abuser, not the victims), many of us choose to "fight back" by decreasing our involvement in the industrial economy, by "living simply so that others may simply live."

So we eat less. We drive less. We do not own a car. We take shorter showers. We live more and more simply. We feel more and more pure.

We're doing what we know we can control.

Living simply is a good thing to do. Sadly, it in no way stops this culture from killing the planet. In no way is it a sufficient response to this culture's destructiveness. In no way is it a substitute for actively and effectively resisting actions and policies that harm our (and others') habitat.

I want to be clear. I'm not saying we shouldn't live simply. I live reasonably simply myself, but that's primarily because I only buy stuff I want, and I don't really want a lot (except I'd love to buy a lot of land to protect it, which would of course be analogous to buying individual slaves to free them, which doesn't alter the fact that I want to do it). But I don't pretend that me not buying much (or me not driving much, or me not having kids) is a powerful political act, or that it's deeply revolutionary. It's not. Personal change doesn't equal social change. It's not a significant threat to those in power, nor to the system itself.

Besides being ineffective at causing the sorts of changes necessary to stop this culture from killing the planet, there are at least five other problems with perceiving simple living as a political act (as opposed to living simply because that's what you want). The first is that it's fundamentally as narcissistic and as much a product of magical thinking as Baring Witness or orgasms for peace in that it substitutes private personal actions that accomplish very little in the real world, and a whole lot of wishing ("But if everybody lived simply . . ." they say, to which we can respond, "If we're going to fantasize about everybody doing something, let's fantasize about them demolishing the oil infrastructure to slow carbon emissions") for organized (or solo) resistance. Once again, I'm not dissing simple living. This book started with me shitting in the forest because it makes food for slugs. But I'm not going to trumpet that act as particularly political. Although it does help those particular slugs and the frogs who eat them, it's not going to slow global warming or stop plastics from being dumped in the ocean. Ultimately it won't even help these slug and frog communities, because unless the industrial economy is stopped, global warming and global poisoning will kill them.

The second is that it's predicated on the flawed notion that humans inevitably harm their landbase, in that it consists solely of harm reduction. The world is still a worse place than had you never been born, only this time it's not quite as bad as it would have been had you not been so pure. But humans can help the earth as well as harm it, and simple living as a political act ignores this. There are other things we can do as well. We can rehabilitate streams, we can get rid of noxious invasives, we can remove dams, we can shut down gold mines that are poisoning water sources, we can destroy the industrial economy that is destroying the real, physical world.

The third problem is that it incorrectly assigns blame to the individual (and most especially to individuals who have no particular power in this system except their ability to consume) instead of to those who actually wield power in this system and to the system itself.

The fourth problem is that it fundamentally accepts capitalism's redefinition of us from citizens to consumers, such that the "political acts" of the simple living "activists" are not the acts of citizens, with all the responsibilities citizenship implies, but are explicitly the acts of consumers. This redefinition is as wrenching, alienating, demeaning, disempowering, and wrong as this culture's previous redefinition of us from human animals in functioning communities to citizens of nation-states. Each of these redefinitions gravely reduces our range of possible forms of resistance. Human animals in functioning communities perceive themselves as having a wider range of forms of resistance to threats (both internal and in the case of functioning communities primarily external) available to them than citizens of nation-states, who perceive themselves as having a wider range available to them than consumers.

The fifth problem is that the endpoint of the logic behind simple living as a political act is suicide. If every act within an industrial economy is destructive; and if we want to stop this destruction; and if we are unwilling (or unable) to question (much less destroy) the intellectual, moral, economic, and at least as importantly physical infrastructures that *cause* every act within an industrial economy to be destructive, then we can easily come to believe that we will cause the least destruction possible if we are dead. Partly because it's true. The world would be better off without humans who do not actively attempt to stop industrial civilization from killing the planet.

No, that's not true. Whether or not we "attempt" to stop this culture is irrelevant. Results matter, in this case. The world would be better off without humans who do not actively and *successfully* stop industrial civilization from killing the planet.

Because the industrial economy is based on omnicide (and you thought it would never get around to consuming you?), to participate in this economy without proactively shutting it down is to be thrust into a double-bind, in fact into the double-bind to end all double-binds (in fact the double-bind to end all life). A double-bind is a situation where you are presented with two (or more) options, and no matter which option you choose, you lose, with the additional constraint that you cannot leave. If we avidly participate in the industrial economy, we may in the short term think we win because we may accumulate wealth, the marker of "success" in this culture. But we lose, because in doing so we give up our empathy, our animal humanity. And we really lose because industrial civilization is killing the planet, which means everyone loses. If we choose the "alternate" option of living more simply, thus causing less harm, but still not stopping the industrial economy from killing the planet, we may in the short term think we win because we get to feel pure and self-righteous, and we haven't even had to give up *all* of our empathy (only enough of it to not stop the horrors), but once again we really lose because industrial civilization is still killing the planet, which means everyone still loses. And unless you've found a way to leave the planet—which would be an odious abrogation of responsibility anyway—you can't leave. Except by dying.

The good news is that there are other options.

One option—the option I perceive as the most real, fundamental, necessary, and most importantly *life-affirming*—is not to die, but to get rid of the industrial economy.

Of course it's not really fair of me to say that this option doesn't include our death, because it does: it involves the death of our socially created selves, a death which, since it's been emotionally and sometimes physically beaten into us that this culture is more important than life—and all indicators are that nearly everyone within this culture acts more or less incessantly on this almost entirely unexamined belief—is far more frightening to most of us than a real, physical death.

Another way to say all of this is that before we can get rid of the industrial economy, we ourselves must die. Not physically, but metaphorically. We must be broken. Our civilized selves must die. Our identification with this culture must be broken. The imbecile with a high IQ inside of us must die. All so that we can remember how to think and feel for ourselves, so our native intelligence—which includes our connections to the land who gives us life and supports us—can begin to return.

Note that I'm not saying we *all* have to go through this death and rebirth before any of us begin the necessary work of bringing down the industrial economy. Not everyone—or even enough people—will wake up, and in any case there isn't enough time ("Damn straight!" say the polar bears, migratory songbirds, large fish in the oceans, small fish in the oceans, the oceans themselves, rivers, icecaps, and indeed life on earth). I'm merely saying that *on an individual basis* undergoing that process can lead to more effective resistance, and of course once the process has passed certain critical points, can lead to greater happiness, joy, self-awareness, and, paradoxically, peace.

For me, I needed to break—to die—because I was still a child, still a slave, still, in fact, an enslaved child. In order for this culture to continue, we must all remain children, we must all remain slaves, we must all remain enslaved children. In order for our souls to survive, and in order for us to meaningfully resist, we must grow up. And we must remove our chains. That's not easy, as so much about this culture infantilizes us, and so much teaches us to be subservient to power. Education, for example. I wrote a book about how schooling teaches us to give ourselves away to those with power over us, to sit in chairs and face forward, to wish our lives away. Advertising certainly infantilizes us. Our romantic relationships are quite often infantile, as so many people in this culture look to their partners as substitute parents, projecting onto their partners all of their unmet childhood needs (I'm not the first to comment that the rush of falling in love could very well be this projection, which is bound to disappoint; I'm sure you've known people who've gone partner to partner, always searching for the one who will complete them). That's what the culture wants us to want (and certainly, for example, most popular music reinforces this: singing about romantic love is fine, but where are the songs for other parts of our lives, songs for snowfall, songs for death and grieving, songs for birth, songs of despair, songs of rage against the machine [oops, I guess those do exist], songs of family, songs of community, songs to welcome the salmon home after their long journey in the sea?). This infantile craving starts early because our family relationships are structured so horribly that we never get the healthy (group/tribal) parenting we need, and then we never grow up emotionally, and forever yearn for someone, anyone in authority to take care of us and make decisions for us.[238] And those in authority are only too eager to make those decisions for—or rather most often against—us. The political and judicial systems pound subservience into us. Or how about film? What do films teach us about activism and resistance? Have you ever noticed that a stock plot point in movies with an activist hero is that usually about two-thirds of the way through the film the activist's romantic partner must become jealous of the activism, and either leave or threaten to leave the relationship? What was the last film you saw with an activist in a loving supportive relationship, perhaps with a partner just as impassioned, one with a partner who encourages the protagonist to become better, more effective, more militant? Contrast the shitty, unsupportive, tension-filled relationships normally foisted upon film activists with the supportiveness and affection nearly always shown by the partners of film sports heroes or film high school teachers. What is the real lesson being imparted here? That activism is lonely. Teach at a tough high school and your partner will have your back (and be all over your front), but become an activist—work to dismantle systems of oppressive power— and you'll never again have sex: you might as well join the conservation convent or priesthood.

Our resistance to the planet's destruction—this resistance is often called environmentalism—is of course servile to its core. Our activism consists almost exclusively of begging those in power to go against the requirements and rewards of this omnicidal economic and political and cultural system and do the right thing, something we know they will never do with any consistency, something we know they *cannot* do with any consistency, because to do so would cause the entire economic system (based as it is *functionally* upon unsustainable and exploitative activities) to implode. We never demand they do the right thing. And we certainly never *force* them to do the right thing. And God forbid we actually cause the right thing to be done using our own power. *That* would be too scary. So, we beg them to remove dams. We don't demand they remove dams (or else, what? What are we going to do? Have an orgasm? Take a naked picture of a woman to try to seduce them into removing the dams? Dress up as salmon and have a die-in? That will surely put the fear of Nature into them and force them to do the right thing! Oh, I know: we'll threaten to vote against them, in favor of someone else just as beholden to large corporations . . .). We don't force them to remove dams. And God forbid we remove the dams ourselves. Similarly, we beg those in power to reduce carbon emissions. We don't demand they reduce emissions (And what are the alternatives? Sign an online petition saying they're mean and nasty? Or would that be too aggressive? Would that be assigning blame? Maybe we'll demand they reduce carbon emissions or we'll meditate and send them waves of pink bubbles of love. *That* will show them! Or how about this: I just read on a Christian pacifist website that to attempt to force those in power to not commit atrocities is to not have faith that the Holy Spirit can reach and convert even them. So there's another course of "action" for us: we can sit on our fucking hands and wait for God to do the work for us. We don't force them to reduce emissions. And God forbid we simply destroy the infrastructure necessary for them to emit this carbon dioxide, depriving them of the ability to heat the planet. We as activists are almost entirely servile. And don't just suggest that this means we need to assume positions of power within this omnicidal system. As I've shown in numerous books, those "in power" are and must be themselves servants to the machine-like social structures called corporations, and beneath that to the machine-like social structure called civilization, to The Machine.

Even more than servile, our activism is infantile. It's characterized by magical thinking, denial, fear, narcissism. Just tonight I received a request to attend a conference on sustainability. The organizer also requested I critique his writeup. I commented that he said nothing about fighting back, to which he predictably responded that fighting back never works (ummm, wasn't there this little thing called the Fall of Rome that was hastened just a tad by those nasty barbarians (read indigenous peoples) "fighting back"? There are plenty of examples of people or groups successfully fighting back, and in any case, I don't perceive our other activism as coming anywhere close to working: if *any* of our activism worked, the world wouldn't be getting dismembered before our eyes). I also commented that throughout this entire document that purported to be about sustainability he never once mentioned concern for the natural world, or indeed mentioned the natural world at all. He responded that he didn't want to express concern for nonhumans because he was afraid the conference would then be perceived as a get-together of "tree-huggers," something he evidently considered a bad thing. So let me get this straight: he's going to put on a conference ostensibly about sustainability that avoids talking about stopping those in power from killing the planet and in fact avoids talking about the planet at all—the real, physical planet, filled with nonhumans— remember that? I shared the exchange with a friend, who said, "This is all so typical of the Left, of the so-called resistance: earnest, useless, and full of fear. This is a big reason we always lose."

I won't be going. Nor will I be baring it all for peace, nor having an orgasm on December 22 (at least not for any reasons other than the obvious).

/ / /

More infantilism. Just today I received this description of a talk by feminist "theorist" (scare quotes around *theorist* because her theories are, well,
you'll see) Sally Miller Gearhart: "Gearhart described herself as a 'recovering activist,' [implying, of course, that activism is an addiction to be
kicked] relating a turning point in her life in which she decided to avoid
negative thinking. . . . Although Gearhart had once fought [sic] to preserve
the California redwood forest, living next door to a logger and helping to
build her own driveway, which required [sic] chopping [sic] down redwood
trees, resulted in a reevaluation. In particular she described an instance of
loading a truck with redwood logs, feeling the power of it and realizing [sic:
read, projecting] this was how men felt. She said she did not know if it was
bad that men felt this power and did not think it was wrong that she felt it.
She suggested that perhaps visualizing and fantasizing might be more productive than demonstrations in the end."

This is of course insane on many levels. Naturally, humans consume
trees and other plants and animals to survive—but there's literally a world
of difference between doing that humbly and well, and turning old growth
into junk mail circulars, or "requiring" the killing of trees for a driveway. I
never cease to be appalled and amazed at the level of idiocy required to con-
flate every real human need with corporate, capitalist rapaciousness: it's the
collapse of all meaningful distinctions. And am I the only person bothered
by her seeming to enjoy the power (that's power *over* for those of you
keeping score at home) she wielded to kill others for her convenience? By
power she's clearly not talking about the joy of doing physical labor (in part
because I'd be willing to bet my life that she didn't load the truck by hand,
but merely ordered and oversaw the mechanical killing and transportation
of the trees, consequent to their presumed sale), but rather the power to
dominate. I must admit I'm disgusted by those not repulsed by subjugating
others. But the real reason I include this reference to her has to do with her
last sentence, where she believes that visualizing and fantasizing might be
more productive than demonstrations. Sure, demonstrations can be just
slightly more physical manifestations of magical thinking (If literally millions of people march against the US invasion of Iraq, that will stop the war,
or at least give the Democrats the backbone to end it, right? Right? Pretty
please, with sugar on top?), but at least demonstrations require action in
the real, physical world. Such is not the case with visualizing. One reason

our "resistance" accomplishes so little is that so much of it consists of a bottomless well of infantile, self-indulgent self-deception. But there's something even more disturbing than all this about her comments. Please note that she does not seem to mind killing, *but only when she gets to kill those lower than her on the hierarchy, those she can kill with impunity.* To subjugate those who cannot fight back is okay, and gives a feeling of power, but to stop those who are systematically destroying life on this planet is "negative thinking," or really, unthinkable.[239]

✦ ✦ ✦

We need to grow up.

GROWING UP

The disappearance of a sense of responsibility is the most far-reaching consequence of submission to authority. —STANLEY MILGRAM

The world is a dangerous place, not because of those who do evil, but because of those who look on and do nothing. —ALBERT EINSTEIN

We have not passed that subtle line between childhood and adulthood until we move from the passive voice to the active voice.
—SYDNEY J. HARRIS

It is easy to dodge our responsibilities, but we cannot dodge the consequences of dodging our responsibilities. —JOSIAH CHARLES STAMP

SOME PEOPLE AREN'T CHILDREN; aren't magical thinkers; aren't so afraid of what Daddy will think that they cannot openly acknowledge (and act on) their love and concern for the natural world; aren't so afraid of the pain of despair that they must avoid looking at the horror that *is* this culture; aren't so afraid that they cannot face this culture and our predicament head on. They recognize that this culture is irredeemable and that having orgasms or taking off one's clothes or putting plants on truck factories or having conferences on sustainability that avoid the two primary preconditions for sustainability (the land comes first; and we need to stop these motherfucking psychopaths no matter what it takes) is utterly insufficient to the crises confronting us. There are not merely *some* people who feel this, but

267

many, with numbers increasing every day. And some of them write to me. These more empowered (and honest) emails outnumber the others by a ratio of probably twenty to one. I want to share one of those honest emails now, in great measure because it so well articulates the pain and confusion of the process of breaking, a process I've been through so many times, a process gone through by so many people I know, an almost necessary part of the questioning and confusion and pain and sorrow crucial to growing up and leaving this culture behind.

I had never heard from this woman before. She wrote: "I got up this morning and worried that my pants were far too wrinkled to wear to work, which happens to be a public school. At that moment I had a flash of who I have become. I wish that I could merely worry about the socially acceptable wrinkliness of my pants. But what I want more than pressed pants is answers. Answers that are detailed, specific, hard-core descriptions of what I need to do to fix things.

"Here's my issue: I work with kids every single day. I feel like there's a giant monster among us, as a matter of fact right in front of us. Our kids live in its shadow each day. I'm beginning to view our kids as the 'shadow of the monster kids' (in the most unconscious and Jungian sense).

"Question: Why do we have a generation of kids who cut, pierce, burn, choke, and do a multitude of self-destructive behaviors? They say they do it to feel something, anything, to help solve problems, to get a high, and to get attention—attention from friends, families, counselors, teachers, anyone willing to give it. Where did our sense of connection, our sense of honoring self—body, mind, and soul—go? When did self-harm become a form of expression and/or blatant and crazed addiction? What's missing? How did we get here? Please tell me how to get out of here.

"Question: Why is it that kids come in talking about seeing spirits, dreaming of the devil and sensing energy? They read Harry Potter, watch Ghost Chasers, and play Zombie video games. But when they approach us with these visions, we wonder if they are schizo. How can I help them frame their spiritual existence in a world that is rampant with messages and images of false or fabricated or mediated spirits, but is void and neutered of real conversations about spiritual experiences? How can I remove this taboo?

"Question: Please tell me why we have laws to protect children [obviously the analogous is true for the natural world], but in the state where I live, I've only met one lawyer who will consult with a child without parental consent. Those giving consent are those who beat, molest, and degrade our children. If I was that nasty and controlling I wouldn't give consent. Would you? I sent him a five dollar gift certificate for coffee and my thanks and gratitude, because I didn't know what else to do. Please tell me how we make what is legal the same as what is right and ethical.

"Question: Why is it that I felt guilty that my daughter had to tell her classmates that she didn't like Hannah Montana, because this was less embarrassing than telling them she doesn't know who Hannah Montana is? (We don't watch TV.) Why is it that after two days in the 'happiest place on earth'—Disneyland—I spend a day weeping uncontrollably? Why is it that I feel a tinge of guilt when I write on my daughter's birthday invitations 'We'd love your presence, not your presents'? And that when I allow her toys (and myself stuff), I suffer with the indigenous peoples and forests who were destroyed in order to create them? Can't I shelter her one more hour, day, year, decade from the realities of this culture? I can choose to ignore this cultural cancer, but if I don't, if I can't any longer, please, please, please, tell me how to treat for it?

"I could go on and on and on with my concerns, but I have to go to bed so that I can get up tomorrow to worry about how wrinkled my pants are.

"For a while I felt proud. I've armed myself against the giant monster in my presence. I have a grain of sand or maybe as much as a pebble to throw at it. I drive my hybrid car to the coffee stand, hand over my steel mug (no disposables for me), and fill it up with organic milk and a shot of fair trade espresso. I hurl my pebble while at the grocery store, where I purchase recycled toilet paper and vinegar and baking soda to clean my kitchen. I arm myself with compost piles, green energy subscriptions, and climate change talk circles. I toss my pebble at the monster with all my strength, but the monster is now so enormous that my pebble doesn't reach the top of its big toe. I pray that David and Goliath is a prophetic metaphor, but where the hell is my sling shot? I pray that Lilliputians were on to something. If there are enough of us, can we tie this sucker down and subdue it? Please give me the answers. I need the reassurance.

"You wrote in your book that you could either write or blow up dams. I don't want to write. I want to blow up dams. Since I'm afraid of explosives and too sensitive to survive in prison, my dams will have to be emotional dams, cultural dams, religious dams, normative dams. And then I'll just have to pray that someone else who isn't so afraid of explosives will do the other necessary work."

And now I want to share this anonymous note I was handed at a talk—and which I soon destroyed—written in a thirty-something woman's handwriting: "I am ready. I have learned self-defense, and how to use a gun. I have gained other necessary offensive and defensive skills. And I'm no longer willing to stand by while this economic system kills the salmon. I love them as much as it seems you do. I am primed and ready, and I'm only waiting for someone else to act first. I am waiting, as I believe so many others are, for a spark to set us all off. But I cannot wait forever for that spark, because the salmon cannot wait forever for that spark. I have set a deadline, and if nobody else has begun by then, I guess it will be up to me."

/ / /

Once again, the fact that so many people go through this process of dying to this culture and being reborn as human beings doesn't mean that this process is not painful and confusing, but merely means that pain and confusion and sorrow can be crucial to growing up and leaving this culture behind.

/ / /

There's a poem I read long ago, of which I remember only a few lines. These lines were very important to me in my twenties, as I—the socially constructed I—fell apart, and the me I was becoming was not yet emerging. These lines are: "Wandering between two worlds, one dead/The other powerless to be born/With nowhere yet to rest my head/Like these, on earth I wait forlorn."

/ / /

The Apostle Paul wrote in his first letter to the Corinthians, "When I was a child, I spake as a child, I understood as a child, I thought as a child: but when I became a man, I put away childish things.

If we're going to grow up enough to stop this culture from killing the planet, we're going to have to put away many childish things.

We must first and foremost put away the childish, narcissistic notion that the world exists for our use. It doesn't. And it doesn't matter whether God told us we have dominion over the earth; or whether, equally plausibly, Elvis came to us in a dream and told us the same thing; or whether capitalists tell us that land is "unused" just because no human is using it for fiscal purposes. All the pretending in the world ain't gonna make it so.

Contrast the Christian/Capitalist/Civilized ethic that God gave man dominion over the earth with what Vine Deloria said to me about the point of life from the perspective of a native North American: "In this moral universe, all activities, events, and entities are related, and so it doesn't matter what kind of existence an entity enjoys—whether it is human or otter or star or rock—because the responsibility is always there for it to participate in the continuing creation of reality. Life is not a predatory jungle, 'red in tooth and claw,' as Westerners like to pretend, but is better understood as a symphony of mutual respect in which each player has a specific part to play. We must be in our proper place and we must play our role at the proper moment. So far as humans are concerned, because we came last, we are the 'younger brothers' of the other life-forms, and therefore have to learn everything from these other creatures. The real interest of old Indians would then be not to discover the abstract structure of physical reality, but rather to find the proper road down which, for the duration of a person's life, that person is supposed to walk."

So, on one hand we have an infantile notion that everything in the world belongs to me, me, precious little me, and if it's not used by me it's wasted, and on the other we have the attitude that we are the younger siblings of our earthly neighbors and must learn from them how to play our proper role in a grand symphony of mutual respect. Gosh, I wonder which of these two perspectives will more likely lead to the murder of the planet, and which will more likely lead to sustainable cultures?

If we wish to live sustainably, which at this point means to continue to live at all, we must put away the childish notion that we have the right to take whatever we want from nonhumans.

We must also put away the childish notion that humans are particularly special, or any more special than flying squirrels, sockeye salmon, red cedar, solitary bees, Steller's jays, or oyster mushrooms. Once again, it doesn't matter whether Christianity teaches us the flattering notion that humans are the only ones with souls, or whether science tells us (again and again, ever more frantically) the equally flattering notion (indeed the same notion) that humans are the most intelligent species on earth. Pretending ain't gonna make this so either.

So many indigenous people have said to me that the fundamental difference between Western and indigenous ways of being is that even the most open-minded westerners generally view listening to the natural world as a metaphor, as opposed to the way the world really is. Trees and rocks and rivers really do have things to say to us.

Today I received this note from an activist in Australia: "Just back to work after a two-month reprieve from wage slavery (enjoyed river restoration and littoral rainforest replanting in Northern New South Wales—penance for working for a TV station). The ocean spoke to me when I was there. She told me she was sick and she needed me to help stop them killing her. While I was foraging for mussels she told me that she wanted her babies back, and that I should only take them if I really needed them, if I was really hungry. I've long known that indigenous peoples talk of listening to the world speak, but I've always felt there must be something wrong with me because I haven't been able to hear or speak to my nonhuman neighbours: I've thought perhaps I'd lost that sensitivity or something. But hearing the ocean speak when I was out there made the answer obvious: it's very difficult, almost impossible, to be a wage slave, work in a fluorescent tomb for eight hours a day, live in a concrete bunker, and then expect to have a relationship with living beings. The question remains: what took me so long to realise that? As you'd say: duh."

Another way to say all this is that we must put away the childish notion that the world consists of resources ("a natural source of wealth or revenue") rather than other beings with lives and concerns as important to them as ours are to us.

We must put away the childish notion that humans are exempt from ecological principles. Most of us understand that if rabbits overshoot carrying capacity, they will destroy their surroundings and undergo a population crash. But humans have clearly overshot carrying capacity, and are clearly destroying their surroundings, yet they continue to breed like, well, like they say that rabbits do. They seem to consider themselves the smartest creatures around, and they incessantly claim they care about human life and that human life is amazingly wondrously sacred and special, yet they are making no reasonable effort to avert or alleviate this crash (which will be the result not merely of overpopulation but the culture as a whole, of which overpopulation is merely one symptom among far too many). In fact, anyone observing this from outside would probably come to the conclusion that these human creatures are doing everything in their power to make this crash as painful and deadly as they can.

We must give up on wishful thinking. *I wish the US would stop invading other countries (but I'm not going to stop it). I wish I had a pony (and maybe I'll get one for Christmas). I wish the industrial economy would stop killing the planet (but I'm not going to dismantle it). I wish I had a bicycle (and I've been so very, very good this year). I wish dams would stop killing salmon (but I'm not going to remove any myself). I wish we would end up at a sustainable population without anyone dying. I wish we would be able to stop those who are killing the planet without harming (or inconveniencing) anyone. I wish we could consume the entire planet without killing it. I wish. I wish. I wish.*

Some children can be notorious for short attention spans (some are not). This culture has enshrined short attention spans in its economic system, which is as short-sighted as it is possible to be: would you rather have a living planet forever, or cheap consumables now? We needn't speak our answer out loud: it's already manifest in our actions, and inactions.

Years ago I knew a woman with several children who briefly dated a man who lived in the country. Their relationship ended when one day he showed her children a bird's nest in a tree near his home. He told them of the fragility of the baby birds, and told them to observe the nest only at a distance. The woman's oldest son, who was six at the time, waited till the man's back was turned, climbed the tree, and intentionally threw the nest to the ground. The man realized he did not want to be in a relationship with such a horrid child (this action was in no way out of character for this child), nor with a parent who would countenance such behavior. The planet will soon say the same to us. We need to grow up such that when the Earth tells us, increasingly sternly, to clean up the messes we've created; to metaphorically and physically stop wantonly throwing birds' nests to the ground; to not ignore (or more to the point, abuse, torture, suffocate, cut, rape, murder) the Earth; to not smirk that horrid smirk we've seen on the faces of so many ill-behaved, aggressive, spoiled-brat children as we continue to destroy anything we feel like destroying for whatever stupid reasons we manufacture; to not destroy the planet; to not continue to create whatever messes we feel like (once again for whatever stupid reasons we manufacture) and to only slightly mitigate those messes (perhaps by putting plants on truck factories), then look around proudly as though everyone else is supposed to somehow approve of us being only slightly— and I mean slightly—less destructive, yet still dreadfully ill-behaved, aggressive, spoiled, and indeed sociopathic. It's all bullshit. People sometimes speak of the Earth as our mother, yet any good mother would see through all this in a heartbeat. And the earth does. As the wonderful philosopher Kathleen Dean Moore wrote to the CEO of an oil company, in response to an ad stating that "Mother Earth is a tough old gal": "If the Earth really were your mother, she would grab you with one rocky hand and hold you under water until you no longer bubbled."

Here's something else we must give up. Have you ever interacted with a child who pretends that covering his or her face conveys invisibility? More or less all of us do this on at least two levels. One is that we pretend that just because we have chosen not to look at the problems we're facing they must not exist. If we switch senses, this is all the equivalent of plugging our ears and singing "La la la la," behavior that describes the bulk of this culture's discourse, politics, "resistance," and philosophy. The second is that many of us seem to believe that if we perform slight actions to make ourselves inconspicuous, the attentions of those in power will never fall on us. So we resist less and less, in less and less effective ways ("free speech zones" anyone?), and still the glaring light of the Panopticon keeps us in view.

Related to this is another childish behavior we must put away. One of the central means by which abused children maintain the pretense of control in uncontrollable, abusive situations, is to pretend—and of course you'll see the similarities here to simple living "activists"—that the problems are caused by *them*, and not by their beyond-their-control abusers. They pretend that if they can only be good enough children, can only clean the dishes well enough (or on the larger cultural scale, perhaps not drive a car), can be quiet enough (become vegetarian or vegan), and so on, that the Dreadful will not happen. But the Dreadful has already happened, and continues to happen, and will continue to happen again and again, incessantly, until this culture has crashed (or been crashed) or killed the planet, no matter what good and pure children we are.

But that's only if we are children, and understand as children, think as children, are powerless and dependent as children. Adults have other options. Grow up, find these options, and follow them.

We must put away the childish notion that the health of our communities is not our responsibility. Children need not take responsibility for the health of their families, and their larger communities. Adults must.

We must put away the childlike passivity manifested by so many activists toward those in power. So many of us seem to think that those in charge are (emotionally, intellectually, and spiritually) adults, and that they have our best interests at heart. We seem to think that the Great White Father will handle everything for us. So many of us seem to believe we are too small to do anything ourselves. We believe the fate of the salmon, or polar bears, or life on earth, are all out of our hands. All of this is false.

We need to grow up enough to recognize that others exist.

We need to grow up. We need to take responsibility for ourselves, and we need to manifest responsibility to our communities.

And that growing up can require a death and rebirth. And this death and rebirth can be painful, and full of despair. Yet some people do it. Some people do it.

/ / /

Others don't. Even many who've made it through earlier stages can get diverted at this stage—perhaps because the pain is too much—into a sort of pseudodespair that does not lead to rebirth, but rather to stagnation. This pseudodespair often arises among people who continue, whether or not they admit it, to identify more with the culture than with life, which means, once again, that in their perspective, the end of the culture would be the death of life. That's pretty scary right there, but the people who've made it this far in the process usually also have realized that this culture kills life. So they're stuck: if they decisively act to stop this culture from killing life, they destroy that which they (incorrectly) identify as life; if they don't decisively act, all life will be destroyed.

So they do nothing. This pseudodespair often leads to a sort of fashionable nihilism that declares there's nothing to be done; or that the culture's destructive momentum is too fierce for us to effectively resist; or that it's too late anyway. Excuse after excuse to do nothing. As with the "enlightenment" we discussed before, this form of "despair" and "nihilism" very well serves both the fiscal needs of those in power, who face less resistance because of this "enlightenment," "despair," or "nihilism" on the part of those who, if they were sane, would surely resist; and it just as well serves the safety needs of, in this case, the "despairing" and oh-so-hip "nihilists" who (of course) not only need not now resist, but, because they are "way past" all of that "simple living nonsense" need not even be inconvenienced in the slightest.

I have absolutely no respect for those who remain in that state.

/ / /

Those who've made it this far, past the pseudodespair and despair, even past this death and rebirth, can still get distracted from the real problems, and the real possibilities. Pretend you no longer (or never did) fall into (and for) the spectacles, the frenetic and meaningless attitudes and activities that keep so many people distracted from the murder of the planet, distracted from their own role (and more importantly acquiescence) in it, and distracted from themselves. Pretend you not only did not attack those who reminded you that the culture is killing the planet, but you also have psychically and physically survived attacks made upon you for this same reason. Pretend you understand that the damage this culture causes really is damage. Pretend you keep your magical thinking to a reasonable minimum. Pretend you understand that the problems are the culture's fault, and not primarily yours. Pretend you do not take on what is not yours. Pretend you have felt and survived the despair that comes to any sane being even remotely honest about this culture and the horrors upon which it is based, the horrors upon which it has always been based. Pretend you've been able to break your identification with this culture. Pretend you understand that the culture won't last much longer.

What do you do?

Some people, at this point, simply walk away, and consider that sufficient. Even someone as brave as Lewis Mumford ended his brilliant work *The Myth of the Machine: The Pentagon of Power* with, "On the terms imposed by tech-nocratic society, there is no hope for mankind except by 'going with' its plans for accelerated technological progress, even though man's vital organs will all be cannibalized in order to prolong the mega-machine's meaningless exis-tence. But for those of us who have thrown off the myth of the machine, the next move is ours: for the gates of the technocratic prison will open auto-matically, despite their rusty hinges, as soon as we choose to walk out."[240]

There are many people who, having gotten this far, "walk out" of the "technocratic prison" and then . . . well, then nothing. That's as far as they take it, and they consider themselves done.

Unfortunately that's not really helpful. There is, remember, a real world being killed, and from the perspective of the salmon, it doesn't matter so much whether you continue to participate in the system or whether you psychologically and emotionally and spiritually "walk away." It's great for you that you're out of the "technocratic prison"—except of course that you're not: since this prison culture has enclosed the entire world, at this point you *can't* walk away; you can only pretend to: when dioxin contami-nates every mother's breast milk, where would you go?—but the salmon need for dams to come down, and they need for oceans not to be mur-dered, and they need for industrial logging, fishing, and agriculture to be stopped, and they need for global warming to stop, which means they need for the oil economy to be stopped.

There are some who take it one step further than simply walking away. They begin to learn survival skills so they may personally live through civ-ilization's endgame. They learn how to make fires with flint and steel; how to catch, kill, and clean animals; how to identify local edible plants. They stock up on beans and rice, put caches of weapons and medicines in the forest so they can find them when trucks no longer bring food into cities. They're attempting to be as ready as they can for this culture's inevitable—and by now impending—crash.

None of these are bad things to do. As the economy—and then this entire civilization—crashes, I want these people on my side.

But preparing yourself and your family for the crash—a common phrase for this these days is "making a lifeboat"—is no more sufficient than walking away. Indeed, it's merely a variant of the same thing. The people who do only this are doing nothing to halt this omnicidal culture.

They're not helping the land. They're not acting such that the world is a better place because they were born. Ultimately, if their concern is primarily for their own safety and the safety of their families (*me* and *mine*) with no significant and *primary* concern for the land that gives them life, they are still, for all the work they've done, stuck in the same narcissistic, anthropocentric mindset that holds our own survival separate from and superior to the survival of the land. Rich people hoard cash, gold, and weapons in their gated compounds, ready to fend off the rabble they know will try to take their riches during the breakdown of civil society; and anti-civ survivalists hoard rice, beans, and weapons, then prepare to fend off the rabble they know will try to take their food—at least these riches are less abstract—during the collapse of civilization. Salmon, polar bears, and redwood trees ask, "What's the fundamental difference?"

It's not sufficient for us to break our identification with this culture, if we merely transfer our primary identification to ourselves: to do so is merely another variant of this culture's narcissism. We must recognize that there exists something larger than ourselves, and it is not the culture. It is the land which gives all of us life. We must recognize that we have obligations—joyous obligations, terrifying obligations, deep and meaningful obligations—to this land. The real, physical, land. The real, physical, water. The real, physical air.

Survivalism isn't enough. Nor is it enough to make a "lifeboat" that you hope will not sink through this culture's chaotic collapse, a lifeboat designed to carry you and your family and your friends.

Far more important than our own personal survival is the survival of the land. Far more important than making a lifeboat for me and mine is doing something to protect the land where I live, to make it so that not merely *I* and *mine* will have a somewhat better chance of surviving, but doing something to make it so everyone—human and nonhuman alike—has a better chance of making it through that imperfect storm.

/ / /

Someday you will die. It is possible to act such that you make the world a better place before you die.

/ / /

This is the pledge I make to the land where I live, to life on earth, and to you: I will make the world glad I was born. I will make it so that my birth, my life, and my death make the world a better place than had I never existed.

/ / /

Here is why I write: if my books help catalyze or cause people to better resist this culture, and if these people help stop this culture from killing the planet, the world will be a better place because I was born. If not, then I will have to do something else. Because I will live and die such that the world is a better place because I was born, because I lived, and because I died.

/ / /

What pledge do you make? How will you make the world a better place because you lived, and because you died?

PART III

THE FUTURE: BUSINESS
AS USUAL

The inhabited buildings slowly extrude their continuous ribbons of compressed garbage and trash. The ribbons fall onto the cargo belts that move steadily toward the high ridges at the city boundary. In these populous continents, each city presses against the next, and so the waste ridges from a network, through which tunnel the intercity roads. Each city posts frontier guards, to prevent a neighbor from tipping its trash over the crest.
—KEVIN LYNCH

Anyone who believes exponential growth can go on forever in a finite world is either a madman or an economist. KENNETH BOULDING

PRIOR TO THIS CULTURE, making the world a better place was dead simple. We only had to be born, live, eat, shit, defer to the land and to the lessons taught to us by our local traditional culture about *how* to defer to the land (do not kill buffalo when mothers may have young, do not kill too many buffalo, and so on), and eventually die to give our bodies back to the land. This is what salmon do. This is what hummingbirds do. This is what all wild beings—including wild humans—do. And it works.

Now it's not so easy. Not only because this culture is killing the planet, and we must stop it; and not only because the industrial economy is inherently destructive, meaning every action involving the industrial economy is destructive; but more intimately because this culture has made our bodies as toxic as the culture itself.

It is popular for writers concerned about the planet, the future, or both to extrapolate current events in attempts to determine just where this culture is leading us. Speculative scenarios about the future can entertain us—and sometimes do little more. But they can also help us make decisions, inspire us, and provoke thought, discussion, and best of all, action.

So far in these pages we've discussed the history of garbage and waste, we've discussed people who work with garbage and waste, and we've delved into some of the deeper implications of what it means to be a living creature on a planet dominated by such an incredibly wasteful culture. As we approach the end of the book, it's time to consider the future. And not just the next decade or two, since most "solutions" to the problems of waste are doomed from the start by a pathological shortsightedness. How will this culture impact those living centuries or millennia from now?

Visions of the future have a tendency to become dated—or even silly-looking—as we actually approach the future. Part of this is due to the inherent complexity of change in the real world, and the difficulty in coming up with a model that can even remotely mimic that complexity. Part of it is that books or films are more dramatic and exciting when subtleties are glossed over, when timescales are compressed, when events are Hollywoodized and made larger-than-life: think *The Day After Tomorrow*. Or sometimes a story about the future is constructed as an allegory or propaganda, where the plausibility of events is not as important as the moral or political point being made. Oftentimes people construct a vision of the future because it's how they'd most like to live (perhaps Star Trek), or how they'd be most afraid to live (perhaps Mad Max), or because they think it would be really, really, neat.

I'd like to lay out a few scenarios of the future, specifically looking at waste, so we can gain further insight into how the path we take will affect those humans and nonhumans who come after us. I'll try to make these scenarios as plausible as possible, based on research rather than on Hollywood-style drama for the sake of drama. I'll construct scenarios, and then draw my points from them, rather than vice versa.

At the same time, I'm not going to insult your intelligence by pretending I'm a "fully objective analyst" (as though such a thing were possible or desirable). All writers are propagandists. If I weren't horrified by the dominant culture laying waste to the planet, I wouldn't be writing this book. My scenarios will be as plausible and well-reasoned as I can make them, but in the end the questions I want them to answer are biased in favor of living creatures: what events now will provide those who come after us with a livable landbase? And always, most importantly, how do we stop this culture from killing the planet?

/ / /

At the beginning of his excellent book on garbage, *Wasting Away: An Exploration of Waste*, the late author Kevin Lynch describes a worst-case scenario of the future: the world has become unimaginably polluted, and cities are surrounded by vast and constantly growing mountains of garbage, each mountain being fed by a ceaseless conveyer belt that dumps an endless flow of garbage out of the city. In fact, in this scenario, there is little between cities except mountain after mountain of garbage.

Lynch's book is insightful in many ways, but this scenario illustrates a lack of plausibility I'd like to avoid. First, there isn't enough cheap energy left to process the entire planet's surface into garbage (at least, we hope not). Second, there probably aren't enough raw materials in the world to cover so much of the planet's surface in garbage. And third, if the world were indeed so polluted, the toxicity of the environment would kill most humans (and other living creatures), leaving very few humans to continue consuming.

I'm sure there are some people who would look at Lynch's scenario as yet another example of "doom and gloom environmentalists" trying to frighten people into giving up their disposable coffee cups by conjuring some fantastically implausible hell-world. I suspect that Lynch's goal was simply to make a point rather than outline a truly realistic future. At the same time, his worst-case forecast seems eerily familiar if we simply expand our scope beyond human cities in the industrialized nations.

The Tijuana city dump we discussed earlier truly is a ceaselessly growing mountain of garbage. There are hundreds or thousands of places like it in India and China and South Africa, where much of the garbage consists either of the by-products of sweatshop industries producing consumer products to be sold in the industrialized nations or of electronic waste exported from industrialized nations once consumers have had their way with it.

Or recall what's happening to the oceans. A 1999 study showed that water in the Pacific Ocean contains six times as much plastic as phytoplankton. Researcher Charles Moore repeated the study in 2002 and found a whopping 10:1 ratio.[241]

For comparison, imagine if this were the case in a terrestrial environment. The amount of biomass in an average temperate deciduous forest is about six pounds per square foot. Multiplying this by the ten to one ratio we just mentioned reveals there'd have to be sixty pounds of plastic per square foot for the forest to be comparably polluted. How much is this? Since, for example, styrofoam has a density of about two-thirds of a pound per cubic foot, for there to be ten times as much styrofoam as biomass on a forest floor, there'd have to be sixty pounds of styrofoam per square foot, which means the forest floor would be covered by styrofoam to a depth of ninety feet.

Truly prodigious amounts of garbage are produced and disposed of somewhere, but what well-heeled middle-class human would want to live in a city surrounded by garbage? Part of Lynch's vision has come true—it's just that the garbage is being exported, like many of this culture's problems, so that the costs are paid by poor humans and nonhumans instead of by those who create the problems in the first place.

/ / /

Throughout this book we've been describing those who are killing the planet and those who rationalize it or who attempt to maintain this culture at the expense of the world (and at the expense of fiscally poor humans) or who promote "solutions" to the problems of this culture that take this culture as a given and the natural world as secondary, as immature, narcissistic, insane (out of touch with physical reality), and/or sociopathological. But maybe those descriptions have been far too kind. Maybe the truth is much simpler, and the people we've been describing are just plain old garden-variety greedy abusive assholes who prefer their own wealth—those who come up with and/or implement "solutions" to, for example, global warming that promote industrial capitalism over the living planet are often extremely well paid for doing so—over the health of the planet. I mean, the problems are not cognitively challenging. Ask any reasonably intelligent six-year-old how to stop global warming caused in great measure by the burning of oil and gas, and the child should be able to give the obvious answer: stop burning oil and gas. Ask an adult who does consulting for high-tech venture capital firms how to stop global warming, and the answer you get will almost undoubtedly be far more protective of high-tech corporations than the planet. And the answer you get won't work.

Bet your life on it.

/ / /

Think about it. Your bathtub is overflowing. The water is running full-blast. What's the first thing you do? Do you begin to carry water outside in a teaspoon? Do you use energy-intensive machines invented at great taxpayer expense by your buddies at huge high-tech corporations and fueled by oil refined at great taxpayer expense by your buddies at huge oil conglomerates to suck up the water, transport it elsewhere, and dump the now-polluted water into what were once salmon-bearing streams? Do you plant native grasses on the bathroom floor, and call it a "new model for sustaining industry"?

Or do you turn off the tap?

If you're not insane, you turn off the tap.

But the problem is that the greedy abusive assholes—okay, so they're sociopaths, too, and immature, and so on—at the top do their damnedest to make sure the bathtub always overflows on someone *else's* floor. And because they don't care about others, they have no real incentive to turn off the tap.

This means someone else has to turn off the tap.

This analogy of an overflowing bathtub is true for carbon dioxide, plastics, and other waste products of this culture.

We need to turn off the fucking tap.

And while we're at it, we should pull the fucking plug.

/ / /

There are three main future scenarios I'd like to explore, as follows.

First, I will extrapolate a future from the continuance of business as usual. What would happen if things pretty much continued the way they're going, with no major social transformations? What kind of world would be left?

Second, I'll explore a "technotopia." What would happen if all the "ecological" high-tech fantasies came true? What if every product manufactured was built with recycling in mind, if every new technology was integrated into what the environmental intelligentsia would label an ecologically sane society, and if maximizing an indefinitely sustainable quality of life was the goal of industrial society?

And last, what if none of the above happened? What if industrial society simply gobbled up what's left of the cheap energy and resources and then collapsed, as so many civilizations have in the past?

/ / /

Let's start with the "business as usual" scenario, since it's the most straightforward. In this scenario, things keep going pretty much the way they have been, with most trends continuing more or less into the foreseeable future. People keep buying and throwing away massive amounts of disposable products because it's easy, cheap, and convenient to do so. Corporations keep manufacturing and selling those products because it makes them piles of money, and those same corporations keep producing vast amounts of industrial waste and pollution as by-products of manufacturing processes. The environmental intelligentsia keep telling us that if we just tweak the processes (putting plants on truck factories, purchasing compact fluorescent light bulbs, and so on) we'll be able to continue this exploitative lifestyle—in other words, the environmental intelligentsia continue to serve those in power by preemptively rationalizing away the necessity of the fundamental changes required to keep this culture from killing the planet, by lulling people into believing the superficial changes suggested by members of the environmental intelligentsia are "revolutionary" and sufficient. Indigenous and poor people keep being displaced by the extensive resource extraction industries necessary for all this manufacturing to take place, and many of them continue to be enlisted into the garbage process. And governments continue to use overwhelming force to facilitate all of the above activities. (What else are governments *for?*)

Since the global economy is predicated on and requires continual growth, the above activities happen more and more, on a larger and larger scale, year after year. The population continues to grow, and economically poorer nations continue to industrialize and consume more. It's reasonable to expect increasing total industrial activity, increasing amounts of waste produced, and increasing toxification of the global environment.

The optimists among you might perhaps point out that even though those trends would continue, we would also see continuing trends of growing environmentalism in governments, corporations, and society at large. Recycling programs are growing, you might point out. There are campaigns afoot to reduce the use of disposable plastic bags and incandescent lightbulbs or even outlaw them completely. There are advertisements in completely mainstream magazines for people to recycle their old cell phone batteries! We've made recycling mainstream in only a few decades!

But people a few decades ago didn't have to worry about pollution from cell phone batteries at all. So it's hardly progress to move "forward" to a point where you've only partly solved a problem that didn't even exist a few decades ago. Consider that in 2006 alone over *1 billion* cellular phones were sold.[242] And those cell phones contained not only batteries, but displays, plastic cases, and microchips, all of which also caused by-product waste during manufacture. It's estimated that manufacturing a two gram computer chip produces some fifty pounds of waste, including toxic waste.[243] All of this hardly marks progress, especially considering that only a century ago almost every item and material was, as we discussed earlier, recycled.

This is really just a continuance of another business-as-usual trend. The number of items recycled, in absolute terms, will continue to increase. But the total number of disposable items produced, in absolute terms, will also continue to increase. So the percentage of items recycled will stagnate, and the total amount of waste will continue to grow even though recycling will be a great success in growth terms.

Let's look at a tangible historical example of this. According to the Environmental Protection Agency, in 1960, the United States produced a total of 88.1 million tons of municipal solid waste. That is, garbage of the sort collected by the curbside, but not industrial, construction, or hazardous wastes. In that same year, 5.6 million tons of waste was recycled. Flash forward to 2005, when 245.7 million tons of garbage was produced. In 2005, 58.4 million tons of the waste was recycled.

We can look at these numbers a few different ways. If we want to be optimistic, we can look at recycling by mass and say "Wow, they recycled ten times as much waste in 2005 as in 1960!" Or we can run the percentages. In 1960, only about 6 percent was recycled, compared to almost 24 percent in 2005. So, great, right? Except that the total amount of nonrecycled waste produced increased dramatically, from 82.5 million to 187.3 million tons. Not so great.

Here's another way we can look at it. Even though the total amount of waste did increase by quite a bit, so did the population. So maybe—even though this isn't the sort of thing that would really matter to phytoplankton, albatrosses, or indigenous people whose land is being turned into garbage dumps—that increase is because of population growth. According to the US Census Bureau, the population of the US increased from 180 million in 1960 to 296 million in 2005. That's an increase of 64 percent. But the amount of waste produced increased by 179 percent, far more than population change can account for.

One important fact we should remember is that these numbers are only for municipal solid waste. They don't include industrial wastes; they don't include wastes which are burned, buried, or otherwise disposed of on site; and they don't include wastes which are released as gases, like carbon dioxide. In fact, the EPA estimates that industrial waste far outweighs municipal waste. They estimate that annual industrial waste production in the US is a whopping 7.6 billion tons.[244] All municipal waste—everything you and your neighbor and the business down the street put out on the curb—weighs less than 3 percent of this industrial waste.

If we divide municipal waste by population, we get an average of 1660 pounds per person per year. But if we include industrial waste, per capita waste production jumps to 26.4 tons per person, or 52,700 pounds.

Let's pause to consider this for a moment. Imagine you live in some average house in some average neighborhood. Let's say you're a die-hard simple living environmental activist. You decide to try to produce zero waste for a year. You recycle everything. You bring cloth bags when you go shopping. You shop for groceries at the farmers market, or in the bulk food store. You get small items and appliances second hand, because they don't come in wasteful packaging. If an item or appliance breaks, you either fix it or give it to someone who can fix it. If you go to a store and any given product is disposable, or if it comes in disposable packaging, you either don't buy it or figure out a way to make use of the disposable parts.

Now, you know that municipal waste accounts for 1660 pounds per person per year. You also know that this number includes not just waste from your residence, but also waste from government offices and businesses, anything that might be collected alongside residential waste. So you march down the street to visit those offices, a sheaf of waste reduction pamphlets in hand, and manage to convince them to cut down on their waste enough to eliminate your share of it. Great job! The local corporate newspaper even features you on the front page!

But there's a problem. Even though you've eliminated all of your household waste; even though your neighbors give you funny looks because you're still wearing those ragged five-year-old tennis shoes you refuse to throw out; even though you have personally done everything you could think of to reduce your waste, short of moving out into the woods and living off the land, there's still a problem. Although you've managed to stop the better part of one ton of garbage from going into the landfill this year, your per capita share of the industrial waste produced in the US is still almost 26 tons. That's 37 times as much waste as you were able to save by eliminating a full 100 percent of your personal waste.

Okay, now you're pissed. You've washed out a lot of plastic sandwich bags this year. You've darned a laundry hamper full of torn socks and fixed your broken toaster three times, but that industrial waste just keeps on flowing like no one even noticed. Maybe you're even starting to feel depressed about it, and who could blame you? The amount of waste produced by industry far outstrips your own personal waste. What can you possibly do about a problem that big, something so out of your control? Hold that thought, because we'll be coming back to it, and we'll have some answers for you.

In the meantime, back to our scenario. Back to those 7.6 billion tons of industrial waste created each year, being churned out as you read this. All these numbers, the millions and billions of tons, are mind-boggling. What the heck does it mean to be producing this kind of garbage? It's hard to get a grip on, especially since municipal waste collection systems are designed to move garbage quickly and efficiently out of cities. A few size comparisons may help us get a better perspective on this situation.

A Boeing 747 can carry a load of about 377,000 pounds. If we wanted to ship every ounce of municipal waste produced in the US over to some "underpolluted" foreign nation, we would have to send 1.3 million Boeing 747s per year (using the 2005 numbers for garbage production). If you went and put your lawn chair down at the end of the runway at our imaginary garbage-shipping airport, you would watch a 747 filled to the brim with garbage roar over your head every 24 seconds, day and night, every single day of the year. But as always, we have to remember that most of the waste an industrial culture produces is made at a factory, not a household. So let's throw those 7.6 billion tons of industrial waste into the mix. If we decide to ship *that* waste on the 747s as well, then we might have to hire some more air traffic controllers at our imaginary airport. We now have to arrange to launch 31.5 million 747s per year. If you're sitting at the end of the runway with your lawn chair and your stopwatch you'd better have a good pair of earplugs. A 747 will be screaming past every 1.3 seconds, twenty-four seven. Picture a nose-to-tail string of 747s launching perpetually.

And all of this is just from one country. We haven't even talked about the rest of the industrialized nations.

I don't know about you, but my head is spinning a little now that we're talking about tens of millions of airplanes. So let's pick something a bit larger for comparison. Let's pick something big and famous and recognizable. A skyscraper. Let's look at something the size of a World Trade Center tower.

Each of the big World Trade Center towers had 110 stories, with a total of 3.8 million square feet of office space. Imagine every office and every corridor and cubicle and elevator filled with garbage, nine feet deep. We can assume garbage has a density of about 1500 pounds per cubic yard. This means we should be able to fit about 32.4 million cubic feet, or 1.9 billion pounds of garbage, into our skyscraper. That's just shy of a million tons per skyscraper.

So let's take annual waste production, both industrial and municipal, and add them together, which gives us about 7.8 billion tons. That would fill 8211 of these skyscrapers per year. That's 22.5 skyscrapers per day. So imagine that we have a construction crew building a new skyscraper every hour of every day—minus a one hour lunch and two fifteen minute breaks, to make the math work out to 22.5 buildings per day (and to meet labor laws). And the first thing they do, each hour, once they've finished the construction, is to fill the skyscraper from the lobby to the observation deck with the previous hour's garbage production.

If you wanted to stick to building these skyscrapers only in American cities with a population of more than half a million, you could put a new skyscraper in every city on just about every working day, Monday to Friday, of the entire year.

Imagine we were building a garbage metropolis filled with garbage skyscrapers. Our WTC-sized tower is 208 by 208 feet. A standard Manhattan block is 264 feet by 900 feet, so we should be able to cram in four per block if we don't care about parking or parks, which is a reasonable assumption since a neighborhood with four million tons of rotting garbage on every block probably won't a be very popular tourist destination. If we put four of these skyscrapers on every block, we'd be building nearly forty new blocks every week. The total land area of all of Manhattan is just shy of twenty-three square miles. Within less than four months, our tireless construction crew would cover every single acre of Manhattan—Central Park included—with garbage-filled skyscrapers. They'll fill three Manhattans per year, with a few hundred million tons leftover.

Now, dramatic as these illustrations may be, we have a caveat. So far we've been looking at total garbage output. But the fact is, of course, not all waste goes into the landfill. We would never actually see all of this waste in one place. According to the EPA, more than a third of municipal waste is recycled or incinerated, and most industrial waste goes into water. Though this would decrease our net waste quantities, it's not as much of a consolation as it could be. It essentially means that while the great majority of the waste is still being wasted, it's just being disposed of in the larger world. Hence, dioxin in every stream. Hence, plastic in the middle of the Pacific Ocean outweighing phytoplankton ten to one.

Back to our scenario. What if business as usual continued? What if current trends were extrapolated into the future? Since we've been talking about a forty-five-year interval, let's use that same time interval for our extrapolation, and we'll look at the a time between 2005 and 2050. It's difficult to predict exactly how things will change, so let's keep things simple and use the same percentage changes over this interval.

Between 1965 and 2005, municipal waste production increased 179 percent. So let's apply that to our combined municipal and industrial waste production, which gives us 21.8 billion tons per year for 2050. What I want to answer is: how much garbage would be produced in the US in forty-five years of business as usual? If we assume we have a constant increase each year between 2005 and 2050, it all adds up to (in a rather devilish coincidence) 666 billion tons over forty-five years.

To go back to our earlier comparison, that's 3.5 billion 747s. That's over seven hundred thousand skyscrapers full.

That's 263 Manhattans.

That's seventy-three Grand Canyons full of garbage.[245]

That's a hell of a lot of garbage.

And that's just for the United States.

/ / /

The nonindustrialized nations—often called the "third world"—are rapidly industrializing. China now exceeds the United States in industrial activity. India is a spinning machine of outsourced technology. The nonindustrialized nations in general are seeing major growth of industry, and often of personal consumption as well.

There are no clear estimates of how much garbage is produced in the entire world each year. Although cities with well-developed municipal waste collection systems keep track of how much garbage they move, it's nearly impossible to keep track of garbage not moved through such systems. That means we have very little information about waste disposed of in areas without well-developed waste collection systems, that is, waste burned or buried onsite, or trucked to an untracked dumping ground. All of this means that if we want to try to figure out how much garbage the global economy is producing we're going to have to make a lot of assumptions. But even a fairly approximate idea of global economic garbage generation will help us get a better grip on our waste problems.

About 20 percent of the world's human population receives about 80 percent of global economic income.[246] Let's assume consumption mirrors income, and waste mirrors consumption.

Now, we calculated that if you included industrial waste, the US economy produces about 26.4 tons per capita. The US is somewhat more wasteful, on average, than the rest of the industrialized nations. So let's assume that the average per capita waste production of the wealthiest twenty percent of humans is a nice round twenty tons per person. The poorest 80 percent have a per capita income of about 0.7 percent of the wealthiest 20 percent. Let's assume their waste production mirrors that, and that they each account for 0.13 tons of waste per year (about 267 pounds).[247]

The human population of the whole planet in 2005 was 6.5 billion. For the poorest 80 percent, population times waste per capita yields just shy of 700 million tons. The richest 20 percent account for 26 billion tons. That's a total of 26.7 billion tons globally.

Again, all this is a bit messy, partly because a lot of industrial waste produced in nonindustrialized nations—like China—is produced in export factories, so it's hard to say in that case who's responsible for what garbage.

Keeping all that in mind, however, let's do another scale comparison: 26.7 billion tons is the equivalent of 142 million 747s per year, or four and a half per second. That's more than 28,000 skyscrapers full. That's just shy of one Manhattan per month.

Let's try to guess what the increase would be by 2050. If we assume the same growth for global garbage production as we did for US garbage production—179 percent—then we get 74.5 billion tons. If we try to add up all of the garbage produced in our forty-five-year period, assuming constant growth, here is what we get: 2,277,000,000,000 tons. That's nearly 2.3 trillion tons, or a bit more than 4.5 quadrillion pounds. This is 900 Manhattans. This is fifty-eight Grand Canyons full. If you were to take all of the garbage and smear it across the state of, say, Kentucky, it would form a solid unbroken layer more than four feet deep. If you wanted to make it just deep enough to block the light and smother all plant life—say, about four inches deep—you could cover an area twice the size of Texas.

*   *   *

If we wanted, we could run some more numbers: how many more mega-tons of plastic dumped per year, how much dioxin in the water until cancer rates double, how long until the plastic in the Pacific Ocean outweighs the plankton by fifty or one hundred or one thousand times. But beyond a certain point the precise numbers don't matter. The trends are clear enough, and their effects sufficiently devastating, that if you're still reading this book you probably don't need more numbers extrapolated to understand the situation. And if you're a politician, CEO, or member of the environmental intelligentsia, the numbers probably won't cause you to act to defend the planet anyway.

′ ′ ′

Fortunately for us and those who come after, the "business as usual" scenario has some deeply problematic assumptions. The largest underlying flaw is the assumption that there are enough raw materials in the world to continue business as usual for more than a few decades. We've already passed peak oil, which means the days are numbered for the cheap energy used to extract, process, and yes, also to recycle synthetic materials. A 2006 study examined the minerals required to bringing the industrial infrastructure in the "developing world" to the scale of the industrialized nations. The authors determined that this would require essentially all of the copper and zinc in the earth's crust (and possibly all the platinum) as well as near-perfect metal recycling.[248] We could go through the industrial demand of other resources as well, but again, that would be an unnecessary diversion. We know that this culture cannot continue with business as usual, because that is what it means to be *unsustainable*.

Since we're making assumptions, we could also go ahead and assume that future energy shortages would be overcome by cold fusion, massively increased coal mining, the discovery of immense quantities of pixie dust far below the surface of Kansas, or by any number of other improbable or dangerous or magical means.[249] But as far as waste is concerned, this only makes the problem worse. Cheap energy makes it easier to extract resources, and to manufacture and ship products. If you knew that one week from now you would be magically transported to the year 3000, what would you do? You could pray for oil to run out as soon as possible, so that the world you appeared in would not be full of the garbage of more decades or centuries of civilization. Hopefully you would do much more, but, once again, we'll come back to that thought later.

This "business as usual" scenario is of interest to us not because we think "business as usual" will last for millennia, but because of the impact every year of it will have on those who do live in the millennia afterwards. How it changes, or what it changes to, is something we deal with in our other scenarios.

Since we were just talking optimism a moment ago, let's move forward to the next scenario and discuss a hypothetical technotopia.

TECHNOTOPIA

Will life be worth living in 2000 AD? . . . Scientists have looked into the future and they can tell you. . . . You will be whisked around in monorail vehicles at 200 miles an hour and you will think nothing of taking a fortnight's holiday in outer space. . . . Doors will open automatically, and clothing will be put away by remote control. . . . In commercial transportation, there will be travel at 1000 mph at a penny a mile. Hypersonic passenger planes, using solid fuels, will reach any part of the world in an hour. . . . And this isn't science fiction. It's science fact—futuristic ideas, conceived by imaginative young men, whose crazy-sounding schemes have got the nod from the scientists.

It's the way they think the world will live in the next century—if there's any world left! —"WILL LIFE BE WORTH LIVING IN 2000 AD?" *WEEKEND MAGAZINE*, JULY 22, 1961

Such vision of endless mechanical progress, such totalitarian utopias, such realistic extrapolations of scientific and technical possibilities all played a more active part in practical day-to-day changes than has usually been realized. These anticipatory subjective promptings were always in advance of actual experience, insistently beckoning, pointing ahead to the next step, breaking down resistance by suggesting that any attempt to reduce the tempo of change or to alter its direction was doomed by the very nature of the universe. . . . —LEWIS MUMFORD

THE WORD "TECHNOTOPIA" is a portmanteau of "technology" and "utopia." It conjures up a vision of a world of advanced technologies in which automation and the use of various advanced, often hypothetical technologies allow (some) human beings to enjoy a high quality of life and large amounts of leisure time. Of course, all human cultures have had technology, and many indigenous cultures have been marked by a high quality of life and large amounts of leisure time as well, but for my scenario I'll focus on a society driven by high levels of industrial technologies. And although some visions of a technological future have been driven by a cornucopian worldview where technology renders a healthy environment somehow obsolete, for the purposes of our discussion let's take a slightly less preposterous approach and assume that this future society makes sustainability a primary social value.

To summarize, this technotopia is an "ideal" society driven primarily by technological innovation and marked by a generally high quality of life. In this scenario, high levels of technology can be maintained indefinitely through near-perfect recycling and manufacturing systems. Population and consumption are firmly constrained in an attempt to maintain this society indefinitely. These constraints could be through market forces (having kids is expensive), by government regulations, or by social mores regarding reproduction. For the purposes of this scenario you can choose whichever option seems most appealing—or least dreadful—based on your own political bent. It doesn't really matter how, just that mass society has somehow come to terms with the fact that it lives on a finite planet where infinite growth simply won't fly.

This hypothetical society has placed limits on growth in an attempt to keep disruption of remaining ecological systems minimal. However, because of land demands for agriculture, and especially for biofuels, most of the terrestrial surface of the planet is farmed to meet human needs for energy, fiber, and other raw resources. Mining, remember, is based on exploiting nonrenewable resources. In our technotopia, all minerals that are economically accessible have already been mined out, leaving metal recycling and the use of biologically-derived materials supply the manufacturing industry.

Remember that human quality of life is paramount in this technotopia—and we will ignore for now the fact that from the beginning, this culture has been based on the gaining of riches by the wealthy at the expense of the labor of the poor and the land (in other words on slavery), so I'll do for a moment what most writers in this culture do, and only concern myself with the quality of life of the beneficiaries of this pyramid scheme—but that as an industrial society it's still based on the activities of machines. Machine manufacturing is usually big, fast, hot, or otherwise hazardous to humans. So, dangerous or toxic industrial jobs are carried about by a vast number of robots which are made mostly out of materials like biodegradable corn plastic. To sustain their construction, maintenance, and fueling, these robots collectively consume more resources than the entire human population. Any toxins that are produced are either remediated or sealed in underground vaults. Now, unsustainable industrial activities like mining have been largely phased out—mines are now used to store toxic by-products of industry in hermetic containers—but landfills and old industrial sites are mined for the resources they contain.

/ / /

This technotopia does not use—or require—fossil fuels. As a finite resource, any significant use of them is unsustainable.[250] In modern society, fossil carbon extracted from the ground is physically made into plastic and many other synthetic materials. This technotopia will have to get that carbon from other sources. It's safe to assume that they will also want to make their synthetic materials as nontoxic and biodegradable as possible. There *are* polymers that exist currently that conform to these requirements. Do those polymers have some potential to help our technotopians?

I decided to talk to an expert, Dr. Tillman Gerngross. Dr. Gerngross is a professor of engineering at Dartmouth College and holds advanced degrees in chemical engineering and molecular biology. Although Dr. Gerngross has studied biologically derived plastics in great detail, he's not employed by a bioplastics manufacturer, and so is able to offer us a more honest perspective (never forget Upton Sinclair's line about how hard it is to make a man understand something when his job depends on him not understanding it).

Bioplastics are plastic materials derived from biological materials, like corn, as opposed to oil, as is the case with most modern plastics. Although they can be biodegradable, bioplastics aren't necessarily so. Many early plastics were derived from biological materials, but were deliberately designed to be non-biodegradable. Dr. Gerngross and I discussed mostly three types of specific plastics: polyhydroxyalkanoates (PHAs), polylactic acids (PLAs), and various starch composites (which are usually rather brittle).

I asked Dr. Gerngross how these plastics are manufactured. He suggested that we focus on the PHAs and PLAs, which are the dominant approaches in the United States. "The starting materials for any of these approaches are typically carbohydrates, mostly sugars. So of course, various agricultural feedstocks lend themselves to providing these sugars," he told me. These sugars can be put through a fermentation process and converted either directly into the desired plastic, or indirectly through different chemical intermediates.

Dr. Gerngross told me that one of the key issues in bioplastics manufacturing is that "you need a source of carbohydrates, and when we look at industrial-scale sources of sugar or carbohydrates, corn-derived dextrose or glucose is *the* primary source." Because we have a lot of information about growing corn, and about the specific processes used to convert those corn sugars into plastic, we can make a very solid analysis of what's involved. "And so then you go back and say, 'How do we make this particular material?' Well we make it by growing corn in large amounts. And then you can say, 'We have to make a ton of glucose, to feed it into a process by which we convert that glucose either into an intermediate, to make polylactic acid, or directly into polymer: what's the environmental impact of generating a ton of glucose?'" We can ask, "'What are the yields in corn farming? What are the fertilizers? What are the pesticides, herbicides, insecticides, all that stuff that goes into corn farming? What's the energy required to make the fertilizer?' And you can do a full analysis of the environmental impact of making a ton of glucose."

Dr. Gerngross also told me that we have "very solid numbers" on the energy in various steps of bioplastics manufacturing, but that "the unfortunate net result is that going through these multiple processing steps requires a lot of energy to get glucose, and requires even more energy to convert that glucose into a polymer. And our analyses in my lab here have shown that it's almost a wash with conventional polymer production. In fact, the energy consumption is significantly higher." Dr. Gerngross pointed out that the energy required to turn oil into a plastic—say, polyethylene—is actually relatively small, and the conversion steps are comparatively energy efficient. In other words, even though a lot of energy goes into plastics manufacturing in the modern industrial economy, a lot *more* energy would have to go into it if they were making plastics out of corn sugar instead of oil.

Dr. Gerngross made another important point. He reminded me that most of the energy used to grow corn and to process bioplastics comes from burning oil and most especially coal. "So really, when you then look at the whole situation from a net environmental impact situation, clearly these biopolymers have a negative net impact on greenhouse gas emissions, clearly they have a net negative impact on land use, because the land you're using to grow corn is not used to sequester carbon through a natural environment or biotope, and in fact is competing with corn for animal feed and for human use."

He told me that people in the bioplastics industry are taking the approach that "it's green because it can be grown again." It's true that source crops can be grown over and over, he told me, "But what is not true is that the energy going into the process is renewable. In fact, it's referred to by many in the industry as 'the dirty secret.'"

Dr. Gerngross does not have a very high opinion of the ecological implications of the bioplastics industry. "My view is, if the objective is to have a positive environmental impact, the impact of these two technologies does not hold up to any meaningful scrutiny. If your desire is to preserve fossil resources, I think there's no case to be made, because they're just going to burn it all."

He then made another important point. Even though conventional plastics manufacturing involves using oil, coal, and so on, much of the carbon in those energy sources ends up incorporated in the final product. "You can make an argument that making polyethylene involves extracting fossil carbon out of the ground, burning part of it, and then ending up with fossil carbon in the form of a plastic bottle or a plastic bag. And when it goes back into a landfill it goes back into the ground. It's basically a very efficient carbon sequestration mechanism." In contrast, bioplastics manufacturing involves "using a lot of agricultural land, and you're using significantly more energy, which currently is derived by burning fossil carbon (therefore, you have a negative impact on air quality) to make something that is biodegradable, which arguably also is not such a great thing."

Why doesn't he think making biodegradable plastics is such a great thing? He told me that if they do produce biodegradable plastics, and "if these materials end up in landfills they will degrade in landfills at a rather slow rate. But because of the anaerobic environment in landfills, the carbon that is released is not released in the form of CO_2, it is mostly released in the form of methane. And methane is a very potent greenhouse gas. It's about twenty times more potent than CO_2." He said that he's not an advocate on either side, but that he wants people to look at the problem from a more rigorous and analytical perspective, "not this fuzzy 'it's green because it comes from a renewable plant' approach." Which, he pointed out, doesn't necessarily mean much, given the unsustainability of current agricultural practices.

I asked him what he thinks is behind the growing popularity of "green plastics," if those plastics aren't actually as green as they are marketed to be. Dr. Gerngross told me that there are several aspects at play. "One is that the USDA has been extremely aggressive in finding new ways of getting stuff that farmers grow to consumers. This is sort of an obvious statement, but look what has happened with corn, the ability of making extracts and intermediates of corn commodities that now go into so many different things. I think of the large amount of corn that's converted to glucose and high-fructose corn syrup: that's what the USDA wants. They want large outlets for the stuff that farmers make. Whether that's good for the environment or for society, that's a completely different question." He pointed out that the corn lobby has been very successful in getting people to eat more high-fructose corn syrup. Americans now eat about sixty-three pounds per person per year, as a sweetener in everything from soda pop to ketchup to yogurt. This, as Dr. Gerngross pointed out, has ultimately resulted in "not only obesity, but also diabetes." But remember, you can only dump so much corn product into food. People can only eat so many corn tortilla chips in a year, only chow down on so many hamburgers made from corn-fed feedlot beef, only slurp down so many Coca-Colas. Eventually, if you are say, Cargill, and you want to keep making more and more profits, you have to find a place to dump all that corn. "I think the USDA is very heavily promoting new ways of using agricultural raw materials to make more than just food. Certainly when you look at their granting pro-grams, they've promoted all those very heavily, with the completely ludicrous argument that it's going to prevent oil imports, et cetera, et cetera."

He said that another reason "green plastics" are becoming popular is that the public doesn't take into account the behind-the-scenes energy costs and ecological impacts of different materials. Materials manufacturers aren't really bothered by this fact. "They're not in the education business," he told me.

Dr. Gerngross outlined a study done by McDonald's in which they examined the environmental impact of Styrofoam fast-food containers and compared them to cardboard containers. What they found was that it actu-ally took significantly more energy, and made a significantly larger environmental impact, to produce cardboard containers. "So, McDonald's was fully aware of that. Now when you go out and ask the public, 'What do you think is more environmentally friendly? A polystyrene cup or a cardboard cup made from renewable resources?' What do you think the answer's going to be? I think the vast majority of people are going to say, 'Well, the cardboard is more environmentally friendly.' So McDonald's made a marketing decision: they said, 'We don't care, really. We would rather do what the public perceives as environmentally friendly than what is really environmentally friendly.' [Or rather, what is slightly less envi-ronmentally hostile.] For them that's a wise marketing decision, but again it's not necessarily the best thing for the environment.

"And the dirty secret, of course, is that the US paper industry *burns* more fossil carbon than the entire chemical industry uses as raw material. Again, this question of whether something is renewable runs into this very misconstrued notion that if that *raw material itself* is renewable, that's a good thing. The problem is that the raw material may be renewable, but the energy required to convert the raw material into something useful . . . typically it's very, very energy intensive. And the paper industry is, I think, a great example of that."

I asked Dr. Gerngross if these plastics—PLAs and PHAs—were actu-ally biodegradable. What would happen if you put them out in the environment? Would they yield toxic by-products? He told me that these bioplastics, at least, are essentially biodegradable. Many microbes have enzymes that can break them down, and the plastics would be a kind of food source. This wouldn't produce toxic by-products, but he reminded me that the methane produced is "natural," but can still have a negative impact if produced in large quantities.

Since he's a molecular biologist, I asked him if it would be possible to selectively breed or engineer bacteria that would be capable of breaking down conventional plastics, and here's what he told me: "It's possible, but again, why would anyone want to do that and release those bugs into the environment? I don't think that, necessarily, would be a very good idea. And in addition to that I would argue that degradability is not a good thing. The reason it's a good thing is because society contains a distribution of different people, and some of them don't care about the environment and are willing to throw out plastic bags or plastic bottles. Now, are we trying to support that behavior by saying, 'The plastic bottle that you are throwing out of your car window, we're now going to design, engineer in a way where it degrades?' We can do that, yes. But is that a good thing? I don't think so. I think we should manage our environment in a more socially responsible way."

/ / /

The specific shortcomings of bioplastics actually outline some of the overarching problems with the technotopia scenario, and with technofixes in general, with technofixes being defined as proposed solutions to societal problems that rely solely on the introduction of a new technology without requiring changes in the structure of industrial society. If the culture is from core to surface, bottom to top, micro to macro based on exploitation, what sane person could think that yet another piece of technology that facilitates this exploitation will make the culture any less exploitative?

As of this writing, there's a headline on a popular technology news site about scientists who have managed to get bacteria, with the help of several intermediate steps, to produce gasoline. Below the article is a series of comments about how great it is that "we" won't be dependent on oil, and about how now "we" can simply make renewable hydrocarbon fuels. The problem with this, and with other biologically synthesized fuels like biodiesel, is that there is only so much photosynthesis that can happen on the earth's surface. British newspaper columnist George Monbiot has noted that to fuel all of the cars in Britain with biodiesel would require farming every single acre of the country just for biodiesel, leaving no room for anything else.

The planet, having a limited surface area and limited growing seasons, has a finite total capacity for photosynthesis, a capacity measured by the net amount of carbon from the atmosphere "fixed" each year and integrated into organic matter: the amount of carbon that is saved in any given area through photosynthesis is called "net primary production" (NPP). A recent study estimated that humans "appropriate" about one quarter of the Earth's net primary production, through direct consumption in the form of food, indirectly by feeding plants to farm animals, or by land use changes like deforestation that affect NPP. In the dry understatement characteristic of published scientific papers, the authors note that "[t]his is a remarkable impact on the biosphere caused by just one species." They also warn that their "results suggest that large-scale schemes to substitute biomass for fossil fuels should be viewed cautiously because massive additional pressures on ecosystems might result from increased biomass harvest."

To some people, one quarter may not seem like a lot. After all, we're still leaving most of the net primary production for the planet, right? Well, not so fast. First of all, this planet is home to millions or tens of millions of different species. This is something like having a city of millions of people where 99.99999 percent of the people live in modest houses and apartments, and one person lives in a gigantic house that covers a quarter of the entire city. In other words, this arrangement is not winning any awards in the fairness department. Especially when you consider that not only do humans make up the tiny, tiny, tiny minority of all species, but also the tiny, tiny, tiny minority of all biomass, being vastly outweighed by other groups of species, like insects, and even other specific species, like giant squid.

Moreover, it's not like this carbon appropriation is happening evenly all over the planet. Human tax collectors aren't dropping down and scooping up a mathematically fair amount of carbon from each square meter on the planet. Rather, some areas are totally destroyed or radically altered for the purposes of resource extraction, which results in major species extinctions.

If the issue of fairness doesn't faze you, here's an argument that the more self-interested can get behind. The Earth's normal "photosynthesis budget" isn't spent on lottery tickets and potato chips (or at least, the portion that isn't appropriated by humans isn't). All of that photosynthesis is spent maintaining healthy ecosystems, replenishing species, nourishing soils, and taking carbon out of the atmosphere so that the planet stays at a reasonable temperature. That photosynthesis is being spent to hold the planet together. Once again, how do you think the planet got to be so beautiful, resilient, and fecund before this culture started to destroy it?

A major cut in the photosynthesis budget can be a big deal, just as a major cut in your household budget can be a big deal. Imagine you make, to choose a nice round number, a thousand dollars every two weeks. During those two weeks, you need to spend two hundred dollars on rent, two hundred dollars on food, two hundred dollars on transportation to work at your job, and three hundred dollars on medical costs, clothing, and general maintenance. One day when you get to work, your boss—he's a real jerk—tells you that he needs to "appropriate" 25 percent of your salary so he can get a new car. That's progress, after all. "Don't worry," he tells you. "You still get to keep most of it." So what do you do? Try to find a cheaper apartment? Start dumpster diving? Wear old clothes? Start farming out your children to relatives who are just as strapped? You can adapt to a small cut in your household budget, but a big cut means trouble with your essentials. The same goes for the planetary photosynthesis budget, and the planet doesn't have any friends it can go to borrow money from if it hits rough times.

Within certain limits, human civilization can appropriate photosynthesis without major short-term consequences. But as that cooptation reaches those limits, the consequences (in terms of climate change, drought, or general ecological collapse) become overwhelming, even locally: prior to the arrival of this culture, what is now Iraq was covered in cedar forests so thick the sunlight never reached the ground, the Arabian peninsula was oak savannah, and so on. This culture destroys landbases wherever it goes. These consequences become a vicious cycle, in which more primary production is appropriated each year to deal with the consequences caused by last year's consumption. Furthermore, and this is a running theme in this book, the benefits of this consumption go to a small group in the short term, and the costs are paid by others in the short term and everyone in the long term.

And then there are matters of scale. This is another shared problem between bioplastics in specific and technofixes in general. An act that's sustainable at one size, clearly, is not necessarily sustainable when grossly enlarged. A very large system has quantitatively *and* qualitatively different impacts on its environment than a smaller-scale one. That's why a water strider weighing one-thousandth of an ounce can move effortlessly across the surface of a lake, resting on the water's surface tension, but a cow weighing one thousand pounds would plunge thrashing through the surface.

And then there are impacts that an environment has on a creature. As evolutionary biologist J. B. S. Haldane famously wrote: "You can drop a mouse down a thousand-yard mine shaft; and, on arriving at the bottom, it gets a slight shock and walks away, provided that the ground is fairly soft. A rat is killed, a man is broken, a horse splashes." The vast majority of animals on the earth are tiny invertebrates like worms and flies. And the vast majority of living creatures are microscopic. Clearly there are many advantages to staying small.

If you grossly enlarge any act to an industrial scale you will see destructive impacts to and from the environment. If you take the backyard sugar cane plot and turn it into a 10,000-acre ethanol farm, you are going to cause major ecological and social destruction. And at the same time, that industrial-scale structure is going to be much more brittle than a smaller structure, and more prone to collapse.

Our last common problems, and possibly our worst, have to do with social impact, power relationships, and control. That's a subject that I'll return to shortly.

/ / /

If I'm going to talk about technology, it's a good idea to have a clear definition of what it is. Often people who critique technology, or civilization, or "progress," get pigeon-holed as "antitechnology" or "luddites." Part of this stems from meanspirited and misleading rhetorical techniques, and part of it stems from muddy ideas of what technology actually is.

Some people would define technology as anything humans make. I think this is too miserly a definition (and of course narcissistic, but why should that surprise us?)—a variety of nonhumans use tools creatively, including, among many others, primates, various birds, and invertebrates like octopi. A more inclusive definition is appropriate. I like to say that technology is any structure a creature employs which is not actually part of that creature's body. This fits (more or less) with the definition of technology we used earlier, when talking about magical thinking.

By this definition, a snail's shell, for example, would not be a technology, because it is part of the snail's body. A bird's song, on the other hand, is a form of communications technology. Birds can adapt songs to different circumstances, and many birds will learn new songs (or other sounds) and potentially employ them in problem solving. A bow and arrow is a technology. A beaver dam definitely qualifies as a technology, and so does a bird's nest. A bent piece of wire used by a crow to retrieve a tasty piece of meat from out of reach is a technology, as would be the knowledge of how to make it. Some might argue that a crow bending a piece of wire into a tool doesn't actually count as technology, because the crow is only making a tool out of something it found. But humans do the same—even advanced composite materials are made out of modified precursors, not conjured out of midair.

Some examples pose fascinating conundrums when considered under this definition. For example, a hermit crab will find empty shells, or even empty cans or containers, and then use one of those as its new home. Does this qualify as technology?

At this point, the lines between what is and is not a technology become blurred (and that, of course, is part of the point). A fox might curl up in a hollow log, and though this is not as sophisticated as the crab's means of attachment, it is in many ways the same. Is that a technology? How is a fox using a hollow tree to keep warm all that different than you sitting in your wood home? The aquatic caddisfly larva will cover itself in a protective coating of available debris, from twigs to pebbles to small fragments of metal. This debris essentially becomes part of the larva's body, which would from one perspective imply that it is not technology. But how is that different than other forms of armor? How is that different than the pacemaker placed in someone's chest, or the pharmaceuticals you ingest and that become a part of you? Or what about those species of cephalopods, like octopus and squid, who communicate through shifting patterns of color on their skin? If a bird's song counts, why not a squid's color patterns? Or a spider's web, secreted out of the spider's own body, but an undeniably sophisticated and adaptable structure which surpasses many examples of human high technology?

One of the things that makes technology so interesting—and so deadly—in this culture is speed. Technology now changes quickly, whereas a few centuries ago technological change was not so stunningly fast. Cell phones. Computers. Garage door openers: anyone over about forty remembers when they were unheard of, and now in the US at least they're everywhere.

This rapid change is central to the issue of accumulating waste and pollution, because humans are creating more and more materials which have no timely or natural means of decomposition. Indeed, often the motivation is to produce materials that resist decay.

For non-technological structures this is generally not an issue. A snail's shell, or a tree's leaf, or a bone, are all built from a shared biochemical "palette." In the billions of years life has been on this planet, living creatures have thoroughly explored the range of molecules that can be produced by living cells. When new molecules were developed that offered a survival advantage, the genes that produced them spread gradually. Sometimes those molecules would have resisted decay. Plants in particular have an incentive to produce indigestible compounds, since that's one of the few means they have to discourage predators. But as those durable molecules became more commonplace, there would also be a greater incentive for predators and scavengers to develop an ability to digest them.

Imagine that a plant develops a waxy coating that can't be digested by any other creature and doesn't break down under normal environmental conditions. If this means the plant has fewer predators, it may produce more offspring and the genes for the waxy coating will spread. But at the same time, that waxy material will begin to accumulate in areas where the plants grow. The more commonplace this plant becomes, the greater the concentration of the wax, and the more incentive another creature has to work out how to eat it.

If it takes a long time for someone to figure out how to eat the wax, and it accumulates in the environment, it may actually start to harm other creatures that share habitat with the wax-producing plant. This may prevent the plant from thriving in areas it already inhabits, but it won't stop the plant from spreading into new ones. Scenarios in which a living creature actually produces enough waste for it to accumulate to a danger point are exceedingly rare in nature.

Okay, fine, but are there actually any examples of a living creature producing enough of one indigestible material for it to accumulate significantly in the environment? Has any metabolic by-product ever built up to the point where it was a threat to other living creatures on a global scale? The answer to both questions is yes. On the surface, garbage apologists can use this as an excuse—"what we're doing is natural"—but the way those scenarios played out in nature shows us yet again why our modern-day waste situation is so unnatural, and so destructive. More on this later.

We talked about an inedible waxy coating building up in our hypothetical example above. But there are real-world examples of materials that accumulate because they aren't very digestible. In fact, close to two thirds of all the nonfossil organic carbon in the world is tied up in just two biological materials—cellulose and lignin.

Cellulose incorporates the majority of that carbon. Cellulose is a crucial building block for plants. Structurally, it has a superficial similarity to synthetic plastics. Cellulose and plastics are both polymers—large molecules made by connecting many identical units. One crucial difference is that cellulose is made out of many sugar molecules. This means it can be broken down into sugar, and then into water and carbon dioxide. In contrast, synthetic plastics are built from nonbiological units, which often include chlorine and fluorine atoms which are decidedly not harmless. In fact, some early plastics were made by modifying natural polymers like cellulose: that's where "cellophane" gets its name.

Living creatures in general are very capable of dealing with sugar molecules, but in cellulose these molecules are linked to each other in a way that is difficult to separate. Because of this, many creatures—humans included—cannot digest cellulose. This isn't necessarily a problem: even though we can't digest cellulose, it's not harmful for us to eat. It *is* good to get a certain amount—that's what dietary fiber is made from—but we can't live off of plants consisting largely of cellulose, such as grasses.

Now, if *no* one could digest cellulose, that would be a problem, because cellulose would continue to accumulate, and (harmless as it is) we—the inclusive we, including all beings, including plants, including the earth—

don't want the entire biosphere to be made out of cellulose. Fortunately, some creatures have evolved that ability. When a new material shows up, if anyone can figure out how to digest it, it's probably going to be bacteria or other microbes. They have a couple of advantages; they have short and rapid generations, allowing them to essentially evolve faster, and bacteria can share genes with other bacteria who aren't their offspring. Larger creatures may eventually work out partnerships with these bacteria. This is the case with cows and other ruminants. Although cows cannot digest the cellulose that makes up much of their food, one of their four stomachs house bacteria who can eat cellulose. (That stomach is called the rumen: hence the name ruminants.) Those bacteria also synthesize vitamins and other beneficial compounds for the cows.

Although we haven't evolved to digest cellulose, it's actually not a tough nut to crack, biochemically speaking, if we compare it to lignin. Lignin is what makes wood, in a word, wood. It's strong, tough, and very difficult to digest. That's why if you're lucky enough to take a stroll in an old growth forest, you can still see fallen logs from decades, or in some cases centuries, ago. Lignin, which makes up about 30 percent of the world's non-fossil organic carbon, takes a long time to break down. And not many living creatures have mastered the trick. Really, only fungi are adept at breaking it down.

Now, again, if you're lucky enough to take a stroll in an old growth forest, you can see that this is not a bad thing. In healthy forests in general, stumps and fallen logs play vital ecological roles. They provide shelter for animals from ants and sow bugs all the way to elk and bears. They act as nurseries for young trees and other plants. Have you ever seen a seedling sprouting from the top of a rotting stump? And perhaps in the same forest, have you seen a mature tree with roots branching out in mid air, where a stump was when that tree was a seedling? The variation in terrain created by fallen trees yields a variety of ecological niches and microclimates, moist lower areas and dryer mounds, which ensures the ecological diversity of the forest's plant, fungi, animal life, and so on.

The carbon in cellulose has a similar role to play, though at a different scale. Fallen leaves on the forest floor retain moisture and nutrients, and feed the living detrivores there. As they're broken down, they're incorporated into the soil, increasing the soil's ability to hold water, air, and nutrients, and feeding the microbes who feed the tree roots.

If cellulose and lignin were magically altered so they broke down overnight, the results would be disastrous. All of the carbon stored in fallen trees and leaves would be released into the atmosphere. The forest floor would become lifeless as it lost the ability to retain water and nutrients. Many plants and animals would lose their habitats and become extinct. Seedling trees would no longer have suitable places to take root, and the forest would die.

This is what makes these durable, natural polymers so different from synthetic polymers. They're not waste. Rather, their evolution laid the basis for a great increase in the volume and diversity of life. Fallen trees and leaves are not just *in* the forest, they *are* the forest. They are its future as well as its past. Who can say that of a plastic fork?

/ / /

There's another big obstacle to an ecologically sustainable industrial tech-notopia: war.

Supposedly green military technologies have been making headlines more and more over the last couple of years. Wind turbines constructed at Guantánamo Bay, programs to develop a biofuel for fighter jets, and a hybrid-electric military jeep are among the most notable examples. It's up for debate whether such initiatives demonstrate simply a series of shallow public-relations moves or a military recognition of peak oil and a looming energy crisis. In either case, most thinking people realize that war and sustainability don't go together.[251]

In the technotopia I'm describing, the people of the future deal with finite supplies of various valuable metals by instituting a perfect cradle-to-cradle approach to industry, similar to that proposed by William McDonough (never mind that it still, several hundred pages later, is not going to happen). All manufactured metals are carefully fabricated, tracked, and recycled when a product reaches the end of its use (I almost wrote *life span*, except of course products aren't alive: I mention this to show how deeply we're all enculturated to identify the products of this culture with life). We can talk about how (im)plausible that is at the best of times, but war is not the best of times. War materiel, from bullets and uniforms, to missiles and bombs, to tanks and fighter jets, is built to be destroyed. There's a good chance that in a time of war almost anything built by or for the military could be torpedoed, sunk, burned, blown up, nuked, or otherwise obliterated. That's not a good way to build a cradle-to-cradle industrial system. In fact, various munitions are built specifically to *be* destroyed and converted into dust and unsalvageable wreckage. They are designed to become waste, and to waste something—or someone—else. The worst munitions don't just physical destroy the target—they irradiate it. They also irradiate the battlefield, and the landscape. Case in point: depleted uranium.

Depleted uranium (DU) is primarily a by-product of the processes that "enrich" uranium used for nuclear warheads and reactor fuel. The military mostly uses depleted uranium for two purposes—armor and ammunition. The reason bullets are commonly made of lead is that lead is a fairly dense metal. Bullets are made out of dense materials so that a small projectile can carry of lot of energy. The less dense a projectile is, the less damage it can cause—think nerf guns. Well, depleted uranium is about twice as dense as lead, which means it can do a lot of damage, and is especially effective in armor-piercing munitions. It's also "self-sharpening," meaning that when a DU projectile strikes a hard object, it fractures into smaller, sharp shards. Most of a projectile is turned in to dust, but the remainder—which may weigh several pounds in larger shells—is a solid lump of uranium left on the battlefield. DU is also pyrophoric, capable of spontaneous ignition. This means when an armor-piercing DU penetrator is fired at an armored vehicle, it may punch through the armor and, by the time it enters the crew compartment, be fragmented into a radioactive dust that will catch fire. Just what you want in an armor-piercing shell. Not what you want if you care about a non-irradiated populace and landscape.

By the 1970s, the US had stockpiled about half a million tons of depleted uranium. Since DU is radioactive, it was proving expensive to properly store such large amounts. However, at that time the Pentagon was searching new materials for armor-piercing munitions, and realized it could, as it were, kill two birds with one stone—make more effective weapons, and get rid of large and expensive stockpiles of DU. Isn't recycling grand?

DU munitions were first used in the 1973 Arab-Israeli War, but the US military has used DU in the first Gulf War, in Bosnia, Kosovo/Serbia, and in the more recent invasion and occupation of Iraq.[252]

Representatives from those countries have attempted to determine what health, safety, and decontamination measures are appropriate, but the US military has been less than helpful, even refusing to give locations where DU munitions have been used. During the first Gulf War, US forces fired close to a million rounds of DU ammunition, with a total mass of DU somewhere in excess of three hundred tons.[253] It's also been used in training operations in the US, despite the fact that DU use is prohibited outside of combat.

Although the existence and use of DU munitions was largely unknown outside of the military, it gained public attention in the early 1990s because of the Gulf War Syndrome. American veterans of the Gulf War began reporting a host of strange symptoms, including immune system disorders and birth defects, and the departments of Defense and Veterans Affairs now recognize brain cancer, amyotrophic lateral sclerosis, and fibromyalgia as potentially connected to the Gulf War.[254] A host of potential causes were suggested, including vaccines administered to soldiers, chemical weapons, and infectious diseases. The use of DU munitions was also suggested, since soldiers handled the munitions and were exposed to dust particles on the battlefield. After the Gulf War ended, Dr. Doug Rokke, who served in the US Army for thirty-six years, became the director of the Depleted Uranium Project for the Department of Defense. It was his job to investigate the health impacts of DU for the army. Starting from the perspective that his job was simply to make sure that DU munitions could still be used, he was shocked by what he found, and the information he found brought him "to one conclusion: uranium munitions must be banned from the planet, for eternity."[255] He has since become an outspoken critic of DU weapons.

The morality and legality of using DU weapons has been, of course, extremely controversial. The US military continues to insist, predictably, that DU is essentially harmless in terms of radioactivity. It has been argued that depleted uranium munitions, though clearly a type of nuclear weapon, do not qualify as such because radioactivity is not their primary effect. Of course, radioactivity is not the primary effect of an atomic bomb, either—the primary effect is an enormous explosion. But that didn't stop thousands of people in Hiroshima and Nagasaki from dying of radiation poisoning.

One report from the former Yugoslavia concluded that the US could not be brought before a war crimes tribunal for using DU munitions in that country, because there is no treaty which specifically and explicitly bans the use of depleted uranium.[256] This seems a little like saying you can't convict someone of murder for beating another person to death with a ball-peen hammer, because there's no specific law against beating someone to death with a ball-peen hammer. There have been more sensible interpretations of international law regarding such weapons. Humanitarian lawyer and UN delegate Karen Parker analyzed international laws that apply to DU and concluded that "DU weaponry cannot be used in military operations without violating [humanitarian law], and therefore must be considered illegal." The use of such weapons "constitutes a violation of humanitarian law . . . its use constitutes a war crime or crime against humanity."[257] Parker writes that there are "four rules derived from the whole of humanitarian law regarding weapons," and hence, four tests we can apply to any weapon to see whether they are at all compatible with international humanitarian law: 1) The Territorial Test: "Weapons may only be used in the legal field of battle, defined as legal military targets of the enemy in the war. Weapons may not have an adverse effect off the legal field of battle." 2) The Temporal Test: "Can only be used for the duration of an armed conflict. A weapon that is used or continues to act after the war is over violates this criterion." 3) The Humaneness Test: "Weapons may not be unduly inhumane." 4) The Environmental Test: "Weapons may not have an unduly negative effect on the natural environment."[258]

Parker concludes that the use of DU munitions fails all four of these tests. It cannot be contained to a given field of battle, since DU dust or DU contaminated soils can easily blow great distances from the original battlefield. And DU, with a half life of 4.5 billion years, will continue to irradiate and toxify long after everyone who even remembers the war is dead. DU also fails the humaneness test, because DU contamination kills by inhumane means (such as cancer and kidney disease) and can also cause birth defects in those not yet born. And clearly, DU fails the environmental test, since contamination is extremely long term, and any cleanup (which almost never takes place) is extremely expensive and limited.

It's estimated that each of us ingests about half a milligram of uranium each year in our food and water—that's for people who aren't living in or near a war zone where DU munitions are actually being used.[259] But there's another industrial by-product that emerged from the nuclear industry which we consume in far greater quantities.

/ / /

One of the cities where I work as a paramedic has a factory complex which manufactures uranium fuel rods for nuclear power plants. Last year we got a call for a patient who "got uranium in his eye." I was a bit worried, but fortunately it wasn't as bad as it sounds: I learned that uranium isn't the most pressing concern. When uranium is refined, it's combined with hydrofluoric acid, an extremely dangerous compound of hydrogen and fluorine dissolved in water. Fluorine wasn't produced commercially in the US until the dawn of the nuclear age, but you likely encounter it on a daily basis now.[260] It's put in most of North America's municipal drinking water.

When I think of opposition to fluoride in drinking water, the first image that comes to mind is the scene in *Dr. Strangelove* where the insane General Ripper proclaims that water fluoridation is masterminded by an "international communist conspiracy to sap and impurify all of our precious bodily fluids."[261] It's not an accident that this comes to mind. Although water fluoridation was a highly contentious and controversial issue in the United States during the 1950s, advocates of fluoridation eventually won in part by ridiculing opponents like members of the John Birch Society, who genuinely believed that fluoridation was part of a conspiracy to poison the American people.[262]

So I have to admit, when someone first suggested to me some time ago that the fluoride in drinking water was a toxic industrial by-product, I wasn't sure whether I should take them seriously. In fact, for years I automatically relegated the idea to the status of urban myth. Sure, it wouldn't be the first time public health organizations and other government agencies deliberately acted against the health of the public. I thought of the *Tuskegee Study of Untreated Syphilis in the Negro Male*, a forty-year-long study conducted by the US Public Health Service in which hundreds of black men were deliberately denied treatment for syphilis even when safe treatments had been readily available for decades. Or Operation Whitecoat, a secret US Army project lasting from 1954 to 1973, in which various biological weapons and vaccines were tested on thousands of conscientious objectors, mostly Seventh-Day Adventists. Or perhaps the Cold War-era Project SHAD, in which the US military tested biological and chemical weapons on unknowing and un-consenting subjects. But deliberately dumping industrial by-products in the drinking water and calling it healthy? It just seemed too outlandish. Too much like the evil plan of a cackling supervillian in a James Bond film.

I read through a dozen articles and reports on water fluoridation and related controversies without finding a single mention of the sources of the fluoride used in municipal drinking water. And then I found a somewhat obscure statement by the Environmental Protection Agency, in which they blithely endorsed the practice. In their own words, the EPA "regards such use as an ideal environmental solution to a long-standing problem. By recovering by-product fluosilicic acid from fertilizer manufacturing, water and air pollution are minimized, and water utilities have a low-cost source of fluoride available to them."[263]

What does this mean? Well, the fluoride used in municipal drinking water is from mineral processing plants.[264] The smokestacks of phosphate fertilizers have scrubbers installed to reduce air pollution by removing pollutants from the outgoing smoke. In this case, however, the pollutants contain a sizeable amount of fluoride. The substances recovered by the scrubbers are repurposed and put into drinking water. A convenient way to minimize pollution, as the EPA put it. And it is, or at least was, a serious source of toxic pollution.

I'm typing this with my computer resting on a very old wooden milk crate. It reads "Dunnville Dairy, Dunnville, Ontario." Dunnville is a small southern Ontario town where my grandfather worked as the milkman in the middle of the last century. In 1960, Dunnville became the home of a brand new monument to progress—the smokestack of their very own phosphate plant. As the Canadian Broadcasting Corporation reported several years later: "Farmers noticed it first. . . . Something mysterious burned the peppers, burned the fruit, dwarfed and shriveled the grains, damaged everything that grew. Something in the air destroyed the crops. Anyone could see it. . . . They noticed it first in 1961. Again in '62. Worse each year. Plants that didn't burn were dwarfed. Grain yields cut in half. . . . Finally, a greater disaster revealed the source of the trouble. A plume from a silver stack, once the symbol of Dunville's progress, spreading for miles around poison—fluorine. It was identified by veterinarians. There was no doubt. What happened to the cattle was unmistakable, and it broke the farmer's hearts. Fluorosis—swollen joints, falling teeth, pain until cattle lie down and die. Hundreds of them. The cause—fluorine poisoning from the air."[265]

The problem wasn't just in Dunnville. Problems were commonly reported near phosphate processors. A 1972 US Department of Agriculture Handbook warned: "Airborne fluorides have caused more worldwide damage to domestic animals than any other air pollutant."[266] Of course, since the 1960s and 1970s, air pollution control measures on phosphate stacks have significantly improved. Scrubbers are now present to capture such hazards so that they can be more sensibly disposed of in our drinking water.

And current industrial sources of fluoride are hardly exceptional. Before phosphate fertilizer by-products were the source of choice, most fluoride for drinking water was a by-product of aluminum refining.[267] And essentially no forms of fluorine were produced in the US before the Manhattan Project. But now two thirds of Americans ingest about a milligram of fluoride with every liter of water they drink, making it the most common drug administered in the US.

Examining the benefits and hazards of water fluoridation are beyond the scope of this book.[268] But there are two lessons to be learned from all this. First, for the US government to deliberately scatter a known, toxic, industrial by-product across someone else's land would hardly be unprecedented—they already do it with their own drinking water. And second, you don't need a dark communist conspiracy to undertake such a venture: capitalist business as usual will more than suffice.

/ / /

A few years ago, at the online Derrick Jensen discussion group, a new poster started saying things like "we don't need to worry about environmental catastrophe. All of this pollution is just a new nutrient. At one point oxygen was a deadly pollutant to most creatures and caused a massive extinction, but look how well that turned out. Really, industrial civilization is just a totally natural step in our planet's evolution."

The poster did get one thing right. When life first appeared on Earth, there was very little oxygen gas in the air or dissolved in water. Photosynthesis had not evolved yet, so only limited sources of energy were available, mostly from digesting various chemicals like hydrogen sulfide.

When photosynthesis did evolve, about 2.7 billion years ago, it was the start of big changes. For anaerobic organisms, oxygen is toxic. And when oxygen became widespread in the atmosphere, it caused the death of those anaerobic species who couldn't hide themselves underground or evolve to tolerate the gas. This event is called the Oxygen Catastrophe, or sometimes, more optimistically, the Oxygen Revolution. Many species did go extinct.

But the poster had a bunch of things wrong, too. There are some big differences between current events and the Oxygen Revolution. First, that event involved the production of really just one chemical. It's pretty conceivable to me that life could evolve to tolerate one new chemical, but modern industry is producing thousands upon thousands of new toxins. And second, the issue of time. Evolution takes time. Between the evolution of photosynthesis and the oxygen-induced extinctions around 300 million years passed. Most modern pollutants have been around for less than a century.

We could outline other differences, too, but I think the main distinction is on the long-term impact on life on this planet. Photosynthesis allowed life on Earth to make use of energy from the sun, instead of relying on small amounts of chemical energy from (mostly) underground. Although some species did go extinct, because of the change the planet can support a far larger and more diverse population of life than it could before. An oxygen rich atmosphere also allowed the evolution of large, fast moving, warm-blooded animals (ourselves included). All that sounds like a net improvement for life.

But industrial civilization isn't offering that at all. In fact, it's decreasing the population and diversity of living creatures in general. In addition, it's shifted human society from relying on solar energy to relying on small (and short-lived) amounts of chemical energy from underground, now in the form of fossil fuels. And civilization is taking oxygen out of the air and replacing it with the gases carbon dioxide and methane, essentially moving the atmosphere slowly back towards what it was like before the Oxygen Revolution.

The poster on the discussion group was arguing that industrial pollution was the same as the Oxygen Revolution. But in every way that matters, it appears to be the opposite.

/ / /

This attempt to naturalize industrialized ecocide by equating it with great ecological changes of the past is something I've seen many times. Of course, it's now generally accepted that human beings—specifically industrial civilization—are causing one of the largest mass extinctions in our planet's history. If you're reading this book, you'll probably agree that's not a good thing. However, it's also been argued that since mass extinctions have happened in the past they are natural, and therefore good, implying that it's okay for industrial civilization to destroy the planet.

This is a bit like saying that because someone fell out of a tree and broke her leg that it's then okay to run over her with a truck. Just because something happened in the past doesn't make it right to replicate it. Especially since some mass extinctions were caused by things like asteroid impacts, which are essentially unpredictable and unplanned accidents in ecological terms.

Moreover, the current mass extinction—what's called the Holocene mass extinction—is markedly different from mass extinctions of the past. Yes, something like an asteroid impact kills a lot of individual creatures, as well as many species, and it changes the planet's ecology. Well, that's happening now, you might say. The difference is that when those mass extinctions pruned the tree of life, they also allowed it to branch out again. Mass extinctions cleared habitat and ecological niches, leaving room for new species to evolve. These extinctions may be followed by an explosion of radiative evolution, in which nature has room to try out new and experimental ways of living. Some evolutionary traits are lost, but many are gained.

However, for this to happen there must be room. The activities of industrial civilization now destroy or reduce habitat in the long term. Cities and roads are paved over and inhospitable. Farms wipe out ecosystems and replace them with a few domesticated species. This kind of habitat reduction and ecological imperialism means there's no room for the kind of explosive evolution we've seen in the past.

That is, of course, unless civilization collapses, and farms and cities begin to return to wild habitat, a subject to which we'll soon return.

In the technotopia we're describing, of course, this does not happen at all. Instead, human impacts continue to grow, with farms for biodiesel and bioplastics expanding rather than contracting. A technotopia would make the Holocene extinction permanent, and leave the tree of life crippled forever.[269]

/ / /

A couple of years ago I was sitting with a friend of mine, talking about peak oil, the collapse of civilizations, and related ideas. This friend worked on community gardens, helped people repair their bicycles, dumpster dove, and generally lived a pretty low-impact life. So it surprised me when partway through the conversation he said, "Well, we don't need to worry about any of that, because we can just grow food in orbital space colonies and ship it down to Earth."

I was dumbstruck. I didn't even know how to answer. *I must have missed the point where we stumbled into bizarro-land,* I thought. He might as well have said that magical pixies were going to end world hunger, or that Xandraxis from the Fifth Dimension was going to teleport in and make new rainforests grow out of discarded coffee cups.

Partly I was surprised because this wasn't a stupid guy. He was pretty smart, as, I presume, are a lot of people who are attracted to the technotopian solution. But a belief in technotopia is essentially a form of magical thinking. And not even magical thinking in that Starhawk-style of think-positive-but-still-do-activism kind of way. The problem is, since the idea of technotopia is disguised in high-tech terminology, people don't think of it as magic.

Magical thinking, if you recall, is sometimes defined as, "The erroneous belief that one's thoughts, words, or actions will cause or prevent a specific outcome in some way that defies commonly understood laws of cause and effect," or more succinctly as "a conviction that thinking is equivalent to doing." This seems to sum up many beliefs about industrial technology, and sometimes activism, especially pacifist activism that proposes things like "meditating for world peace." The die-hard pacifist, and the die-hard technotopian both believe that just *thinking* hard enough will solve any problem, either through cosmic vibrations or new technologies.

When I was writing this section, the amazing activist and writer Lierre Keith told us she couldn't wait for the section to be finished. "Technotopia is where progressives have gone to die," she told me, "and the idea is in serious need of debunking." It's true, and I think it's fairly evident why that is. Anyone who cares about human welfare, social justice, or ecology can see that the planet is in a lot of trouble. There are more people every day, consuming more every day. Those in power are getting more powerful every day, and the gap between them and everyone else continues to widen. Ecological limits are being trampled. And these basic trends have been at work, with a few interruptions and collapses, essentially since the beginning of civilization. Environmentalism has failed to stop or reverse global destruction, and many of the social justice gains that have been made are now dependent on the goodwill of governments or on the surplus production of a system which is itself unsustainable and based on exploitation. In other words, it's clear to intelligent people, and painful to sensitive people, that we're in a lot of trouble and that our efforts to deal with that have so far been pretty ineffective.

/ / /

So what do you do if the odds are against you, if the situation is incredibly complex, and you don't see a way out that doesn't involve a lot more pain and suffering for a lot of people? Sometimes you cope by pretending that everything is going to be all right in the end, because that makes it a bit easier to get through the day, and maybe that helps you to do work that is important and valuable. If you've been raised by a society of God-fearing men and women, maybe that belief is in heaven or the Second Coming. On the other hand, if you've been raised by a society of gadget-loving *Star Trek* watchers constantly shown a future where technology has ended poverty and where even the most dire problems can be solved by reversing the polarity on the deflector dish, maybe you have a different belief.

There's no question that industrial technology is good at solving certain problems. But drawbacks include: industrial technology depends on a large-scale and centralized society; those in power choose the problems it will address; and every problem it does solve creates a cascade of new problems.

In any case, those who believe that orbital space colonies or the like will feed us in the future won't be moved by technical arguments to the contrary. Although the belief is ostensibly one of science, they won't be impressed by discussion of energy return on energy invested, or the technicalities of carrying capacity or nutrient recycling. The belief that technology will solve all of our problems is a comforting article of faith.

In the end, technotopia offers a pacifying false hope. It used to be that the discontented masses were promised "pie in the sky when you die," mediated by a class of theocrats. In the more "rational" modern age many people won't believe religious promises of the afterlife. So a new promise, predicated on the same model, promises a future of technological bounty, mediated by a class of technocrats. In both cases, the promise serves to lull and distract potential dissidents, and prevent them from taking responsibility into their own hands.

TECHNOTOPIA: PRODUCING WASTE

Most creatures produce wastes that may be poisonous to themselves or even to others, and so can disrupt the wasting cycle. Man [sic] is unique, because he makes substances that are poisonous to all living things [sic], including himself. —KEVIN LYNCH

WHY DOES THIS SOCIETY produce so much waste?

Imagine that you are one of the chief architects of a technotopia. If you and your team want to design a wasteless society, answering this question will surely be at the top of your to-do list. You can't create a society without waste if you don't understand why the dominant culture is so incredibly wasteful.

You could simply say that members of this society are lazy. They value convenience over sustainability. They're greedy for material possessions. They're addicted to consumption. They're pathologically short-sighted.

And these things may very well be true. However, as an ultimately rational technotopian, these answers aren't good enough. They don't go deep enough. They aren't precise enough. And they're not especially actionable. It's kind of hard to build a good, proactive waste-elimination plan around the fact that people are just plain lazy. (Unless you want to make wastefulness a lot of work, of course.)

As a rational technotopian, you'll probably want to task some other rational person with the job of answering this question. Let's say you choose an economist.

An economist might start by looking at the "incentives" people are offered to be wasteful: What are people getting out of it? What are the rewards for being wasteful? And conversely, what are the penalties?

At an individual level, in modern society, there are a great many incentives for making wasteful decisions. Some of these incentives are financial, some of them are social, and some of them are psychological.

I went to the store recently to buy some ink refills for a pen I like to use. When I got there, I found that it was more expensive to buy refills than it was to simply buy the entire pens new. This is a financial incentive, and financial incentives are usually pretty obvious. Inkjet printer cartridges are an even more egregious example. Printer manufacturers will often inflate cartridge costs, and subsidize new printer costs, because they make their profits on cartridges. This price-gouging sometimes extends to the point at which it can be cheaper to buy a new printer than to replace a few cartridges, thus encouraging people to simply buy an entire printer. A shopper thus has the choice of making an ecologically bad choice (buying a new printer to replace their perfectly good one) or a financially bad one (padding the pockets of a sociopathic corporation).

For people who have very little money, disposable items are usually cheaper than their more durable cousins. Thus, if you live on a fixed income like social assistance, you again have an incentive to buy a more disposable product. You may be unable to afford a more durable equivalent.

Not all incentives are so obvious. Let's look at so-called "feminine hygiene" products. The term "feminine hygiene" has seen a lot of use in the past century. During the 1920s and later, the term "feminine hygiene" was primarily a euphemism for birth control.[270] A casual survey of historical documents of the time might lead you to believe that "feminine" hygiene apparently had a lot in common with hygiene in the kitchen, bathroom, and hospital. That's because from around 1930 to around 1960 the single most common method of birth control in the United States was a *post-coital Lysol douche*. Yes, Lysol. That Lysol. Household cleaning product Lysol.

Beginning in the 1920s, the makers of Lysol ran a series of advertising campaigns promising that douching with Lysol would eliminate the "feminine odor" that might cause a woman to "spend the evenings alone" or even lose her marriage: "Yes, the proved germicidal efficiency of 'Lysol' requires only a small quantity in a proper solution to destroy germs and odors, give a fresh, clean, wholesome feeling, restore every woman's confidence in her power to please."[271]

Others gave instructions on how to use Lysol as a contraceptive: "The douche should follow married relations as a cleansing and antiseptic agent . . . Lysol has that rare quality of penetrating into every crevice and furrow of the membranes, destroying germ-life even in the presence of organic matter."[272]

Many of these advertisements featured endorsements by European "physicians" who, upon later investigation by the American Medical Association, did not actually seem to exist. Lysol was not even an effective contraceptive, a fact known in the 1930s. In their 1936 book *Facts and Frauds in Woman's Hygiene*, Rachel Lynn Palmer and Sarah K. Greenberg, M.D., remarked of the Lysol ad campaigns that, "Such sentimental trash would be laughable were it not for the tragedy of the many women who have become pregnant because they have relied upon antiseptic douches."[273] Unfortunately, many women were also burned by the strong antiseptic.[274]

Around the same time, another company was at work in the nascent "feminine hygiene" industry. Kimberly-Clark had been a large producer of "cellucotton," a material the company had developed for use in dressings during the First World War. After the war ended, the company was left with warehouses full of the stuff, and plenty of money invested in the infrastructure and factories for its manufacturing. They had a lot to lose if they couldn't figure out a new market for their cellucotton.

Up until around that point, women commonly made their own reusable menstrual pads from sewing scraps, rags, and similar materials. But in 1920, Kimberly-Clark introduced the Kotex sanitary napkin, a product which conveniently incorporated a large amount of cellucotton. The advertising campaign emphasized the high-class nature of the napkins. One advertisement asserted that "*80% or more better-class* women have discarded ordinary ways for Kotex."[275] Kotex ads also equated disposability with modernity and progress: "Just as the coming of telephones and electric lights changed old habits of living, so too Kotex warrants the forming of a new sanitary habit." "Study lamps instead of pine torches. Printed books instead of written parchments . . . a new sanitary habit made possible by Kotex."

The advertising account for Kotex became a hot property, in part because of its disposability. As the head of an advertising agency who courted Kimberly-Clark said, "The products I like to advertise most are those *that are only used once!*"[276]

Women found that they could customize Kotex pads for their own needs by changing the shape or opening up the pad to change the amount of stuffing. This customizability was a major source of popularity. At the same time, it was the beginning of the end for homemade pads in the mainstream. Beginning with businesswomen and women at school, more and more women found themselves too busy to make their own, or lived in small urban spaces with little room to store cloth, rags, and sewing supplies. Although wealthier women were certainly the first adopters, extensive advertising campaigns eventually (and successfully) targeted women of all classes.

Though Kotex is an especially notable example, nearly every modern disposable product had a reusable, or reused, predecessor. And each disposable replacement product had an advertisement and propaganda campaign designed to discredit that predecessor and sell more disposable items. Toilet paper, from around the 1870s, was often just reused newspaper or catalogue pages (although sometimes softer papers were saved for use by guests). Advertising campaigns insisted that reused papers directly caused hemorrhoids (which is untrue) and clogged the plumbing. In the 1910s, a campaign was undertaken to convince people to use disposable paper drinking cups when out of the home, starting with trains. This campaign was led by health authorities concerned that disease would be spread through public fountains, and also of course led by paper manufacturers. More "respectable" travelers generally carried their own collapsible or folding cups, and some paper companies made folding paper cups that could be reused. But of course, there was more money to be made by designing cups to do the opposite. The Individual Drinking Cup Company designed the "Health Kup" to be "destroyed if you try to fold it for a second use." The Health Kup became widely adopted, and soon had a name change to Dixie Cup.

Products like Kleenex and cardboard cereal boxes shared much of the same history. Disposable packaging in general was advertised with an emphasis on sanitation and cleanliness, since beforehand many products—even things like toothbrushes—were sold from piles or bulk bins at stores which many customers might handle. So products like toothbrushes, for example, were marketed with their sanitary boxes as the main selling point. It's ironic that a cultural obsession with cleanliness has led to the production of so much garbage.

So what are our individual incentives to be wasteful? Our historical examples show some common threads. Certainly convenience. Hygiene and cleanliness. An association with affluence or being higher-class. The association with progress and modernity (since disposability is the future, what choice do we have?). The price tag also plays a role, since disposable and cheaply made products are usually much cheaper than durable and long-lasting versions.

But let's not get too bogged down at an individual scale. As we've already discussed, personal waste accounts for only a tiny percentage of total waste production. Industrial-scale waste is a much larger issue. Corporations have many incentives to be wasteful. Or perhaps it would be more accurate to say that corporations exist to make profit, and that waste-less industrial activities would be extremely expensive, if not impossible. Waste is an externality. And from the beginning corporations have been social machines and legal fictions set up for the sole purpose of privatizing profits and externalizing costs: what do you think "limited liability" really means?

More to the point, a corporation makes money by selling products, and it makes more money by selling more products. You can sell more products by making the useful lifetime of a product as short as possible, through either obsolescence or breakage. Of course, in an industrial age, the most important thing a company manufactures is not a physical product. A company grows by manufacturing *demand* for its products. And so companies are constantly encouraging people to throw away what they have to buy something new, regardless of whether the propagandized benefits are true or not.

In the years following the Second World War, the rapid rate of plastic production began to cause a major problem for plastics manufacturers. By churning out economically inexpensive items so rapidly, they had begun to saturate the market. The solution? As plastics manufacturers were told at a 1956 industry conference: "Your future is in the garbage wagon!" The production of huge amounts of waste became not merely a by-product of capitalism, not accidental, but an explicitly stated goal of industry. The priorities of manufacturers should be "low cost, big volume, practicability, and *expendability*."[277] And so it was. The subsequent production of vast quantities of disposable items like plastic plates, diapers, and sunglasses was combined with marketing campaigns emphasizing fashion and a need to throw away unfashionable old items and purchase new ones.

Consumers were unaccustomed to such behavior, and the memory of a long Depression was still fresh for many. Plastics companies found that in the late 1950s customers were still saving and reusing their "disposable" cups. So they created major "educational" campaigns to convince consumers that it was appropriate (as well as classy and hygienic) to promptly dispose of their plastic consumables.[278] As a major advertising journal had previously observed, "The future of business lay in its ability to manufacture consumers as well as products."[279] Instead of advertising the virtues of their particular product compared to other similar products, they increasingly began to glorify consumption itself.

And when they can't convince people to buy their products, they simply destroy the alternatives. Clive Ponting describes one example in his book *A Green History of the World*: "In 1936 three corporations connected with the car industry (General Motors, Standard Oil of California and the tyre company Firestone) formed a new company called National City Lines whose purpose was to buy up alternative transport systems and close them down. By 1956 over 100 electric surface rail systems in forty-five cities had been purchased and then closed. Their biggest operation was the acquisition, in 1940, of the Pacific Electric system, which carried 110 million passengers a year in fifty-six communities. Over 1100 miles of track were ripped up, and by 1961 the whole network was closed."[280]

Governments also have incentives to encourage wastefulness. Governments rely on tax money, tax money comes from economic activity, and economic activity usually consists of turning parts of the natural world into garbage, directly or indirectly. And governments themselves buy and use huge amounts of disposable goods. Governments benefit by encouraging production, and by encouraging corporations, which (in theory at least) creates jobs. Governments want to keep the economy going. As president Herbert Hoover proclaimed in 1931, a time of dangerous frugality, "The sole function of government is to bring about a condition of affairs favorable to the beneficial development of private enterprise."

Author Richardson Wright summed up a lot of this in 1930 when he wrote (in a glowing endorsement of industrial-scale disposability): "We live in a machine age. To maintain prosperity we must keep the machines working, for when machines are functioning men can labor and earn wages. The good citizen does not repair the old; he buys anew. The shoes that crack are to be thrown away. Don't patch them. When the car gets crotchety, haul it to the town's dump. Give to the ashman's oblivion the leaky pot, the broken umbrella, the clock that doesn't tick. To maintain prosperity we must keep those machines going."[281]

Beyond a certain point, of course, it becomes obvious that all this might not be such a good idea. Eventually all of the garbage starts to pile up, the pollution starts to accumulate, raw materials become in shorter and shorter supply. But that doesn't change the behavior of those who run governments or corporations, because governments and corporations are based on short-term thinking, competition, exploitation, and the externalization of costs. A government that doesn't keep the economy going gets overthrown. A CEO who doesn't turn a profit gets fired. A corporation that can't sell more disposable junk than competitors goes bankrupt. The whole system is based on making decisions in the very short term, and then pushing the consequences off into the future, or out to other places.

Yes, governments and corporations occasionally make decisions that seem to break this pattern, but those instances are the exception rather than the rule. And more often than not, those supposed exceptions don't hold up to much scrutiny.

Even if they did it wouldn't help us very much. If the government of any given country mandated a 10 percent reduction in disposable packaging, that country would simply take eleven years to produce all the disposable packaging garbage it would previously have made in ten. (And of course, as we've previously discussed, consumer waste is only a tiny percentage of total waste: it really wouldn't make much difference at all.)

One problem with this kind of short-term decision making is that by the time a situation is really bad it's too late to reverse the damage.

I began this section with economists and incentives, and an economist might say that the way to change someone's behavior for the good is to change or remove their incentives for bad behavior, or offer incentives for good behavior. Incentives often suggested include things like tax breaks for polluting less, or carbon credits, or increasing the cost of garbage disposal. It's been suggested, rather absurdly, that the capitalist economy could be made completely sustainable simply by adding a system of incentives.

Sounds nice in theory if you really like the status quo, but there are a few problems. First, there must actually be non-wasteful options available for people to be encouraged to choose. Capitalism is supposedly all about choice, and if I walk into the grocery store I can find tens of thousands of different products. Of course, not a single one of these products will be ecologically sustainable, so really we're only offered the illusion of choice.

A second problem is that most suggested incentives target consumers. We've already talked about the problem of getting sidetracked by household waste when the vast majority of waste is produced on an industrial scale. But even beyond that, consumer-based incentives tend to be especially ineffective. Who is going to impose them? Corporations don't want to impose "artificial" incentives by making less wasteful products less expensive, because this cuts into their profits. Yes, less wasteful products could be less expensive to manufacture, but that's already reflected in the cost. Governments can impose taxes on especially wasteful products, but those have their own pitfalls. If you significantly increase the price of a commonly used but wasteful product, rich people simply absorb the extra cost, and poor people (who may simply be buying the cheapest version) struggle with it.

/ / /

Sure, there are exceptions to this, too. There's no real reason people couldn't use cloth bags instead of plastic or paper ones when shopping, for example. Cloth bags don't break easily, they last for decades and then can be composted, and they have a negligible ecological impact compared to plastic or paper bags. The only reason people use paper or plastic bags is convenience, because they don't want to be bothered bringing their own bags to the store, and because most stores subsidize the waste by giving bags away for free.

You can argue that plastic shopping bags can, at least in theory, be recycled, and therefore wouldn't be so bad if people disposed of them responsibly. But even that isn't really true, because most plastics aren't so much *recycled* as they are *downcycled*. Carol Misseldine, sustainability coordinator for Oakland, California, notes; "We're not recycling plastic bags into plastic bags. They're being downcycled, meaning that they're being put into another product that itself can never be recycled."[282] And frankly, even downcycling would be optimistic: in 2000, only 5 percent of US plastics were recycled.[283]

Most cities don't even have facilities to recycle plastic grocery bags, so some grocery stores have offered in-store recycling of plastic bags. And if you're going to bring your old plastic bags back to the store, why not just use a cloth one?

The fact that plastic bags are still the status quo shows how dire our situation is really is. If you can't get people do something easy and simple to reduce waste, how can you get them to do something more challenging?

/ / /

Part of the problem is that incentives that apply at a personal level aren't determined at a personal level. The reason it's so convenient to create waste by, say, using disposable plastic bags, is because of decisions made at corporate and governmental levels. And those decisions are largely determined by the larger system—industrial civilization—that corporations and governments are embedded in. Specific programs and laws that reduce waste can—and generally should—be put into effect. But these tend to have a small impact in the scheme of things, partly because they focus on consumer-level issues.

Moreover, these programs can act as distractions and diversions to activists and citizens. Here's an example. My career as an activist started around age ten, when I helped to start an environment club at my school. We sold Tupperware and reusable bags, encouraged household composting, and generally worked on waste reduction issues. In high school, we set up separate containers next to the garbage bins for students to put their recyclable plastic and glass bottles in. Then we would collect them, sort out all of the garbage that people had thrown into the clearly marked bins (they also threw recyclables in the adjacent garbage), wash the bottles off with a hose, and arrange to have someone pick up the bottles for recycling (which was rather more difficult to find at the time).

You could argue that instead of doing this, we should have spent our time lobbying government for mandated recycling programs, and then we wouldn't have to go to all the trouble of washing bottles ourselves. But that's not where I'm going with this, because neither option addresses our fundamental underlying problem. Sure, a small reduction in waste is better than nothing, but it's not good enough. And if we spend our time doing only "better than nothing" when we could be stopping all waste, the eventual outcome is the same. We're like Skinner's pigeons—bobbing our heads and turning in circles instead of trying to get out of the cage.

All this brings to mind Thoreau's observation that "There are a thousand hacking at the branches of evil to one who is striking at the root." Most incentives that motivate people aren't from programs, they're systemic. If you want to change a system that's profoundly broken, you don't tinker with programs, you get a different system.

If you really want to remove the incentives for being wasteful, you have to change this culture at the deepest level you can find, and then deeper than that, and deeper still. How deep can we go? Yes, individual people are wasteful because it's cheap, convenient, and they don't know any better. Corporations make a profit because you make more money by selling things that shortly thereafter become garbage. Governments benefit when corporations do this because it keeps the economy going and offers taxes.

These things only persist because the consequences of wastefulness are exported, like so many trash-laden barges, to seemingly distant places and times. If you want to stop the incentives for wastefulness, you have to stop the systems that export and postpone the disincentives. You have to bring the consequences home. And most of all, you have to stop the industries that manufacture waste in the first place.

/ / /

We could fantasize about a technotopia where government simply banned waste. Where it made corporations responsible for the eventual disposal of everything those corporations produced. But such a fantasy contravenes the deepest tendencies of civilization, and the structural rules that reward short-term gains over long term considerations (including life on this planet), and that reward people for taking more than they give back.

We could also fantasize that guns can give birth and that stealth bombers convert carbon dioxide to oxygen and pencil lead, and generate energy to boot. Governments and corporations are social machines invented to accomplish specific purposes—working in tandem to centralize power (and privatize profits) and to externalize costs; in other words, to facilitate and magnify exploitation; in other words to wage more effective war upon the poor and upon the earth—and it makes as little sense to expect these social machines to accomplish goals other than those for which they were created as it does to expect other instruments of war to perform life-fulfilling activities.

While we're at it, why don't we fantasize that you can convert a living planet to machines without killing it. And we can fantasize that dams don't kill salmon, that burning oil doesn't release carbon dioxide, that clearcutting forests improves their health, that you can suffocate oceans without killing them.

/ / /

Hell, most people in this culture already live according to these fantasies. These fantasties are printed in nearly every newspaper, trumpeted from every television. They form the basic for the legal and economic systems. They *are* this culture.

/ / /

This culture is based, once again, on a straight line. Its mythologies are linear and progressive. They depend on a beginning, middle, and end. The natural world is based on circles and cycles. A linear, progressive culture is not compatible with a finite world.

/ / /

Living creatures carry their own mechanisms for decay. The lysosomes in our cells rupture and begin to digest our cells when we die. Microorganisms on our skin and in our digestive tracts help to break us down into small, digestible molecules.

This—our flesh—is the most intimate, fundamental gift we can and must and do give each other. Our bodies are the most ancient, most vital of all gifts.

We have violated this most ancient of all agreements, and the cost for doing so is extremely high.

TECHNOTOPIA: INDUSTRY

The machine is antisocial. It tends, by reason of its progressive character, to the most acute forms of human exploitation. —LEWIS MUMFORD

So far from presenting its utopia as a beautiful dream, its effort terminates all too often in a kakotopia or realizable nightmare. —LEWIS MUMFORD

THE MAIN REASON A TECHNOTOPIA won't happen is because it is necessarily based on an industrial society. So if I want to critique this technotopia, or industrial society in general, I need to define what it means to be "industrial." First and foremost, an industrial society is one that depends on machines for the essentials of daily human life.

Machine production maximizes labor efficiency: generally, one person operating a machine can produce much more than many people without machines. This is well illustrated by industrial agriculture. With the use of a tractor, one person can plow and plant hundreds of acres. Using only hand tools, one person would be hard-pressed to plow and plant even one acre.

But maximizing labor efficiency doesn't mean maximizing other kinds of efficiency. Industrial agriculture produces an immense amount of food *per farmer*, but less food *per acre* than is possible under human-directed agricultural methods like permaculture.[284] Industrial agriculture also produces much less food *per calorie* than other methods. Industrial agriculture requires, on average, upwards of ten calories of *fossil fuel energy* per calorie of *food energy* produced. Of course, forms of agriculture and horticulture that predate industrial agriculture all yield more energy than must be invested in them, or there would be no point engaging in them.

These specific points about industrial agriculture illustrate a great deal about industrial society in general. An industrial society is based on access to cheap supplies of energy, which is why industrial agriculture can afford to be so energy-inefficient. It also depends on large areas of land into which it can expand for resource exploitation: it requires that there always be somewhere "beyond these current mountains" for it to take trees, fish, fur, all life. And the malignant expansion of industrial civilization has taken place in an era where both of those preconditions were made available (often by the use of a great deal of violence and genocide).

One reason industrial activities are so efficient in some ways and so incredibly inefficient in others, is that intelligence is selectively employed. A tractor magnifies the effort of one farmer, but it does so in a way that is stupid and clumsy compared to the precision of a pair of hands. The same applies to the vast majority of machines. The result of this is that machines require uniform resources to work with in order to be more efficient than human beings.

Think about the massive Central American garden terraces that have nourished people for millennia. Pick up a tractor from a grain farm on the wide, flat prairie and drop it onto those terraces and see how well it does with the not-very-uniform terrain. Next, take a piece of birch bark and try to put it through a printing press. Third, mass produce buildings using locally available timbers instead of machine-milled plywood and two-by-fours.

This is not to illustrate that humans are "better" than machines— although of course in a moral sense they are, as all living beings are—but to make a specific point. Humans are more adaptable and intelligent, meaning we can usually make more efficient use of available materials than a machine.

/ / /

Industrial society has grown so much because people are systematically rewarded within it by increasing production. Within society that means more money and power. Between societies, this means that a society that produces more can outcompete or even conquer another. Labor efficiency is central to maximizing production, but land and energy are only important to the extent that they boost or interfere with production in the short term. All decisions are made on the short term, because in such a highly competitive society, any group that can't compete in the short term will not exist to make decisions later on. This is strangely appropriate, seeing as a society that makes all its decisions in the short term will not exist in the long term.

/ / /

The brilliant writer, thinker, and historian Lewis Mumford described differences between what he called "polytechnic" and "monotechnic" approaches. Polytechnic approaches involve using many different technologies to meet human needs. Monotechnic approaches, on the other hand, prioritize technology for the sake of technology, to the exclusion of other options, regardless of the impacts on human beings or the planet. Mumford's favorite example of the monotechnic approach was the automobile, because automobile-based transportation systems thrive at great human and ecological cost, and grow at the expense of other modes of transport like walking or bicycling.

It would be fair to say that agriculture is the first example of a truly monotechnic approach; an approach that set a pattern for all of civilization's future technologies. If that's confusing, think about what agriculture *is*: you take a piece of land and destroy all visible plant and animal life on it; use plows to destroy the structure of the soil underneath; replace them all with one monocultural species; kill any competitors or predators; harvest that monocultural species; repeat. As Lierre Keith writes, "Agriculture is carnivorous: what it eats is ecosystems, and it swallows them whole."[285]

Agriculture is monotechnic in the sense that it eliminates biological diversity, to be sure, but there's much more to it. Indigenous societies are generally quite mobile, and can move to make use of the many different foods available in different seasons in healthy ecosystems. The same goes for other material gathered or hunted, such as firewood, furs, or medicinal plants. Though early agricultural societies certainly gathered food initially, it doesn't take long to deplete what is available around a village, which would have made those societies even more dependent on agriculture. Agriculture also eliminated many of the birth control methods that were intrinsic to hunter-gatherer life.[286] That, along with other changes, like the ability to replace breast-feeding in young children with foods made from stored grains, led to a trend toward constant population growth that worsened local ecological destruction. Agriculture grew at the expense of technologies, skills, and social structures used by indigenous peoples.

/ / /

For a technotopia to succeed, it needs to meet two main conditions. First, sustainable alternative technologies need to be possible—and they have to be developed—to replace essentially all current (unsustainable) industrial processes. And second, those sustainable technologies actually have to be implemented within that society's social conditions, power structures, and economic system.

The first condition requires a profound leap of faith. The problem is that industrial civilization itself is based on unsustainable technologies, like annual grain agriculture and the use of fossil fuels. If you invent a new technology that depends on unsustainable infrastructure, your technology is simply not going to be sustainable. Proven sustainable technologies of the past—like beaver dams or cheese fermentation—start from the context of a healthy living biome. It's safe to say that proven sustainable technologies of the future will start there as well.

The term "green technology" has been thrown around so much these days that it is essentially meaningless. Most supposedly green industrial technologies—like bioplastics—fall apart under scrutiny, and only look promising when compared to industrial society's abysmal ecological track record. Industrial green technologies are more promise than substance, and in essentially every case green technologies are essentially prospective or theoretical. If we rely on them for our future, we are once again giving up our agency, once again making ourselves powerless, once again falling for a distraction when we should be looking for the way out of the cage.

But it's the second condition that's really the final nail in the coffin of technotopias. Belief in a technotopia requires not only that sustainable industrial technologies exist, but that those technologies be implemented at the expense of current unsustainable technologies. And the history of garbage we've discussed in this book has been the complete opposite. Over the past century, low waste systems have been deliberately dismantled by corporations and governments, specifically *because* they were low waste systems. We've discussed some of their motivations for doing this, but there are deeper and subtler reasons that we can tease out.

I think the idea that "technology is neutral" is one of the most dangerous myths of our time. If we fall for the myth, it blinds us to the many ways that various technologies determine social structures and influence power relationships. We're often told that technology is "simply technology" and can be used for "good or for evil" but is fundamentally amoral. In other words, that whether a technology is harmful or beneficial depends on the intent of whoever is using it. Hence, we should ignore issues around technology and focus on more productive approaches to changing the world, like choosing which wealthy capitalist to vote for at the ballot box.

This belief probably has more to do with constant repetition than with deliberate misdirection. If we can see through the myth, however, we can see a whole new layer to the human world, and a whole new history. One of the first and most brilliant people to explore and document this history was Lewis Mumford. Throughout the vast majority of the twentieth century, Mumford wrote extensively, beautifully, and insightfully on the interplay between technology, history, hierarchy, social mythologies, and systems of control. And he didn't buy into the "technology is neutral" myth. Mumford made a clear distinction between two classes of technologies: those which were democratic, and those which were authoritarian. Moreover, Mumford explicitly recognized that society structure and technology went hand in hand, and used the term "technics" to encompass both the cultural and industrial aspects of technologies.

According to Mumford, democratic technics are comprised of some of the earliest human technologies. These technics are human and community scale, "resting mainly on human skill and animal energy but always, even when employing machines, remaining under the active direction" of autonomous human communities.[287] Such technologies, Mumford notes, are widely available, require minimal equipment or resources, and both robust and highly adaptive. These technologies, like gardening or pottery, "underpinned and firmly supported every historic culture until our own day." In addition, they had something of a moderating effect on the impact of tyranny. "Even when paying tribute to the most oppressive authoritarian regimes, there yet remained within the workshop or the farmyard some degree of autonomy, selectivity, creativity."

Authoritarian technics are another story. Rather than being ancient, Mumford identifies them as being more recent, specifically equating them with the origin of civilization. Authoritarian technics are based on centralized, top-down control, on a much larger scale. Mumford writes: "The new authoritarian technology was not limited by village custom or human sentiment: its herculean feats of mechanical organization rested on ruthless physical coercion, forced labor and slavery, which brought into existence machines that were capable of exerting thousands of horsepower centuries before horses were harnessed or wheels invented."

He continued, ". . . above all, it created complex human machines composed of specialized, standardized, replaceable, interdependent parts—the work army, the military army, the bureaucracy. These work armies and military armies raised the ceiling of human achievement: the first in mass construction, the second in mass destruction, both on a scale hitherto inconceivable."

The development of authoritarian technics climaxes in what Mumford called megamachines, massive centralized social machines with human beings as their components. For Mumford, these megamachines could certainly include industrial technologies, but that aspect was secondary, since megamachines like the Roman Empire predated the industrial revolution.

Obviously, Mumford's concepts of authoritarian and democratic technics overlap with his ideas of polytechnic and monotechnic approaches.

Mumford's categories are a useful starting point for debunking the myth of technological neutrality. There are also specific questions we can ask if we want to evaluate any particular technology for its neutrality.

First, there are questions about the prerequisites for the technology. How many people are required to make and run the infrastructure? Does the use of the technology presuppose the existence of other infrastructure like roads or energy distribution grids? Can the materials be obtained sustainably? Can human-scale communities with minimal hierarchy implement the technology? Or is a large, hierarchal society required to utilize it?

And second, there are questions about who can direct and make decisions about the technology. Is the infrastructure distributed so that it is under the control of many people, or is it centralized under the control of an elite? Does the technology allow people to become more autonomous and fulfilled, or do people surrender their autonomy to meet the needs of the technology? And who gets veto power? Can the technology be repealed or stopped?

And third, there are questions about the inherent effects of the technology. Are any benefits widely available? Are there barriers to accessing them? Is the technology useful to all, or only to an elite? And are the costs paid by the same people who get the benefits, or are the costs exported? Does the technology affect the degree of stratification and hierarchy in social?

Many of the technologies that we take for granted clearly fail the neutrality test and fall into the camp of authoritarian technics. Nuclear power and nuclear weapons are especially strong examples of this.

For nuclear technologies even to be developed requires the existence of a megamachine. Nothing less could marshal together the large numbers of people and raw materials required for an endeavor like the Manhattan Project. For example, as one of several massive complexes created for the Manhattan Project, the US government built an entire secret city called Oak Ridge to manufacture enriched uranium. Oak Ridge was home to the largest building in the world (which housed the first nuclear reactor), and by itself used *one sixth* of all electricity generated in the US.[288] Constructing the infrastructure for nuclear power is simply beyond the capability (and likely even further beyond the desires) of decentralized democratic communities.

Clearly all decision-making about the technology is highly centralized. That certainly goes for today, when much of the nuclear energy in the world is under military control. And it definitely went for the Manhattan Project—anything that is a complete state secret can hardly be called democratic. Furthermore, nuclear infrastructure requires a regimented hierarchy to function, and requires the people involved to sacrifice their autonomy to meet the needs of the technology and the social structure implementing it. In one illuminating case, a World War II-era Oak Ridge employee wondered, "Why on earth did they have all these high-school girls running this machinery? We could have blown up the whole of Tennessee!" Visiting the still active facility six decades after the bombs were dropped on Hiroshima and Nagasaki, she got her answer. "I was told that they wanted young women who would do what they were told and not ask questions. Really, we were just robots."[289]

What about the effects of the technology, the distribution of benefits and costs? It's clear that the purpose of nuclear weaponry, from the beginning, has been to cement and increase the power of the countries that created them—more accurately, to cement and increase the power of megamachines, a primary goal of all megamachines being to increase their own power and extent. In that regard, nuclear weaponry has so far been a success. So it certainly has benefits for those in power, though I think few people would say that the world is better off with nuclear weapons than it would be without them. And the costs, as usual, have been imposed on those with relatively little power. The Japanese civilians who were killed or injured at Hiroshima and Nagasaki. The Micronesian islanders (and other indigenous peoples) who were displaced or poisoned by nuclear testing on their territory. The many people in Iraq and other countries who have been poisoned by the use of non-fissile nuclear weapons like depleted uranium munitions. And, of course, all of the (human and nonhuman) people displaced, oppressed, exploited and killed by the megamachine so that it could create such weapons in the first place.

/ / /

Many people would probably admit that nuclear weapons don't fall into the category of neutral technology, but might insist that it's an exception to the general rule. So let's choose another, less obvious example.

In the interests of fairness, I'll simply go to the Technology portal on Wikipedia and choose a random page. (Oddly enough, the current featured article is "Nuclear Weapon.") And our next example? The steam locomotive.

All right, prerequisites. Metal is the single greatest infrastructural requirement for the steam locomotive. And metal means mining, both for ore and for fuels like coal to operate smelters and forges. In a number of very real ways, the steam locomotive—and all rail transit—is a product of mines. The metal that forms a locomotive, the steam engine that drives it, and the rails it travels on all originate from mines.

The first practical steam engines were invented only about three hundred years ago. Because of their large size and weight, they were stationary. The first and most important application of the steam engine was in mines. Since deep mineshafts are essentially gigantic wells, large machines were needed to continuously pump out water to allow miners to work. It was in this application that early steam engines were refined and developed. Steam engines weren't used in self-propelled vehicles until some seventy years later.

Rails themselves, originally made of wood, also originate in mining. Because metals are usually extracted from much larger amounts of ore, miners must find a way to move the ore to a place where it can be melted and refined. Almost five hundred years ago, miners in Europe started to use rails for this, though carts were hauled by hand or draft animals rather than by machines. Again, it was for mining that the railway was gradually developed—to carry ever-heavier loads of ore—until it reached the solid metal tracks we see today.

These different mining technologies converged to create the steam locomotive only two hundred years ago.

The steam locomotive also has social prerequisites, especially because of the extensive systems of railroads put in place for locomotives. Railroad construction in the North American West in the late nineteenth century involved tens of thousands of laborers at a time, most of them poorly paid and poorly treated Chinese immigrants, who had horrendous injury and mortality rates because of the dangerous work. Those railroads also presupposed the dispossession, displacement, and genocide of the many indigenous peoples whose territories they were built on. It seems fairly clear that building railway networks is the job of a large, hierarchical, and expansionistic society.

Question two, decision-making. Who controls steam locomotives? As a massive machine requiring extensive mining (smelting, etc) and transportation infrastructures to build and maintain equipment, the steam locomotive is not a community-scale device. In North America a century ago, steam locomotive manufacturing was essentially limited to three specialized companies.[290] At the same time, it's not as centralized as our previous nuclear example. Even though locomotives and trains are themselves under the control of an elite, the infrastructure is very spread out. Individual communities cannot use that infrastructure for themselves, except by buying passage on a train. However, they can use—and have used—the distributed nature of railroads as a means of political leverage in disputes with empire. It's pretty easy to block the tracks. And when the going gets rough, railways—especially bridges—have been a favorite target of resistance and guerilla groups, since they are often unprotected and fairly easy to disable.

Category three, effects. Steam locomotives—and railroads—are the basis of an industrial empire. In a sense, they shrink the world, and allow governments and corporations to extend their influence, especially in military and industrial terms. They make possible and speed up the transport of large volumes of raw materials and other goods across long distances over land, for which railroads are still used today. And they allow the movement of heavy equipment and machinery over great distances, and to places that previously could only be reached by hauling over dirt roads or wagon trails.

Railways allow those in power to increase and concentrate their power ever more, both materially and socially. A railway brings distant resources within reach of the engines of industry, and can ship the engines of industry to those resources. By dramatically increasing the geographic scale and convenience of resource extraction, railways make possible the creation of much larger and more powerful megamachines.

Let's talk briefly about two other consequences of steam locomotives, to show how deep and wide can—and will, and often must—be the consequences of a piece of technology. The invention and use of any piece of technology will by necessity affect many other parts of any culture, because for better or worse, technology and culture are intertwined: certain attitudes and social mores and reward systems will lead to certain technologies; and certain technologies will lead ineluctably to certain changes in perception, experience, behavior, and personal and social (cultural) mores and systems of rewards. The first of the two social consequences I want to mention here is that railroads led to the standardization of time zones. Prior to railroads, different communities kept their own time. Standardization wasn't necessary because by the time someone walked three days from one town to another, whether the time was now 8:15 or 8:34 in the evening rarely was of consequence. But with railroads came railroad timetables, and with timetables came the necessity for one community's 8:15 to be another community's 8:15. Now, superficially, this may not seem important to you, but the point is that the existence of railroad locomotives requires—as do so many technologies—standardization, even in parts of our lives where we might not expect.

The second consequence to mention here is that railroads really were the prototypical modern corporation. As I wrote in *The Culture of Make*

Believe, "Corporations are a legal device invented in the eighteenth and nineteenth centuries to deal with the myriad of limits exceeded by this culture's social and economic system: the railroads and other early corporations were too big and too technological to be built or insured by the incorporators' investments alone; when corporations failed or caused gross public damage, as they often did, the incorporators did not have the wealth to cover the damage. No one did. Thus, a limit was placed on the investors' liability, on the amount of damage for which they could be held liable. Because of limited liability, corporations have allowed several generations of owners to economically, psychologically, and legally ignore the limits of toxics, fisheries depletion, debt, and so on that have been transgressed by the workings of the economic system.

"By now we should have learned. To expect corporations to function differently than they do is to engage in magical thinking. We may as well expect a clock to cook, a car to give birth, or a gun to do other than that for which it was created. The specific and explicit function of for-profit corporations is to amass wealth. The function is not to guarantee that children are raised in environments free of toxic chemicals, nor to respect the autonomy or existence of indigenous peoples, nor to protect the vocational or personal integrity of workers, nor to design safe modes of transportation, nor to support life on this planet. Nor is the function to serve communities. It never has been and never will be. To expect corporations to do other than to amass wealth at any (externalized) cost is to ignore the system of rewards that has been set up, to ignore everything we know about behavior modification: if you reward someone—those investing in or running corporations, in this case—for doing something, you can expect them to do it again. To expect corporations to do other than they do is at the very least poor judgment, and at the very worst delusional. Corporations are institutions created explicitly to separate humans from the effects of their actions, making them by definition inhuman and inhumane. To the degree that we desire to live in a human and humane world—and really, to the degree that we wish to survive—corporations need to be eliminated."[291]

The invention of steam locomotives was tightly tied to the rise of corporations. Does anyone still want to argue that technology is inherently neutral?

A common thread in both the nuclear and locomotive examples is that both require a large, hierarchical society in order to develop. And in both cases, the effects of the technology are to benefit those at the top in their efforts to make megamachines even larger, to gain more power. Now, we can get into an argument about whether making bigger megamachines is good or bad (though if you're still reading, that's probably not going to be an issue), but it's clear that such an act is far from amoral.

The issue is complicated slightly by the fact that the development of technology is an iterative process: new technologies are built upon old ones, often stacked so deep that the foundational technologies become invisible and unquestioned. When people talk about technology they almost invariably mean something from the past few decades, not the technological bases of industry or agriculture.

Because of this stacking, we're sometimes put in confusing and morally complex situations—technics that could be somewhat democratic may be based on deeper authoritarian technics. While I like the idea that the Internet can be used for rapid, long-distance communication between equals, the Internet and all electronic communications are based on deeper, authoritarian technics like mining and industrial manufacturing. And despite the tendency of some academics to refer to this age as "postindustrial," the character of modern civilization is determined much more by its clanking industrial foundation than its digital veneer. Not only did the manufacture of my computer require the ecological and social damage caused by mining and manufacturing, the social conditions required for its manufacture emerge from and lead to the dominance of authoritarian technics. Because of this, even the most theoretically liberatory technologies, like the Internet, end up being dominated by corporations and governments that use them to advertise to, propagandize, spy on, and track people. We should also never forget that although we may be able to use computers to quickly communicate with others of those who oppose the destruction of life on Earth, computers are used more efficiently by those who are killing the planet. Indeed, much modern commerce and war (and surveillance) would be impossible without computers.

None of this is to say we should shun potentially democratic—or even potentially useful—technics that are based on authoritarian ones. We can use them judiciously while keeping our end goals in mind. We can disillusion ourselves of myths of a magical and painless conversion to a nonauthoritarian system using identical technologies. We can rid ourselves of the narcissistic notion that our personal purity (spiritual or otherwise) is more important than conditions in the real, physical world, including life on earth. And we can use the most effective tools and techniques to which we have access in order to systemically dismantle the underlying authoritarian system.

/ / /

A central story of the dominant culture is that the past few centuries have been characterized by constant technological and democratic progress. In fact, even the people who insist that technology is inherently amoral will often insist that technological and democratic progress are inextricably linked. Mumford didn't buy it. Instead, he thought that recent technological progress was characterized primarily by a growth in authoritarian technics: "[W]hat we have interpreted as the new freedom now turns out to be a much more sophisticated version of the old slavery: for the rise of political democracy during the last few centuries has been increasingly nullified by the successful resurrection of a centralized authoritarian technics."[292]

Lewis Mumford doesn't deny the many struggles toward democracy, but asserts that overall, modern authoritarian systems are winning, and with less resistance than cruder authoritarian systems of the past: "Why has our age surrendered so easily to the controllers, the manipulators, the conditioners of an authoritarian technics? The answer to this question is both paradoxical and ironic. Present day technics differs from that of the overtly brutal, half-baked authoritarian systems of the past in one highly favorable particular: it has accepted the basic principle of democracy, that every member of society should have a share in its goods. By progressively fulfilling this part of the democratic promise, our system has achieved a hold over the whole community that threatens to wipe out every other vestige of democracy.

"The bargain we are being asked to ratify takes the form of a magnificent bribe."[293]

Mumford, writing close to fifty years ago, couldn't possibly have anticipated all of the new "magnificent bribes" that high-tech industrial culture has offered us. And technotopia *is* one of these bribes.

/ / /

One of the earliest articulators of the technotopian ideal was Buckminster Fuller. A writer, lecturer, designer, and architect with a career spanning five decades, Fuller gained public prominence in the fifties and sixties, and left his mark on many different fields. Fuller's renown has waned somewhat since his death in 1983, but even people who haven't heard his name will recognize some of his iconic inventions, like the geodesic dome.

Although Fuller lacked a university degree and was not wealthy, he was a driven and prolific inventor and innovator from the 1920s onward. An early advocate for sustainability, Fuller was driven by this question: "Does humanity have a chance to survive lastingly and successfully on planet Earth, and if so, how?" Fuller had a genuine desire to improve conditions for humans using technology, a goal stemming in part from his young daughter's death from polio and meningitis. He wrote that industry should be converted from manufacturing weaponry to making "livingry": technology that benefits humans (Fuller was fond of inventing new words and terms). He was aware that the planet had finite resources, and popularized the term "Spaceship Earth" to emphasize the idea. Fuller was the prototypical technotopian—whose aim was to apply "the principles of science to solving the problems of humanity"—and his influence has been enduring, if not widely recognized.

Fuller's goal of learning about the world and employing technology to benefit everyone—at least many humans—was certainly laudable and authentic. But his specific ideology, like the idea of technotopia more broadly, falls down in a few ways. His doctrine was profoundly anthropocentric. He preferred to call humans "earthians," as though humans were the only species living on planet Earth rather than one of millions. His definition of "technology" is essentially the same as "industrial technology." His inventions, though clever, were plagued by implementation problems and most never entered production, leading some to criticize him as a hopeless utopian dreamer. Furthermore, many of his ideas were themselves inconsistent with his stated philosophy.

Take his concept of "ephemeralization." Fuller observed that in his lifetime the size of machines continued to shrink, even as their functionality continued to increase, becoming ever more "ephemeral." Think of computers, getting smaller and smaller so that the cell phone in your pocket has more computing power than the early, room-sized computers. Or compare a 4000-pound 1950 Buick to a 1600-pound Smart Car. Or compare wood and stone buildings to those built with lightweight space-age materials. As knowledge and technology increased, Fuller believed, material items could become physically less substantial. Fuller presumed that since these smaller machines were constructed out of less material, production and population could continue to increase even with finite resources. In fact, Fuller went so far as to say that this could continue indefinitely, and that there was no upper bound on production. It's surprising that he would say this, especially because of his stated recognition of Earth's finite limits, but he isn't the first technologist to fall prey to magical thinking. Fuller believed that because of ephemeralization, knowledge and information would continue to increase toward infinity, as material technology continued to shrink away to become more and more ephemeral.

Yes, some nanotechnologists are predicting the same thing, and we already discussed why the human population won't continue to grow indefinitely. In addition, Fuller ignores the fact that if you really had the technology to supply an infinite number of humans, you wouldn't *need* humans. I mean, if you were a government with the ability to synthesize infinite quantities of iPods, or cruise missiles, or robots out of next to nothing, why would you even keep humans around? You don't need them to be labourers or manufacturers, and if you can manufacture anything for free, there is no profit to be made from having consumers. If your viewpoint is completely utilitarian, and your production near-infinite, then humans just get in the way. If the ultimate Fullerian future did exist, it wouldn't include humans.

Furthermore, Fuller's analysis of ephemeralization in the twentieth century isn't even accurate. Yes, gadgets are smaller now than they were a year, or a decade, or a century ago. But just because devices themselves are smaller doesn't mean they use fewer resources. As discussed earlier in the book, tiny devices like computer chips are incredibly wasteful. And when talking about technology, the infrastructure can be much more important than the manufactured product. The amount of industrial infrastructure required to produce that pocket-sized cell phone is immense, and that must be factored into an analysis of whether or not ephemeralization is actually taking place.

And of course, they don't make things like they used to. Newer devices may be smaller and lighter, but they're often more fragile and built to be disposable. A new plastic-built electric drill, for example, might be half the weight of an older, metal-built one. But if the older drill has a lifetime—there's that conflation of the mechanical with the living, so we really should say, "the older drill will last"—three or four times longer than the new drill, then we aren't seeing true ephemeralization, just a manifestation of a trend toward profitable disposability.

Fortunately, if you really want to believe in the feasibility of a more "ephemeral" society, you're in luck. The technology has already been invented. Some clever and persistent people have already developed a way (many ways, actually, literally thousands of ways) to replace big, industrial infrastructure with knowledge. They're called indigenous people. Ultimately, hunter-gatherers, with their portable lifestyle, lack of industrial infrastructure, minimal physical goods, and extensive knowledge of the land and its nonhuman inhabitants, have been far more successful at ephemeralization than industrial society ever will be.

Of course, such an approach requires a healthy landbase (and a sane people who value physical reality over social constructions; or perhaps saying this another way, a sane people whose social constructs facilitate and don't hinder their perception of, engagement with, and love for physical reality). And technotopians, like other industrialists, treat all nonindustrial technology as useless, and healthy living communities as little more than impediments to resource extraction.

In short, technotopians are insane: out of touch with physical reality.

/ / /

There's a logical endpoint to this belief, in the minds of some futurists, and it goes far beyond turning silicon into computers. Here's how the argument goes, according to some futurists who call themselves "extropians" (from the word "extropy," the opposite of entropy). Some of the first digital computer parts were vacuum tubes, also called thermionic valves, which we used to modify electrical signals. They were about the size of lightbulbs, and were invented in the late nineteenth century. They were replaced by the transistor, a solid-state device the size of your thumbnail, in the middle of the twentieth century. Standalone transistors were largely superseded by microchips, and a row of microchips fifty long would be smaller than the diameter of a human hair. According to extropians, there's no reason this couldn't go on forever ("The Law of Increasing Returns"), with computation taking place in components the size of a single molecule, a single atom, even using hypothetical structures within atoms themselves. According to some, matter itself could eventually be reconfigured at a subatomic level to be optimized for computing, turned into a material dubbed "computronium."

What would be the purpose of such a material? Some futurists have suggested you could build something called a "Jupiter brain," a single planet-sized hunk of computronium (getting enough matter for such a project would likely require dismantling an existing planet and manipulating it at the atomic level). Not to be outdone, others have suggested building a much larger "matrioshka brain," a set of concentric computronium spheres as large across as the earth's orbit, which would monopolize all of the energy output of the sun and use it for computation.

And how would such vast computation potential be employed? It's been suggested that a planet-sized computer could be used to run simulations of reality—kind of like in *The Matrix*—including simulations of human beings who had their brains scanned and "uploaded." In this simulated environment, the argument goes, human intelligence would no longer be constrained by the biological limits of the brain, and would develop into God-like beings. Generally speaking, those who describe such a future don't have a specific goal in mind for massive amounts of computation, except that sufficiently developed artificial intelligences would be able to develop more advanced artificial intelligences than themselves, and so on, in a runaway cycle dubbed "the Singularity."

Computer scientist and science fiction writer Rudy Rucker casts doubt on the plausibility—and desirability—of computronium: "Although it's a cute idea, I think computronium is a fundamentally spurious concept, an unnecessary detour. Matter, just as it is, carries out outlandishly complex chaotic quantum computations just by sitting around. Matter isn't dumb. Every particle everywhere everywhen is computing at the maximum possible rate. I think we tend to very seriously undervalue quotidian reality.

"In an extreme vision . . . Earth is turned into a cloud of computronium which is supposedly going to compute a virtual Earth—a "Vearth"—even better than the one we started with.

"This would be like filling in wetlands to make a multiplex theater showing nature movies, clear-cutting a rainforest to make a destination eco-resort, or killing an elephant to whittle its teeth into religious icons of an elephant god."[294]

Rucker goes on to point out that there are no shortcuts for the work nature is already doing: its complexity is irreducible.

Others have asked, if you *were* living in a poor simulation of reality, how would you know? What would you compare it to? Especially if your psyche were only a poor imitation of an actual psyche?

It sounds like science fiction—it *is* science fiction—and I hope it stays that way. But the extreme extropian future doesn't interest me because I think it's a literally probable outcome—it's obviously not—but instead it interests me because it's such a remarkably (and inadvertently) clear articulation of the pathology of civilization. It's a digital manifest destiny. It encapsulates civilization's willful denial of the laws of ecology and of physics (the very name is in contravention of thermodynamics), as well as its myths of inevitability and immortality. The dream of continual expansion and "progress." It takes civilization's drive for control to a new level—the vision of tearing apart the entire solar system and remaking it from the subatomic level. And all that to support an imaginary world where they can become as gods, the pantheon of their own mythology.

High-tech trappings aside, the story is strikingly familiar. For millennia, those in power have been dismantling the real world to shore up an imaginary one defined by control instead of diversity—a toxic mimic of the real world. Quarrying mountains to make artificial mountains—monuments to themselves—wrapping themselves in cocoons of civilized culture and infrastructure, attempting to project and imprint their thoughts and words and propaganda on the entire world.

But unlike the fantasy, their work is rough and incomplete. There are still wild places and wild people left; there are cracks in their toxic mythology; we are still born with feelings and knowledge that they (and then we) can't completely suppress. So again, we can ask the question, how do you know if you are living in a poor simulation of reality, a toxic mimic of reality? Well, if you're reading this, it's safe to say you do know, that you've seen through the mythology.

The question is, what are you going to do about it?

/ / /

Empires and civilizations—to whatever extent there's a difference—are fundamentally expansionist. They're driven to expand by their own internal competitiveness and a desperate drive to outrun their intrinsic unsustainability. Those who identify with civilization enjoy the benefits that come when you live beyond your means, when you take more than you give back. But people who are paying attention recognize the finite limits of a planet. So in order to try to reconcile the desire to keep benefiting from living beyond their means, and the finiteness of Earth, some people make up a fantasy of internal expansion. Like Buckminister Fuller, they try so hard to believe that civilization can expand indefinitely without actually taking up any more space.

Some technofix advocates, like James Lovelock of Gaia theory fame, argue that their goal is more a kind of damage control. They agree the Earth is a small place, that the ecological damage is extensive, that the social inertia is tremendous. And they argue that pursuing advanced industrial technology is necessary to reduce the damage.

I'm all for damage control. I'm all for harm reduction. Sure, do whatever can be done to reduce the destructiveness of industrial culture. But I think these advocates are often being disingenuous. For the most part, they don't want to save the landbase if it means making real or substantial changes; they don't want to save the landbase if it means inconveniencing themselves. What they want isn't to kick the habit—to deal with their addiction—it's just to find a superficially different way to get their fix. It's Mumford's slavery, all over again, and these slaves—these addicted slaves—are being bought off cheap.

I'm happy to have a vision and plan for the future. But a vision of shiny toys and glamorous tech that ignores the root issues does us a disservice. It serves a distraction instead of a path, makes waste of our time as well as this planet.

/ / /

The problem with the technotopia scenario isn't really the goal of employing technology to make a better society. Of course a good society should be materially sustainable. Of course it should be more equitable. Of course it should run on solar energy, as all societies of the past have. No, the problem with technotopia is the false assumptions entangled with it, and the almost deliberate ignorance. The assumption that technology means industrial technology; and that industrial technology is neutral. The assumption that technology alone could make an unsustainable society sustainable. The ethnocentrism that assumes indigenous ways of living lack technology, social structure, or well-being. The presumption that progress is always good, and that the centralization of power and resources is progress. The assumption that those in power are simply misguided about ecology. A willful ignorance of the way power and coercion underlie the dominant culture. That it is possible for most industrial technologies to become sustainable. That governments and corporations are capable of going against their own underlying systems. That governments, corporations, scientists, and engineers are going to solve our problems for us. That it is possible to maintain this current way of living with only minor adjustments to our daily habits, even though we know that the root problems go to the core of civilization. That maintaining current power structures with green technologies would actually be a good idea. That maintaining this current way of living is desirable for humans, let alone the other inhabitants of this planet.

Even if you agree with these assumptions—even if your dearest wish is that in fifty years you could go to a sod-roofed Wal-Mart powered by windmills and buy cheap biodegradable soy-sneakers—that doesn't make the scenario more plausible. Industrial civilization is still based on centralizing control and externalizing consequences. The ability to externalize consequences requires a constant flow of cheap energy, new raw materials, and new lands, all of which civilization is now exhausting. Which brings us to our final scenario: collapse.

COLLAPSE

As a result of that deep-rutted heritage, the very survival of civilization, or indeed of any large and unmutilated portion of the human race, is now in doubt. . . . Each historic civilization . . . begins with a living urban core, the polis, and ends in a common graveyard of dust and bones, a Necropolis, or city of the dead: fire-scorched ruins, shattered buildings, empty workshops, heaps of meaningless refuse, the population massacred or driven into slavery. —LEWIS MUMFORD

WE'VE COVERED A LOT OF GROUND in this book. We've looked at what happens to many different things when they break down; shit, plastic, and so on. We've talked about the large-scale systems that create particular kinds of waste. Now it's time to look at the really big picture: what happens when *societies* break down?

A main theme in the scenarios has been that civilization functions by externalizing consequences. The first scenario, business as usual, was about the fantasy of a system that could continue to grow and externalize consequences forever. The second scenario was about the fantasy of maintaining the same social and economic structure, retaining the same privileges, *without* exporting consequences. The third is by far the most realistic. Collapse is about what happens when the consequence-externalizing system *fails.*

Collapse is different from our previous two examples in another important way. Both of those scenarios were about continuing to develop a larger, more centralized civilization, a bigger megamachine, which means more uniformity and predictability. Collapse is the opposite of this. By definition, collapse is the *breaking down* of one big system, which makes room for many different smaller approaches. So collapse is not monolithic, it's not a single event or trend, but many overlapping events and trends with common themes.

The idea of collapse has gained a lot of recognition in recent years, partly because of growing awareness about peak oil and climate change, and partly because more and more people are realizing that collapse is, in historical terms, a rather common outcome for civilizations. In fact, civilizations of the past seem to have two main options in regards to their own demise: collapse, or be conquered and assimilated into a larger civilization which subsequently collapses.

You don't need us to tell you that the dominant culture, civilization, is unsustainable. And you probably recognize that civilizations of the past were also unsustainable, in different particulars but for largely similar reasons. When your society is unsustainable, eventually you have to make a choice: either change or collapse. Actually, that's blurring the truth, but hold that thought because we'll go back to it in a minute.

Probably the best academic study of why and how civilizations fall apart is *The Collapse of Complex Societies* by Joseph A. Tainter. An anthropologist, Tainter wanted to understand why so many "complex societies" in history failed. Complex societies, in his definition, are societies with a high degree of social specialization and stratification; which is to say they have many different kinds of specific social and occupational roles (for example, we have not just healers, but gastroenterologists, oncologists, proctologists, and so on) and many different kinds of artifacts (for example, we have not just spears, but many different calibers and types of bullets). Complex societies are an interesting category, because all civilizations are complex societies, but not all complex societies are civilizations. So research by Tainter and others in that field can give us a lot of insight into the differences between civilizations and other societies that have decided to use complexity to solve some of their problems.

So what is collapse? In his book, Tainter writes that "A society has collapsed when it displays a rapid, significant loss of an established level of sociopolitical complexity." He adds: "Collapse is manifest in such things as a lower degree of stratification and social differentiation; less economic and occupational specialization, of individuals, groups, and territories; less centralized control; that is, less regulation and integration of diverse economic and political groups by elites; less behavioural control and regimentation; less investment in . . . those elements that define the concept of 'civilization': monumental architecture, artistic and literary achievements, and the like; less flow of information between individuals, between political and economic groups, and between a center and its periphery; less sharing, trading, and redistribution of resources; less overall coordination and organization of individuals and groups; a smaller territory integrated within a single political unit."[295]

I think we can probably agree right off the bat that many of these changes would be welcome. "Less behavioural control and regimentation" for one. I would welcome "a lower degree of stratification," "less centralized control . . . less regulation . . . by elites." And we should also remember that these are changes within the collapsing society, not necessarily within smaller communities. So even though there might be "less flow of information" in the collapsing civilization (for example, fewer television channels) there might be *more* flow of information in actual human-scale communities (for example, I might talk to my neighbours). Collapse on the large scale can make room for a blossoming on the community scale.

But all of this is a little vague, a little abstract, a little academic. What would collapse actually look like in the here and now? In part that depends on where you are.

There are many factors likely to drive industrial collapse. Two of the largest are energy decline and rapid climate change. It almost goes without saying that cheap oil undergirds virtually every aspect of industrial society, starting with agriculture. For every calorie of food we eat, something like ten calories of fossil energy are expended, both in farming and food transportation. It is cheap oil that allows global shipping and travel, that permits large-scale manufacturing and construction, that builds and runs the global energy and communications infrastructures.

As peak oil researchers, authors, and activists have made abundantly clear, cheap oil is on the way out. Global oil production has almost certainly peaked and will decline from here on out, and there are no suitable replacements at industrial levels of energy consumption.

It's this change that's likely to set off a long series of cascading effects. Without cheap oil, essentially everything we buy will get more expensive. This will certainly happen with food, a process which has already begun in part because of competition between food crops and biofuels.[296] But it will also happen with other products, many of which (like plastics) are literally made out of oil, and virtually all of which are processed and shipped long distances using fossil energy. Increasing costs mean increasing prices for practically all commodities and consumer items, which will cause a decrease in consumption as people literally can't afford to buy as much. Because of this, people will shift their limited funds toward more essential items, like food, transport, housing, and utilities. Purchasing of nonessential services and items will fall off, and the people who manufacture or sell them will be out of their jobs.[297] Increased unemployment will drive down wages across the board, and the growing numbers of unemployed or poorly paid people will spend even less money.

This will drive a vicious circle of economic contraction. It will be especially evident in places like the housing market. Most Americans have debt, and most Americans who own a home have a mortgage on it. But rampant unemployment would render many unable to scrape together their mortgage payments, and houses would be seized by banks, thus worsening the situation for individuals and families.

In terms of trash production, though, things are looking up. It takes a lot of energy to produce disposable junk. And it requires a lot of disposable income to buy that junk. Industrial collapse means a significant and rapid decrease in the production of waste, both because of a shift toward more reusable items and because of a general economic slowdown. (There's a lot of talk in some quarters these days about how waste costs money, and therefore capitalism will act to decrease waste. But as long as energy is incredibly cheap, we aren't going to see that happen.)

A reciprocal change to this is a dramatic increase in the value of waste. A recycled aluminum can requires 95 percent less energy to manufacture than one made from new materials. Recycled paper uses 60 percent less energy. Recycled glass uses 50 percent less. A decrease in overall manufacturing, and an increase in costs of new materials, means that recycled materials will be more economically appealing and make up a greater proportion of the raw materials used by factories.

And remember the rag collectors who were common on this continent barely more than a century ago? Remember the repair workhouses of the Salvation Army? Heck, remember the garbage pickers and cardboard collectors currently common in the majority world? Well, expect them to make a comeback. A new depression, a major economic contraction, means decreases in both employment and wages. That means it will be more economically feasible for relatively labour-intensive activities like repairing broken items that would otherwise be disposed of, and for the manual collection and sorting of virtually all refuse.

*   *   *

A little while ago I wrote that unsustainable societies have the choice to change or collapse, and then said that was blurring the truth. Well, here's what I meant: societies don't choose to collapse; they choose to continue doing whatever it was they thought made them successful in the first place. If a society knows it's in trouble, it sticks to its main problem-solving method, its main indicator of success, whether that's building stone heads, increasing production, developing new technology, or raising the GDP. A society, or more specifically those in power in a society, never say, "Well, I guess that's it then, boys; we might as well collapse." They always insist on doing what they wanted to do all along, even though, more likely than not, that's what brought them to the brink of collapse in the first place.

Joseph Tainter hypothesized that something a bit like this is the general model for historical collapses. He wrote that complex societies solve their problems by increasing complexity, whether that means trying to develop new technologies, making government bigger, or what have you. Now, initially this has certain benefits. If you and I are making a lot of clay pots, for example, it might make sense for one of us to specialize in forming the pot shapes, and the other on decorating them. This way we could make pots more quickly than we could before. But as complexity increases, the costs of supporting that process start to increase as well. Our pottery operation has expanded, and now we have apprentices, and special rooms for different operations, and plenty of specialized clays and pigments for very fancy pots. And the price of that is incorporated into our very nice, but now quite expensive, pots. Around this point, the costs of complexity start to outstrip the benefits. This means that, as nice as our pottery is, people aren't sure they want to spend that much on a vase.

So what do we do with our pottery business? We've come so far, but we're in trouble: our customers are discontented! Well, obviously the solution is to make a larger number of more elaborate pots! Hire clay moisture technicians, make larger kilns, bring in scientists to develop new levels of pottery glaze technology! After all, that's what made this pottery company great. Some nay-saying doom-mongers say we should go back to making simpler, tastefully designed pots since that's what people "really need." But we can't go back to making simpler pots. That's absurd! We must go forward and continue making ever more expensive pots that are so fancy people will buy them without question! Besides, if we made simpler pots and sold them for less, how could we afford payroll for our large staff of moisture technicians and kiln-tologists?

If our pottery business were a society, this is around the point it would probably begin to collapse. The costs of complexity have far outstripped the benefits, but complexity itself has become entrenched, self-perpetuating, and an end to itself. The same is true with our pottery corporation. At this point, the customers would reasonably abandon us, our company would run out of cash, the staff would leave when they didn't get their paychecks, and we would go bankrupt. But we, like some of the more "successful" complex societies, have a few tricks up our sleeves. We're not going down the tubes yet.

First, we're going to need some money so we can make payroll this month. We'll just dip into the pension fund a bit. And we'll get a loan from the bank and go into a little bit of debt. This is a very popular strategy for complex societies, especially in ecological terms. Whenever a society starts using something faster than it can grow back—cutting down too many trees, or drawing down an aquifer faster than it can recharge—it's going into ecological debt.

Second, there are other small businesses out there, on the fringe of our vast pottery empire, that are making inexpensive but functional pottery. The nerve! We'll put them out of business, and make it a monopoly. That way people will have to buy from us if they want pottery. That's another big difference between a business and a civilization. If you're dissatisfied with a company you can, in theory, take your business elsewhere. A civilization does its best to make itself the only game in town, removing all other options.

If people don't like the monopoly, well, don't they realize how much choice *we* offer? Heck, we have fancy pink pottery, and fancy red pottery, and even some fancy pink pottery. Next year we're coming out with some fancy green pottery; we promise. And besides, you don't really want to use that other pottery, do you? It doesn't have our patented On-Glaze™ technology, which is far too fancy for you to make yourself. And don't forget, our Center for Clay Moisture Research has determined that using pottery not made by us can lead to excessive pottery mold, as well as many lonely nights: scientific studies have shown that only *our* pottery will gain you friends, and, ultimately, help you find your soulmate™. Best of all, we've been making pottery for a long time now, and people don't really remember what pots were like before—all they remember is that pots in the last few years have definitely gotten fancier. We'll just tell them that pottery before us was nasty, brutish, and short.

Now, whether in our pottery business or in real life, none of this can go on forever. Societies can put off collapse by expanding, by going into ecological debt, by limiting people's options for escape. But none of that changes the underlying fact that they're still unsustainable, and that they're unsustainable on a finite planet. Eventually all of these efforts do fail, but the longer they go on, the worse the collapse is when it does happen because there's so little left of the environment, and because people have become so dependent on the system.

In historical collapses, people were not as dependent on the system as they are now, and the system did not have such a monopoly. Many complex societies were formed because they offered real benefits to their members, whether in terms of information exchange, or limited specialization and barter. And many of those societies did not become civilizations, which is to say that they didn't control the food supply and other essentials of their members. Which is to say that they weren't capable of effectively forcing people to continue participating in that society if it no longer benefited them. That's a crucial difference between a complex society and a civilization. It's the difference between a healthy relationship and an abusive one. In a healthy relationship, both partners are getting something out of it, and both are free to leave the relationship if they want. In an abusive relationship, this is not the case.

When historical societies collapsed, their members had real alternative options. There were still healthy communities they could rejoin, communities that had skills like growing their own food, and building shelters—essentially, those healthy communities possessed a full gamut of democratic technics. It was the existence of options that made societal collapse fairly painless for most people. They just had to walk away and return home, to what their people had been doing for thousands of years.

Civilization does everything it can to make sure that's not possible.

/ / /

Back to the real-world collapse scenario: economic depression and energy decline. Aside from voluntary recycling of refuse, there's bound to be an upswing in *involuntary* recycling as well, especially because of rising prices for metals as easily available ore deposits become mined out. Prices for metals like copper and aluminum have been pushed toward historic highs. The past few years have seen a sharp increase in the theft of standing infrastructure to be sold on the black market for recycling. This has caused some bizarre headlines. In Germany last year, someone dismantled and removed five kilometers of disused railroad track to sell for scrap metal.[298] In British Columbia, law enforcement agencies have been trying to stop people from stealing thousands of pounds of aluminum in the diverse forms of goalposts, aluminum sockets on scoreboards, seating, and pipes.[299] The city of Toronto found that aluminum cans were being taken from recycling bins in such large quantities that it asked that "bylaw-enforcement officers be moved to the overnight shift to monitor blue boxes."[300]

In the United States, both aluminum and copper are being targeted in the forms of pipes, radiators, air conditioners, electrical wiring and cables, and even aluminum siding from houses. In Alabama, someone stole an eight-mile-long stretch of power line. That kind of scale sounds like "organized crime," but according to a detective tracking metal thefts, that's not the case: "There's no one or two, there's no six people that are specifically responsible for this. Everybody's doing it. People know the prices are up."[301]

The poorer that people are, and the greater the wealth and class disparity in a region, the more likely people are to engage in forms of involuntary recycling. People may also be willing to engage in more dangerous forms of resource repurposing, mostly out of self-interest, but in ways that may inadvertently accelerate collapse.

Electricity tapping is surprisingly common in the Majority World, especially in slums. Illegal connections to the power grid are common enough in South Africa that utility companies have hotlines they encourage people to call to report "electricity abuse" (and it's interesting to note that by "electricity abuse" they don't mean such abuses of electricity as LED video display billboards, retractable stadium roofs, and bug-zappers; but rather the unpaid use of electricity by the desperately poor). One pamphlet on electricity thefts proclaims that "People who continue putting others at risk through these criminal and negligent acts are snakes and must be reported." Modifications to power lines and substations sometimes cause blackouts, injuries, and death. In South Africa between 2000 and 2003, there were at least 260 deaths caused by electrocution from power line tapping.[302] Dangerous as it may be, the very poor have few choices. "There is no other way," noted one slum-dweller in India, because poor people have to choose between spending their limited funds on "either food or electricity, and food is a natural choice."[303] In Delhi, about 42 percent of the electricity is lost to tapping, and as much as half is lost is lost in the rest of the country. Power cuts are already a regular event in Indian cities, and generating capacity would need to increase seven-fold in the next twenty-five years to meet current demands, while many Indian utility companies are on the verge of bankruptcy.[304]

In some areas in Sri Lanka, the amount of electricity used illegally is actually greater than the legal electrical usage, and "sudden power fluctuations are causing damage to domestic electrical equipment." Those tapping the power are usually poor, of course, but they are sometimes militant as well. A representative of the Electricity Board in Sri Lanka complained that when Internally Displaced Persons (IDPs) tapped electricity "it was difficult to arrest them, as IDP camp members attacked officers of the board whenever they went to check the connections."[305]

The Niger Delta offers a different and even more sobering example. Nigeria is the most populous country in Africa, as well as being the single largest exporter of oil. Much of that oil is extracted from the Niger Delta, where industrialization and frequent spills have degraded living conditions. Years of military rule and the refusal of (mostly Western) oil companies to share oil revenues with the people have led to increasing social unrest in the area. Writer and activist Ken Saro-Wiwa was famously executed—read, murdered—by the Nigerian government in order to silence his opposition to oil interests, Shell in particular.

Oil extraction processes in the Delta are even worse than the oil industry usually is, which is saying a lot. In order to concentrate on extracting crude, most of the natural gas found in the same well as crude oil is simply flared—burned off at the top of smokestacks. The amount burned is equivalent to 40 percent of the natural gas used on the entire African continent. The gas flaring in the Niger Delta is the single largest source of greenhouse gases on Earth.[306] The burning also creates toxic gases and soot that affect those living in the Delta. Ironically, Nigeria imports many petroleum products, such as gasoline, as the county has very few functional refineries.

The inhabitants of the Delta, largely indigenous people who subsist on fishing and horticulture in the ecologically rich region, have become understandably angered by the situation. Living next to such ostentatious wealth and waste, which despoils their land while giving them nothing, it's become common practice for those living in the region to tap oil pipelines passing through their land to gather fuel for cooking or for resale on the black market. This can be a dangerous undertaking. Tragically, many have died when tapped pipelines have caught fire and exploded upon being accidentally ignited by a cigarette or motorcycle engine. The flaming blasts can be enormous, and the (human) death toll equally so. It's not unheard of for hundreds of people to die in a single explosion—on one occasion in 1998, one blast killed more than a thousand people.[307]

Those were terrible accidents, but militant indigenous groups like the Movement for the Emancipation of the Niger Delta (MEND) have taken an organized approach to breaking apart the oil infrastructure. MEND has destroyed major oil pipelines, razed oil industry facilities, skirmished with Nigerian soldiers, and kidnapped foreign oil workers. In a 2006 communiqué to foreign oil companies, MEND warned, "It must be clear that the Nigerian government cannot protect your workers or assets. Leave our land while you can or die in it. . . . Our aim is to totally destroy the capacity of the Nigerian government to export oil."[308]

There's no reason to think actions like these will wane in the coming years, in South Africa, in the Niger Delta, in the world. As the oil supply dries up, as the gap between rich and poor continues to grow, tapping the grid may be more and more rewarding. And attacks on the grid, whether those attacks are masterminded by freedom fighters, warlords, or foreign nations, will offer greater and greater impact.

/ / /

Everything we're talking about here has to do with energy and complexity. Modern day industrial civilization is, in material terms, the largest and most complex society ever to exist. That complexity depends on easily available energy. That energy is required to build factories and machines, and to ship them and their products around the world. With enough cheap energy, you only need one factory in the world to build a given product. And you can ship that product everywhere. Because that factory is the only one in the world, and because making that one product is the only thing it has to focus on, it's going to be pretty efficient at making that product (regardless of social and ecological costs). And you can put that factory wherever it's going to be cheapest for it to make that one product. And because, in economic terms, that product is so cheap per unit, you can afford to make many different items in similar factories. That's where complexity and specialization pay off.

Of course, the natural world doesn't take this to such an extreme. Imagine if the world had, for example, only a single gigantic tree, that specialized in doing all of the "tree" things really well. That sounds kind of stupid, doesn't it? Partly because it would look so silly, and partly because there isn't only one "tree" thing, and partly because relationships in the natural world can't be FedExed or packed into container ships. Trees do many things, have many relationships, some of them awe-inspiringly grand, some invisibly subtle. Complexity in civilization is usually built at the expense of much greater complexity, subtlety, and sophistication in the real world.

As energy decline begins, it will be impossible to maintain that global level of industrial complexity. And the way this plays out will vary greatly from place to place. We've seen some contrasting real-world examples of this in Cuba, North Korea, and the USSR. Communities that are less entangled in the global industrial system, that maintain skills for community sufficiency, and that have a healthy landbase, will fare better. Populations that are thoroughly entangled in that system, that have lost their community sufficiency skills—their democratic technics—and that have permitted or been unable to stop the destruction of their landbases will not do as well. Societies in this position—societies that have accepted Mumford's "magnificent bribe"—will find themselves squabbling with other industrial societies over dwindling supplies of cheap oil.

This brings into the picture another force driving collapse: war.

Large industrial societies require vast quantities of cheap oil to persist. Those in power don't have misapprehensions about this fact. An industrial society without large amounts of energy will collapse. So as the total supply dwindles, industrial countries have to make choices: maximize industrial potential, and conquer or colonize other regions to ensure a supply of raw resources; or minimize industrial activities, and risk being conquered or colonized by others seeking raw resources. It's a poor choice, either way. Who wants to be conquered? Who wants to be a colony? And even if you're one of the conquerors, dwindling supplies mean eventually you'll be fighting other conquerors just to keep your piece of the pie.

Either way, you may be better off if you don't have resources. Studies in Africa have shown that the amount of armed conflict in any given country correlates with the quantity of industrially valuable resources (like minerals and ore) present in that country.[309]

In either case, war has the potential to drastically accelerate collapse. Yes, war has brought countries out of economic tailspins before. The start of the Second World War helped end the Great Depression in the United States. But we aren't talking just about economics here—economics and finances are largely imaginary, since money itself is just something we all agree to pretend is real. We're talking about actual, material, real-world limitations that (to the disappointment of magical thinkers) can't be solved by simply pretending something else.

War machines tend to be very effective at concentrating resources and externalizing consequences, because of their proficiency in using physical and psychological violence to coerce and dominate others. In theory, shouldn't a little war stave off collapse by simply being such an effective megamachine? I mean, if the US is running out of oil, invading a few oil-rich countries should solve the problem for now, right?

But of course we all know that the United States would never invade a country to take its resources. The United States only invades countries which harbor terrorists and that have weapons of mass destruction. Speaking of which, I've been thinking a lot about relations between the United States and Canada. If communities in Canada stop transnational corporations from stealing water and turning much of Alberta and Saskatoon into moonscapes for oil—oh, I'm sorry, I guess the polite phrasing would be "developing water resources and extracting necessary oil from tar sands"—how long do you think it would take the US to invade Canada? Of course the United States wouldn't invade to take those resources, but would only invade to bring Freedom and Democracy™ to that country. So in the interests of the greater public good, we are right here and right now going to start the process of demonizing Canada. Canada clearly harbors terrorists: I've long had suspicions about the Toronto Maple Leaf/al-Qaeda connection, and even more damning, have you ever noticed that William Shatner, a known Canadian, has never been seen in the same room as Osama bin Laden, a known terrorist? Further, William Shatner is an actor, and actors are able to assume accents, and to modify their appearance. That's what they *do*. So how do we really know that William Shatner and Osama bin Laden are not the same person? Think about it. And of course if Mike Harris (the former Premier of Ontario who spent most of the 1990s slashing health care and social programs to fund corporate tax cuts) doesn't qualify as a WMD, I'm not sure what would. The United States must no longer ignore these obvious connections, and must act quickly or risk invasion by the power-mad Canadians.

Now that we've started that campaign, expect *Fox News* to pick up this war cry soon, and for the *New York Times* to quickly follow suit.

But even invading Canada would not allow the megamachine to continue much longer. War can only perpetuate (and perpetrate) a megamachine for so long. The first problem is that since total resources available are finite, more effective megamachines make the problem worse by using up that finite supply of resources faster. Delaying collapse requires consuming less, the opposite of what is happening now.

The second problem is that war machines have a lot of overhead expenses. The more resources are consumed by military in the act of acquiring those resources in the first place, the fewer resources remain for other uses. If a country uses a million barrels a day, and they need to spend two thirds of that to supply their army's overseas oil wars, they aren't going to have much left over. Superficially, this may seem to prolong collapse, but for everyday people it's going to accelerate it by drastically reducing the amount of oil available for civilian applications like growing food, heating and lighting buildings, and moving things and people around. (Of course, if a society is spending most of its resources on the military, we can have a discussion about the degree to which *any* aspect of that society is civilian. Oh, and you did know that the US already spends most of its discretionary funding on the military, right?) This is also why "war" doesn't just mean battles and saber-rattling. War also includes the societal changes required to facilitate on ongoing conflict: militarism, increasing social controls to criminalize dissent, a general trend toward martial law and fascism, massive propaganda campaigns to support all of these, and of course the physical infrastructure necessary to support the military-industrial complex.

The third problem is that even though individual war machines may be good at concentrating resources, they each have the ultimate aim of destroying the other war machines, and capturing or destroying their materiel. But they have a goal beyond that, which is the destruction of "enemy" infrastructures. Destroying "enemy" infrastructures is much more effective than focusing on troops and weapons alone, and for industrial countries with bombers and cruise missiles it's comparatively easy. Factories, pipelines, bridges, and the like are crucial bottlenecks which have long been recognized by the military as effective targets. In a sense, warfare is about collapse—about inducing collapse in the enemy's systems, rather than simply trying to destroy troops and armament.

None of this is new. George Washington was known to Indians as "Town Destroyer" because that's what he did. And of course he didn't invent that strategy for subduing an enemy. We can go back to the Roman defeat of Carthage, and the salting of Carthaginian fields. And we can go further back then that, to the first cities, which had walls not around the outside of the city to protect from supposed marauders, but around the granaries so that those in power could control food supplies. Those in power have *always* known that controlling food supplies (and by extension infrastructure) is crucial to your enslavement of the people thus made dependent.

Inducing collapse is also a goal of guerilla or "asymmetric" warfare, conflict between a complex and centralized military megamachine and smaller, decentralized forces. Asymmetric warfare is interesting for a few reasons. First, it's becoming very common around the world, especially in high-profile wars like those in Iraq and Afghanistan. Furthermore, asymmetric warfare can be highly effective in terms of damage-per-fighter, since successful guerillas can inflict significant damages to their enemy while being less vulnerable to attack. Asymmetric warfare can target the infrastructure required by a large, complex military, even though it requires very little of that infrastructure itself. And asymmetric warfare is likely to become more and more popular in the scenario we're talking about.

Here's what I mean: Imagine that two large, oil-dependent, militarized countries come into conflict over a dwindling supply of oil from a shared supply. There just isn't enough left in the oil fields for both of them. Presume, now, that those in control of those countries aren't completely stupid. They have information about peak oil. They have information about resource depletion, about climate collapse, and about the (in)feasibility of renewable energy supplies. Whether they get into hot war or a cold war, both of them know that they want control of those oil fields. They know that the oil age won't last forever, and whoever can keep their industrial system intact the longest will be able to dominate the other, gain territory, exploit distant resources, and build electric turbines or other infrastructure that will allow them to extend their reign. If they have a military stalemate they both lose, because they spend their energy on armament, and both collapse sooner. However, if one country takes a decisive military victory, the other will collapse, and the now dominant country can delay collapse.

That's where asymmetric warfare comes into the picture. Guerillas require comparatively few resources, and can make very effective use of the materiel to which they do have access. They can continue to operate even when large-scale infrastructure is inaccessible, inoperative, or destroyed. They can work independently, and with minimal communications or centralized control. And of course, asymmetric or guerilla warfare in the past has focused on ambushes, hit-and-run attacks, and the destruction of unguarded infrastructure. By forcing larger military forces to spend limited resources to guard supply lines and infrastructure, and by targeting those lines and infrastructure when they're unguarded, guerillas can defeat much larger forces at a fraction of the cost. This is especially true when they work with a supportive populace.

War has certainly played a part in collapses of the past, and the collapse of the Roman Empire is particularly illustrative. The cause of the collapse of the Roman Empire has been an obsession of historians for centuries, but more recent explanations have moved away from mystical causes like "moral decay" and toward more rigorous understandings.

One of the most exhaustive analyses in the twentieth century was undertaken by influential historian Arnold J. Toynbee, who spent three decades writing his twelve-volume *A Study of History*, which traced the expansion and decay of more than twenty historical civilizations.

Although some previous historians had argued that the decline and fall of Rome was caused by some moral or political change, Toynbee did not agree. In his analysis, the problems had been there all along, and the empire was marching toward its own demise from the very beginning. The problem, according to Toynbee, was that the Roman Empire (like all empires, I would add) was not based on sustainable economics. Instead, he characterized the Roman system as *Raubwirtschaft*, German for a "plunder economy" or "robber economy," a word which deserves more use in our everyday discussion of modern economics. Instead of actually producing things, Rome was externally based on continuing conquest and systematic plunder of colonies and subjugated territories, and internally based on the extensive use of slave labor. This of course recalls anthropologist Stanley Diamond's famous opening to his book *In Search of the Primitive*, "Civilization originates in conquest abroad and repression at home."

Writing decades after Toynbee, Joseph Tainter came to conclusions that share similarities with Toynbee's. In *The Collapse of Complex Societies*, Tainter observes that the Roman Empire "was paid for largely by the monetary subsidy of successive conquests. Captive peoples financed further subjugations, until the Empire grew to the point where further expansion was exceedingly costly and decreasingly profitable."[310] Eventually the Empire had successfully conquered most of its known world, and its series of expansionist wars came to an end in a period remembered as *Pax Romana*, the Roman Peace. Superficially, *Pax Romana* was for some an era of relative prosperity, commerce, and peace (since the Empire had essentially expanded as much as it could). Underneath, though, the Empire was in trouble. Since the Empire was funded by continuing plunder, the fact that it could no longer expand meant that its *Pax Romana* was not a genuine peace, but simply a plateau, like the moment a ball tossed in the air seems to pause at the apex of its trajectory before it plummets back to earth. Also, even though the interior provinces were comparatively tranquil, discontent and rebellions in recently occupied regions on the frontier still made plenty of work for soldiers of the Empire.

Stop me if any of this seems familiar.

Tainter writes that the as the Empire expanded, it required more and more money and resources to maintain its extensive infrastructure and armed forces. Without expansion, the Empire had to increase the exploitation of those already conquered in order to maintain itself. It increased taxation, forced peasant children into slavery, and devalued its own currency in order to artificial inflate the budget. As Tainter writes, the effect of this was to repeatedly pass on current expenses to future generations—again, stop me if you're feeling a sense of déjà vu.

Eventually, of course, the Empire's strategy of displacing costs and consequences caught up with it. Those living in the Empire became apathetic or disaffected: when they had directly or indirectly reaped the benefits of external conquests and exploitation, the Empire was able to buy their loyalty, but things changed when *they* were the ones being (openly) exploited. As treasuries emptied, Rome could no longer maintain its military, the outlying territories crumbled, and the "barbarians" moved in. The people were no longer sympathetic towards Rome, and many rose up and joined the invaders. Eventually they succeeded, Rome was sacked, and the Empire was split apart into declining remnants. Some historians mourn the fall of Rome as a descent into "The Dark Ages," but for the vast majority of those who weren't part of the elite, the removal of the Empire meant a notable improvement in their daily lives. Despite the mythology of toil developed about the so-called Dark Ages following the Roman collapse, many historians believe that medieval peasants actually worked fewer hours than Americans do today, in part because they didn't have to support a parasitic ruler class.[311]

The collapse of the Roman Empire also underscores the effectiveness of asymmetric warfare. Although the Visigoths and Vandals who sacked Rome weren't organized along modern guerilla lines, those Germanic groups shared many of the same advantages. They were smaller groups with support from the citizenry, they had short supply lines, they made use of local resources, and many of their military actions served to disrupt and destabilize Rome rather than to engage in an occupation. Though the smaller Germanic kingdoms that succeeded the Western Roman Empire were not ideal societies, they proved more robust and resistant to invasion than that monolithic empire had been.

/ / /

When I started to write about collapse, I thought it would be the easiest section in the entire book to write. My first book, *Peak Oil Survival: Preparing for Life After Gridcrash*, was about collapse. My website InThe-Wake.org is about collapse. I've written hundreds of thousands of words on the subject in various venues. So I thought that a book chapter on the subject would be a cinch: there's so much to talk about. I could examine historical examples of collapse, starting with the earliest civilizations. I could discuss underlying mechanisms like the concentration of power, resource depletion, and the destruction of agricultural lands that have been common in collapse across millennia. I could bring it up to the modern day, and compare and contrast oil shocks in Cuba and North Korea and examine the dissolution of the USSR. I could outline current trends toward collapse, from Peak Oil to Peak Soil, and catalog the ways that technological and industrial responses to those crises are fundamentally inadequate. I could even talk about how responses to crises like oil depletion—the promotion of biofuels, for example—are making the original crises far, far worse than they were.

And yet, with such a smorgasbord of fodder, I got part way into the chapter and stalled completely, for weeks. At first I thought there was simply too much to cover—it was an issue of paralysis caused by choice. And then I came to realize it was something very different.

When I first started taking issues like peak oil seriously, I began to scour newspapers for clues about how it might be taking effect. When news websites on energy decline and ecological collapse started popping up, I read those, too, often checking several times a day. Part of it was curiosity, and part of it was me looking for cracks in the dominant culture's façade of invulnerability, which would mean cause for hope (or at least, a decreased level of pessimism). Somewhere between the time I wrote *Peak Oil Sur-*

vival and the time it was published, Hurricane Katrina blew through and obliterated most of New Orleans. The pictures and video in the news looked like a civilization that had already collapsed. I followed the news and analysis, but about eight months later, I stopped reading those news sites. It wasn't that the information was frightening or depressing (at least, not any more than before), but that there was actually very little that really seemed new or unexpected anymore, and going to such effort to follow the subject seemed, well, maladaptive. It reminds me of a scene in Neal Stephenson's classic sci-fi novel *Snow Crash*, in which the protagonist, practically dripping with high-tech electronics, sensors, and Heads Up Display info, finds himself the target of some angry and well-armed antagonists: "He turns off all of the techno-shit in his goggles. All it does is confuse him; he stands there reading statistics about his own death even as it's happening to him. Very post-modern."[312]

What stalled my progress here wasn't a lack of data or analysis, it was the dearth of action. If you want to know what happens when authoritarian societies run low on oil, look at North Korea. If you want to know the aftermath of imperial resource exploitation on a massive scale, look at most of Sub-Saharan Africa. There's not a lot we can discuss about collapse that hasn't already happened, or at least begun, somewhere in the real world. That's not to say you shouldn't learn about why civilizations collapse, or how, or what is happening now. But eventually a purely analytical approach reaches a point of diminishing returns. It's long past time to stand around reading statistics about our own demise, and about the demise of the planet. If there's one thing I've learned from everything I've read or written about collapse, it's this: we have a choice, a real choice.

When people are confronted with the likelihood of collapse, I've seen their reactions fall into three broad categories. Most people deny that collapse is even possible: having been born after the Second World War, most living Westerners can only remember the constant growth of industry, and other possibilities are almost literally unimaginable. A rather smaller number of people recognize that collapse is likely, but disclaim any responsibility for action: boggled by the scale of the issue they either insist that government or industry will address the problem, or take the "if we're going to die we might as well party" route. And a much smaller group responds with the resolve, even optimism, that can lead to action.

We do not have to be, and should not be, passive, regardless of when or how we expect things to change. Collapse, by breaking down large-scale hierarchal structures and reducing their ability to project power, proportionally increases our own power on the smaller scale. We can make the choice to take an active role in collapse, and change the outcome.

There's a quotation that's been sticking with me for a long time. It's something that Dietrich Bonhoeffer wrote while he was imprisoned by the Nazis and awaiting execution for his role in the resistance: "We have spent too much time in thinking, supposing that if we weigh in advance the possibilities of any action, it will happen automatically. We have learnt, rather too late, that action comes, not from thought, but from a readiness for responsibility. For you, thought and action will enter on a new relationship; your thinking will be confined to your responsibilities in action."

It's time to talk about action. And it's time to talk about fighting back.

FIGHTING BACK

Suppose that was an awful big snake down there, on the floor. He bite you.
Folks all scared, because you die. You send for a doctor to cut the bite; but
the snake, he rolled up there, and while the doctor doing it, he bite you
again. The doctor dug out that bite; but while the doctor doing it, the
snake, he spring up and bite you again; so he keep doing it, till you kill
him. —HARRIET TUBMAN[313]

A GREAT MANY BOOKS ON ENVIRONMENTALISM written in the last few decades conclude with a listing of things that you, as a citizen, or more likely as a consumer, can do to address the problems. In your capacity as a consumer, you can reduce your consumption or buy products that are allegedly more eco friendly in order to convince corporations to enact change. As a citizen, you can write to "your" congresspeople or other governmental "representatives," and ask *them* to enact change. If you want to do one better, you can donate to a nonprofit organization that will lobby governments and corporations on your behalf. Authors may offer a plethora of different vicarious solutions involving various ways to try to persuade large, entrenched institutions to act against their underlying drives.

We're not going to do that.

If you've gotten to this point and are yearning for a way to reduce your personal use of disposable packaging or compost more of your household waste, there are already hundreds of books with tips on exactly those subjects. It's been done. And we're not saying you shouldn't try to reduce your production of household waste. Minimizing waste is certainly a good thing to do, and we don't want to insult that group of people (which includes ourselves) who have taken steps to reduce their waste. And obviously we aren't saying you shouldn't donate to nonprofit organizations, especially local ones. We *will* say that in general you shouldn't give a dime to big corporate "green" organizations. One example why: Jay Hair, former head of the National Wildlife Federation, immediately went from there to becoming a spokesperson for Plum Creek Timber Company, a timber company so nasty even a Republican called it the Darth Vader of the timber industry. Such is business as usual among the big corporate "green" organizations.)

What we are saying is this: we aren't going to insult your intelligence by asserting that such solutions are even remotely sufficient to address the problem. Removing an extra few dump-truck loads from those seventy-three Grand Canyons is good, but it's a drop in the plastic-suffocated ocean in terms of real change. We don't have the time or patience to immerse ourselves in a fantasy world where corporations and governments act in ways that contradict their own fundamental imperatives and immediate self-interest because we send them politely worded and well-researched letters. And we aren't going to blindly swallow the premise that you, the reader, are a mere consumer, taxpayer, or even citizen. Your identity, your being, is not limited to your economic function in relation to some vast bureaucracy. You are a human being, an animal: whether you recognize it or not, you are a living creature embedded in a network of trillions and trillions of other living creatures, all interdependent.

In a way, that's really what this entire book is about. Is your identity that of a consumer, or a person? When you say "the real world" are you talking about wage slave capitalism, or are you talking about a living breathing world of trees and rivers and lakes and deserts and forests and mountains and seas? And if you identify as a living member of that much larger community, a community that is being systematically destroyed by a toxic mimic of the real world, what are you going to do about it? How are you going to defend your community?

/ / /

There are those who disagree that the world needs defending. I'm not talking about the people who think everything is fine, or those who wouldn't understand or care if every fish in the ocean were killed. I'm talking about people who recognize the scope and seriousness of our predicament, who recognize that the problems we face are deeply rooted and systemic, but who don't think that they in particular need to do anything other than "walk away." Of course, as was already discussed, this is not nearly enough, and we should not pretend it is. Those who would truly walk away, who aim to abandon the dominant culture completely, clearly recognize that this culture is fundamentally irredeemable. And they presumably recognize that civilization's voracious industrial appetite is eating up the planet at an ever increasing rate. So why do they view walking away as an adequate strategy? If this culture is not going to change, where do they expect to be safe? What do they expect will happen as industrial society exhausts its last remaining resources? If this monstrosity is not stopped, the carefully tended permaculture gardens and groves of lifeboat ecovillages will be nothing more than after-dinner snacks for civilization.

I think the problem is partly a lack of historical perspective. It's not as though living outside of civilization is new, after all. A little more than five centuries ago, North and South America (and Africa, Oceania, and Asia; if we go back 2000 years we can add Europe, and if we go back 6000 years we can add the Near and Middle East) were filled with tens of thousands of uncivilized communities. Any five-year-old child among them would have been better at finding wild edible foods than I will ever be. The Confederacy of the Haudenosaunee was a proven and effective participatory democracy that I can only dream of emulating in my community. These continents abounded with warriors who were skilled and courageous beyond my conception. And yet they were all but wiped out by the insane civilized (whom they significantly outnumbered, at least at the beginning) using technologies that are hopelessly crude by modern standards of conquest and genocide. Do civilized people who walk away, many of whom are essentially novices both to living in a healthy community and living with the land, believe they can survive so much better than entire indigenous nations with countless millennia of uninterrupted experience?

I don't mean to sound overly pessimistic. I don't think we're completely doomed. We do have some advantages that those living centuries ago did not. We recognize how deeply pathological this culture is, we recognize the need for it to be dismantled entirely, and we can understand that in a way for which indigenous Americans of the fifteenth through nineteenth centuries (and others of the indigenous) simply didn't have the context.

They were taken off-guard by an enemy more hideous, insatiable, and cruel than they had ever encountered, could reasonably anticipate, or even imagine. We no longer have to imagine. We need merely pay attention. We also have the benefit that modern society is more monolithic, more dependent on a small number of centralized industrial and economic systems, more brittle, and more vulnerable to collapse. And we can use those systems to our own (and the world's) advantage, both by employing and disrupting them.

But if we choose solely to walk away, we give up these and other advantages. If we choose not to fight back, we concede most of the few slim possibilities we have for success, let alone survival.

Like most decisions people make, especially life decisions based on complex or unpredictable situations, the decision to just walk away is not one based on reasoned analysis. And because of this, I think the motivation has much to do not just with identity, but with what Dietrich Bonhoeffer wrote: that action comes from a readiness for responsibility. To succeed in stopping the destruction of the planet, you have to be ready to take responsibility. Not to belittle responsibility by pretending that solely personal actions aimed primarily at protecting you and yours can solve vast problems. Not to surrender responsibility to governments and businesses which claim to act on your behalf or in your best interest. Not to renounce responsibility by pretending that walking away from the destruction will somehow cause it to stop.

You have to be ready to take responsibility to defend your community. And when your community (by which we mean your landbase, the living Earth, your human community, and your own body) is in danger—no, not just in danger but actively under attack—that means fighting back.

/ / /

Fighting back, in the broadest sense, means a great deal. It means giving up on the fairy tale that those in power act in the best interest of us or the planet, or that they are systematically capable of thinking in the long term. It means no longer pretending that industrial progress will bring us to some bright new beautiful tomorrow. It means stepping outside of the carefully circumscribed limits that keep us ineffective. It means deliberate and strategic opposition to those in power, instead of attempts to lobby or convince them to please stop exploiting people and destroying the world.

Fighting back means doing what is appropriate. It means seeking solutions appropriate to the scale of a problem. It means not ruling out actions just because those in power (or Gandhian activists, or the Bible, or those who think buying recycled toilet paper is sufficient, or liberal members of the "loyal opposition") say they shouldn't be used. And fighting back may or may not look like fighting: it doesn't have to look like violence (although it may). It means not using violence when it's appropriate to not use violence. It means using violence when it *is* appropriate to use violence. It means using industrial technology when it's appropriate, and not using it when it's not appropriate. It means being strategic, and being smart, and remembering our allegiances and our end goals.

I've learned many things from the brilliant writer and activist Lierre Keith, and one of them is this: effective organized political resistance, whether it's devoutly nonviolent or espouses a diversity of tactics, is based on force. Those in power understand this, which is why they want a monopoly not just on violence but on force, and why they have gone to such lengths (through the use of COINTELPRO and the like) to sabotage movements that understand force. Radical movements that disavow the use of directed political violence can, in certain circumstances, succeed, so long as they have very large numbers of dedicated people who understand that their goal is not to *ask* for the intolerable to stop, but to force it to stop.

Another person who understood this was abolitionist Frederick Douglass, who said as much in a speech in 1857: "Let me give you a word of the philosophy of reform. The whole history of the progress of human liberty shows that all concessions yet made to her august claims, have been born of earnest struggle. The conflict has been exciting, agitating, all-absorbing, and for the time being, putting all other tumults to silence. It must do this or it does nothing. If there is no struggle there is no progress. Those who profess to favor freedom and yet depreciate agitation, are men [sic] who want crops without plowing up the ground, they want rain without thunder and lightning. They want the ocean without the awful roar of its many waters.[314]

"This struggle may be a moral one, or it may be a physical one, and it may be both moral and physical, but it must be a struggle. *Power concedes nothing without a demand.* It never did and it never will. Find out just what any people will quietly submit to and you have found out the exact measure of injustice and wrong which will be imposed upon them, and these will continue till they are resisted with either words or blows, or with both. The limits of tyrants are prescribed by the endurance of those whom they oppress."[315]

When he said that power concedes nothing without a demand, he wasn't talking about a sharply worded request. He meant, as my dictionary puts it, "to ask for with proper authority; claim as a right." Not to ask authority—to ask *with* authority.

When I interviewed Lierre several years ago, she discussed the basis of political effectiveness in force, and strongly recommended Gene Sharp's book *The Politics of Nonviolent Action*: "His books are profoundly important. His ideas have been used in liberation struggles all over the world, from South Africa to Eastern Europe. He points out that power depends on obedience, and we don't have to obey. The moment the oppressed withdraw our consent, the powerful are left with nothing. Sharp identifies a range of tactical approaches, but they break down into two categories: acts of omission and acts of commission.

"Omission includes things like boycotts, strikes, nonparticipation in illegitimate governments. Acts of commission would include sit-ins, obstructions, and occupations like the forest defense elves in the trees. But either way, nonviolent action is an attempt to coerce an institution that holds power to change.

"There's a tremendous misconception, particularly in the US, that nonviolent action is about somehow trying to educate or convert those in power. It's not. That's pacifism, not nonviolent action. I mean, does anybody really think the owners of the bus company in Montgomery, Alabama had a sudden epiphany? 'We've been so terrible to Black people, oh my god, segregation must end!' Of course not. The boycott brought them to their knees. There may or may not have been individuals whose consciences were awakened, but that wasn't the point. People withheld their economic power until the institution—in this case, the bus company—caved in.

"I think this is so important because the main divide isn't between violence and nonviolence. It's between action and inaction. Properly understood, both militancy and nonviolence are direct confrontations with power, confrontations backed by the threat of force. Both strategies require planning, discipline, and sacrifice. Both kinds of activism will bring the full weight of the wrath of the powerful down upon the actionists. The moment you're successful, the moment power is threatened, you will pay, sometimes with your lives."[316]

/ / /

There's no doubt, we're in a serious situation that requires a serious response. And though we shouldn't unnecessarily provoke those in power, we must recognize that effective political strategy will meet with reprisals from those in power regardless of the specific tactics used. It can be frightening to think of those reprisals targeting you—that's the whole point after all: it's a form of government-sponsored terrorism (and one could easily argue that civilized governments *are* a form of terrorism)—but in the long term (and by now the short term), those in power are destroying the world. What do we have to lose? If we make them really mad, what are they going to do, destroy the earth twice?

Some people will not resist. Some will actively collaborate. Perhaps they benefit, at least temporarily, from civilization's hierarchy. Or perhaps those in power have determined the precise measure of empty promises that most people will tolerate, calculated the "exact measure of injustice and wrong" that they can get away with. But for the rest of us, our job is to devise and enact a plan—many plans, actually—that will make those in power afraid, not just for themselves, but for the entire wretched system that keeps them in power.

So what do we need in order to do that? First, to again quote Lierre Keith: "What any movement needs is an effective strategy. That means identifying two things: where is power weak and where are you strong? The overlap is where you strike. One problem with nonviolence is that it depends on huge numbers of people to be effective. Rosa Parks on her own ended up in jail. Rosa Parks plus the whole Black community of Montgomery ended segregation on the public transportation system. Without a mass movement, the technique doesn't work."

So that's actually two things we need so far: an effective strategy, and an effective strategy that's actually congruent with the numbers of people we have, the resources we have, and the time we have. That helps to narrow things down. Though a wholly nonviolent mass movement bent on systematically uprooting the fundamental causes of human exploitation and ecocide would be wonderful, it falls short in our case as a valid strategy. Currently, there simply aren't enough people willing to address the issue. And worse, there isn't enough time to build that movement—each day that passes means hundreds of species wiped out forever, means more land-based cultures destroyed by the industrial onslaught, means global fossil energy consumption brings us closer and closer to a runaway greenhouse effect, means more plastic and fewer fish in the oceans, means fewer amphibians, and so on, ad omnicidium.

A third asset we need is a collective recognition of the real systemic roots of the problem. As discussed earlier, the garbage problem will not go away because you or I stop producing garbage. And global warming will not go away because you or I stop using gasoline. Those problems will not go away as long as there is a global industrial system that produces waste and burns gasoline. In fact, it's conceivable that in the coming decades a focus on reducing personal consumption could even make things worse. I'm not talking about worsening caused by a focus on more symbolic action at the expense of more effective action, although that's certainly a valid argument. Rather, I'm talking about the fact that we're entering a post-peak period for oil, and for many other commodities. If demand for oil far outstrips supply, and we all decide to get together to reduce our consumption of gas for altruistic reasons, we will reduce demand for the finite supply of oil. The net effect of this "green" action will simply be to make the remaining oil cheaper and more readily available for militaries, corporations, and other institutions which lack our scruples. Which, again, isn't to say we shouldn't reduce our consumption, but we should do it because it's the right thing and not because we expect it to topple those in power.

A fourth prerequisite for effectiveness is a culture of resistance. This should not be confused with an "alternative culture." Instead, a culture of resistance is an explicitly oppositional culture. An effective culture of resistance does not seek a "cultural revolution": cultural change is not the objective, but material change, accomplished though the organized work of a large and diverse group. A culture of resistance is, collectively, that group of people with an understanding of the root causes of their predicament, and a willingness to work together in opposition to authority to address those causes.

A culture of resistance is not the same as an organized resistance movement, but is necessary for the success and growth of such a movement. In every country where a successful revolution has taken place, there has been a culture of resistance. In every occupied nation with an ongoing resistance movement, there is a culture of resistance. If a resistance movement is a sturdy tree, a culture of resistance is the soil from which it grows, a soil itself enriched by the growth of the tree.

There is a story about a member of an Irish resistance group in the early twentieth century. One night, while carrying out resistance activities, he was discovered by the British and shot as he escaped. The man was wounded, but managed to hide in an alley and avoid discovery by the British. Later that night, a group of men on their way home passed through the alley and found the injured man. Though not active members of the resistance, they immediately recognized what had happened and brought the man to a doctor and safehouse. They didn't need to be told to do this—they knew, because their culture was a culture of resistance.

Such a culture benefits from shared goals and group norms that allow the culture to propagate and persist, and gives rise to effective tactics and strategies. Solidarity and mutual aid, such as in the above example, are one important characteristic. Further, those in the culture acknowledge and support a broad diversity of tactics and involvement, with the understanding that they're all working toward the same goals with the same general strategy. This permits individuals and small groups to focus on the projects and tasks they're best suited for, as well as limiting risk to the entire group while supporting those in the most high risk positions.

Here's what I mean by that. In any given army, only a tiny percentage of the army is actually involved in fighting. (For example, in 1918, just before the Irish War of Independence, the IRA had about 100,000 enlisted members, but only about 3000 of them were actually fighters at any given time.) The rest of those involved participate in logistical and support roles, doing recruitment and training; communication; logistics like obtaining, manufacturing, and moving materiel; medical support; and even things as basic as feeding the troops and maintaining equipment. And many of those who fight aren't professional soldiers. An army commonly consists of a core or skeleton of professional officers and noncommissioned officers. When war is declared, that skeleton is fleshed out by conscripts, reservists, or civilian militia.

Guerilla or resistance movements are likely have a similar disproportion between the number of people who actively carry out operations, and the people who support them. Relatively few members of an armed resistance movement actually take up arms as active guerillas. But in order for them to succeed they require a much larger support network of sympathizers, fundraisers, above-ground political agitators, reconnaissance workers, and those who offer direct material support such as food and shelter. Even nonviolent movements are likely to have a parallel structure. Those people who put themselves in harm's way through civil disobedience or direct action—be they forest defenders in tree sits, Project Ploughshares activists smashing military hardware, or indigenous people blockading loggers or miners in their homeland—ultimately rely on those who can offer support for prisoners and their families, medical aid, awareness raising and material support. (For example, when people think of treesitters, how many of them think about the people who bring them food and water, and carry away their shit buckets, people without whom the treesitters would not be able to last more than a few days.)

Governments of occupation realize that cultures of resistance are very dangerous things. They constantly try to undermine group solidarity and a diversity of tactics to split movements and quash actions that might prove effective. The FBI's Counter Intelligence Program, COINTELPRO, is an example of a highly effective program designed to destroy cultures of resistance. In his book *War at Home: Covert Action Against US Activists and What We Can Do About It*, attorney Brian Glick identified four main methods used to target and disrupt everyone from the Black Panther Party to the National Association for the Advancement of Colored People. Those four categories were: "*Infiltration:* Agents and informers did not merely spy on political activists. Their main purpose was to discredit and disrupt. Their very presence served to undermine trust and scare off potential supporters. The FBI and police exploited this fear to smear genuine activists as agents. *Psychological Warfare from the Outside:* The FBI and police used myriad other 'dirty tricks' to undermine progressive movements. They planted false media stories and published bogus leaflets and other publications in the name of targeted groups. They forged correspondence, sent anonymous letters, and made anonymous telephone calls. They spread misinformation about meetings and events, set up pseudo movement groups run by government agents, and manipulated or strong-armed parents, employers, landlords, school officials, and others to cause trouble for activists. *Harassment Through the Legal System:* The FBI and police abused the legal system to harass dissidents and make them appear to be criminals. Officers of the law gave perjured testimony and presented fabricated evidence as a pretext for false arrests and wrongful imprisonment. They discriminatorily enforced tax laws and other government regulations and used conspicuous surveillance, 'investigative' interviews, and grand jury subpoenas in an effort to intimidate activists and silence their supporters. *Extralegal Force and Violence:* The FBI and police threatened, instigated, and themselves conducted break-ins, vandalism, assaults, and beatings. The object was to frighten dissidents and disrupt their movements. In the case of radical Black and Puerto Rican activists (and later Native Americans), these attacks—including political assassinations—were so extensive, vicious, and calculated that they can accurately be termed a form of official 'terrorism.'"[317]

Though COINTELPRO operations were only officially active between 1956 and 1971, of course COINTELPRO-like operations have continued.[318] And furthermore, the success of COINTELPRO means that their strategy has served as a template mimicked by many of those in power attempting to quash resistance without unleashing the public sympathy that comes with more overt fascism. We hear echoes of that same strategy when protestors are described as "good protestors" and "bad protesters" (usually based on who is willing to follow police orders), and when protesters internalize these messages from above and use them to label other protesters as "good" or "bad" based on this same criteria. Any successful culture of resistance must learn from COINTELPRO and programs like it in order to become more robust, and any serious culture of resistance must come up with its own measures of "good" and "bad," "successful" and "unsuccessful."

When the issue of fighting back comes up, I sometimes hear people argue that we mustn't *fight* back, because those in power will only rebuild, or because they'll only increase their repression and violence. That this concern is considered by some to be a valid reason for inaction tells me many things. One of the most important things it tells me is that the people asking that question do not live in a culture of resistance. The question of how those in power will respond to different actions is certainly strategically valid. But in a culture of resistance, it's not a reason to not resist by whatever means are most appropriate and effective. Of course those in power will inflict reprisals on those who resist them. Of course they will try to frighten and terrorize dissidents into accepting their authority. Of course they will try to harm even those who do not directly participate in actions against power. *This is not a reason to hold back—this is why we fight them.* In a culture of resistance, reprisals and state terrorism are certainly not trivial, and they are not ignored. Instead, they underscore the importance of resistance, and strengthen the resolve of those who fight back.

/ / /

Successful resistance movements of all times have recognized that one extremely effective way to counter state reprisals is that each time the state raises the stakes against the resistance, the resistance raises the stakes back against state repression. Many successful resistance movements have recognized that not only does the state not have a monopoly on violence,[319] but that it also does not have a monopoly on either reprisals or upping the stakes.

/ / /

Even if you believe that the dominant culture is flawed but redeemable, even if you believe that government trespasses could be righted by electing a new president, or that corporate excesses could be solved by purchasing a different product, here is a reason to support fighting back and a culture of resistance.

Remember the "Roaring Twenties"? Remember the technological innovations, the growing affordability of new machines like automobiles, the economic growth, the growing freedom for different lifestyles? Remember how it collapsed suddenly and segued virtually overnight into a Great

Depression? Remember how that economic downturn made many countries into breeding grounds for overt fascism, a fascism that had itself been taking root and then growing all through the twenties?

That's why.

It's true that, as we discussed earlier, the collapse of large, centralized organizations can offer great opportunities for community-scale resurgence and resistance. Unfortunately, it's also true that a partial failure of a state or economy *without* the thorough dismantlement of oppressive power structures also provides an opportunity for more ruthless power-mongers.

Fascism and other authoritarian systems do not *originate* from economic collapses. Throughout the 1920s, fascists were gaining ground in many countries. But the Great Depression was a major factor in moving them from marginalized upstarts to ruling governments. Authoritarians have been able to take advantage of social and economic disruptions through the stories they tell, through the myths and propaganda they promulgate.

No government or party gains power by offering a cogent analysis of the underlying flaws of a civilization. A cunning fascist would not claim that the capitalist economy collapsed because it was intrinsically unstable and based on the imaginary inflation of capital. No clever authoritarian would say that energy shortages are caused by the exhaustion of finite supplies of fossil fuels. No, those sociopathologically sly would-be dictators always find a scapegoat and a reason to gather more power, more authority. The economy is weakened because foreign powers are conspiring against us, because there are bloodsuckers parasitizing our society. Energy is short because environmentalists are standing in the way of progress, because terrorists overseas are attacking us and hate our freedoms. Give me your vote. Let me be your voice. Give me the power, and I will do what it takes. I will crush these enemies of our glorious nation, our homeland, our fatherland.

The Nazis, in particular, were experts at this kind of manipulation (although one could argue that they were neophytes compared to the US). Hitler and other Nazi propagandists blamed Germany's economic woes on the Treaty of Versailles and countries which signed it, whipping good Germans into a nationalistic frenzy against those external enemies. This propaganda had at least a partial basis in fact—the Treaty of Versailles required Germany to pay exacting reparations for losses suffered by the

Allies. On the other hand, finding internal scapegoats required what Hitler called the Big Lie, the use of repeated falsehoods so colossal that listeners would not believe that the liar would have the gall make such preposterous statements if they were not true.

When Nazis targeted and demonized internal enemies, such as Jews and Roma, they started by capitalizing on existing ethnocentrism, racism and anti-Semitism. Germans were suffering, they said, because Jews were bloodsuckers who leeched money from hardworking German people while doing no work themselves. Propaganda posters showed Jews depicted as worms with dollar signs and the hammer-and-sickle for pupils. From there the Nazis moved onto even more audacious lies: Jews were part of a conspiracy to control the world and wipe out the Aryan race; at Passover, Jews would try to kidnap and kill Christian infants to mix the blood with their matzoh.

As absurd and clearly irrational as much of their propaganda was, the Nazis were motivated by a sociopathic internal logic, social Darwinism, and a perverse understanding of carrying capacity. German expansionism in the Nazi doctrine was internally rationalized by the German need for more *Lebensraum*, the German word for "living space" or "habitat," a concept which included both land and raw resources. In the Nazi doctrine, other "inferior races" took up *Lebensraum* that rightfully belonged to the superior Germans. It was therefore not just acceptable, but a moral imperative for Germany to enslave, deport, or kill those "inferior races" and colonize their *Lebensraum* with Germans.

Of course, the Nazis didn't claim to have invented this idea. Hitler repeatedly pointed out that his stated goal was to emulate what white colonists had done to the indigenous people of the Americas.

Let's bring this back to the modern day and the concept of collapse. Currently human carrying capacity is vastly overshot through the use of unsustainable practices like the extensive use of fossil energy for agriculture, the erosion of soils with intensive cultivation, and the drawing down of aquifers around the world. When peak oil really hits home, when global trade starts to unravel in earnest, that ghost carrying capacity will evaporate, and food and basic resources will be in globally short supply. People who do not have jobs, people who cannot afford to fuel their cars, people who are hungry, do not want to hear that they're experiencing that privation because generations of industrial humans lived unsustainably and stripped away much of the planet's surface in an orgy of pointless consumerism. They do not want to hear that they will never again experience the levels of material opulence to which they've grown to feel entitled. They certainly do not want to hear that *they* were the ones who used up all of those resources. They would be much more receptive to hearing that some particular scapegoat—Arabs, Mexicans, illegal immigrants, those strange people living across the river—is to blame, and if they'll just give the government a little more power, sacrifice a little more of their freedom, the government will be sure to solve that problem. And heck, if all else fails they can always displace the different people across the river and grow more food on their land. Or they can enslave them and force them to work in labor camps, because those fossil-fueled factories aren't quite so productive since the oil dried up. And if those scapegoats don't cooperate—or even if they do—the government can always kill them. They aren't fully human after all.

The Nazis did all of these things. When Jews, Roma, Slavs, and others were targeted by the Nazis, the German government could claim their land, homes, money, and even personal effects and bodies (including hair, eyeglasses, fat for soap, and gold from their teeth), which could be given to Germans or used in the war effort. And of course, the concentration camps weren't solely extermination camps. Almost all of them were also work camps, commonly with adjacent corporate factories getting slave labor from the inmates.

Although carrying capacity pressures and economic or social collapses are never the sole causes of genocide, many historians and analysts believe they play a major contributing role. According to some observers, even the relatively recent and tragic genocide in Rwanda may have been worsened or partly caused by carrying capacity pressures.[320] But that by itself is never enough to spawn a full-scale genocide. Such atrocities require some preexisting racism or other prejudice accompanied by an authoritarian leadership willing to manipulate the situation for power gain.

Regardless of the exact politics of those fascists, authoritarians, or totalitarians, resistance and dissent represents a threat. A culture of resistance represents a particular threat. In fascism, the identity of the people is subsumed to that of the nation, which is personified by the authoritarian leader, be it the führer or some variant. That leader cannot tolerate dissent, or, as Erich Fromm put it in his essential book *The Anatomy of Human Destructiveness*, self-assertion: "For all irrational and exploitative forms of authority, self assertion—the pursuit by another of his real goals—is the arch sin because it is a threat to the power of the authority; the person subject to it is indoctrinated to believe that the aims of authority are also his, and that obedience offers the optimal chance for fulfilling oneself."[321]

Even modern states officially running under democracy—or, depending on your perspective, a form of Friendly Fascism—work under much the same principal. Those who are not able to break this indoctrination, those who are not part of the culture of resistance, often fail to question authority even when it is utterly clear that they should. Psychologist Bruno Bettelheim, survivor of both Dachau and Buchenwald concentration camps, had a keen understanding of how this played out in the camps: "*Non-political middle class* prisoners (a minority group in the concentration camps) were those least able to withstand the initial shock. They were utterly unable to understand what had happened to them and why. More than ever they clung to what had given them self respect up to that moment. Even while being abused, they would assure the SS they had never opposed Nazism. They could not understand why they, who had always obeyed the law without question, were being persecuted. Even now, though unjustly imprisoned, they dared not oppose their oppressors even in thought, though it would have given them a self-respect they were badly in need of. All they could do was plead and many groveled. Since law and police had to remain beyond reproach, they accepted as just whatever the Gestapo did. Their only objection was that *they* had become objects of a persecution which in itself must be just, since the authorities imposed it. They rationalized their difficulty by insisting it was all a 'mistake.' The SS made fun of them, mistreated them badly, while at the same time enjoying scenes that emphasized their position of superiority. The [middle class prisoner] group as a whole was especially anxious that their middle class status should be respected in some way. What upset them most was being treated 'like ordinary criminals.'

"Their behaviour showed how little the apolitical German middle class was able to hold its own against National Socialism. No consistent philosophy, either moral, political, or social, protected their integrity or gave them strength for an inner stand against Nazism. They had little or no resources to fall back on when subject to the shock of imprisonment. Their self esteem had rested on a status and respect that came with their positions, depended on their jobs, on being head of a family, or similar external factors. . . ."[322]

Bettelheim goes on to discuss how the apolitical middle-class prisoners almost universally failed to adopt the more dignified and effective behaviour patterns of political prisoners, which embodied solidarity and mutual aid. Instead they exhibited "pettiness, quarrelsomeness, self-pity" and took to stealing from the other prisoners. Many of the apolitical prisoners actively collaborated with the guards and informed on other prisoners. This hurt other prisoners, but did not help the collaborators. As Bettelheim observed, "the Gestapo liked the betrayal but despised the traitor."

Clearly, those apolitical middle-class prisoners failed to break their identification with those in power, even when those in power were not merely in error but actually evil, even when those prisoners were directly confronted with this fact, even when those prisoners would clearly have benefitted from breaking that identification. This fact cries for a question: if apolitical middle-class people do not break their identification even when they personally bear the brunt of state violence, how will they come around when the state's violence is primarily directed at others, when the effects of corporate ecocide are displaced years or decades into the future?

The answer is, by and large, they won't. And we should not fool ourselves into thinking that they will. But that does not lessen the importance of building a culture of resistance, only underscores it. Some of the apolitical middle-class prisoners did join with the political prisoners. And even in a place as dismal as the concentration camps, the culture of resistance helped to improve conditions and sometimes permitted organized escapes. Whatever trials we face, a healthy culture of resistance can only help us to keep going.

/ / /

All this is why building a culture of resistance should never be confused with building a mass movement, or making our politics mainstream. Again, it would be lovely if those things were to happen. But when the situation is as urgent as it is, we do not have time to wait for a majority consensus. You would not, after all, expect WWII resistance members in

Germany to wait for the majority of Germans to oppose the Nazi government before they started to take action. You would not expect Tecumseh to wait until all Indians agreed to fight before he confronted colonizing soldiers. You would not expect antebellum slaves to refrain from attempting to escape or revolt until they could gather endorsement from society at large.

Some observers critique those who take action without a mass movement, or outside of the context of a mass movement. Such action, they argue, constitutes a form of vanguardism, a strategy (most notably used by Lenin) in which dedicated revolutionaries attempt to put themselves at the center of a movement in order to trigger a revolution and steer its direction. In a way, this criticism is paradoxical—how can you try to seize control of a mass movement if you aren't actually participating in one? The difference between vanguardism and what we're talking about is that of self-defense, of community defense. A mass movement would be wonderful, but you can't reasonably expect people to stand by and watch the destruction of the planet while waiting for a "mass enlightenment," which, as Bettelheim's example shows, is counter to the (socially created) nature of many people.

Of course, when people do fight back, their example may inspire others to act. Which is why governments are often so afraid of even small acts of resistance.

Here's another reason why people who don't want to fight back themselves should support a culture of resistance. Simply by being more radical, or by taking action, members of serious resistance movements can shift the entire political spectrum. For example, let's say that a group of reformists are attempting to draw attention to a certain issue, perhaps attempting to convince a municipal government to limit suburban sprawl. They issue letters of protest, they sign petitions, and eventually, out of desperation, they blockade a highway in a subdivision under construction. Maybe someone even spraypaints a bulldozer. The police come to clear them out, and they are criticized by newspapers and letters to the editor. The next week, some anonymous party burns down three unfinished houses in that same subdivision. All of a sudden, the newspapers don't seem to care so much about the occasional blockade; they have their hands full criticizing bigger events. Whether or not you agree with burning down unfinished subdivisions, more radical actions can shift the political spectrum, and make more room for actions that aren't as radical.

This kind of dynamic between reformists and revolutionaries is common in history, and often leads to notable progress. Do you think that Martin Luther King, Jr., would have been so successful if the government hadn't been afraid of Malcolm X? Of course, reformists rarely acknowledge this relationship in public. But fighting back can lead to serious results, even if it does not immediately result in the ultimate success of those fighting.

/ / /

Here's a fifth requirement for our effectiveness: long-term thinking.

In the long term, what happens to us here and now is not as important as what we leave behind. Harriet Tubman knew this, and it shows in her story about snakebite at the beginning of this chapter. She knew that healing could not occur until the cause of injury was neutralized, and that meant fighting.[323] And hell, Harriet Tubman knew what it was to be wounded. She bore the metaphorical and physical scars of a life spent under and fighting slavery.

I wish I could say that people with more privilege than Tubman would be able to muster more resources with which to fight back. I wish I could say they would be a fraction as radical—as willing to look for root problems. But I'm not sure that's the case. The more privilege you have, the more you have to lose by opposing those in power. As the Last Poets said, speak not of revolution until you are ready to eat rats to survive.

Regardless of where we come from personally, we each have a role to play in the culture of resistance. And in that context it is our obligation, both morally and strategically, to engage in the most radical action we are capable of doing. To go for the root with as much drive and courage as we can muster.

/ / /

Here are some questions that anyone contemplating serious action should ask themselves.[324] What are the risks if you take action? (Loss of status?

State reprisals? Prison? Torture? Murder by the state?) What are the risks if you don't? (A freefall slide into fascist dystopia? Runaway global warming? The collapse of the biosphere? Loss of self-respect?) What would you need from yourself, from your friends, your family, your community, your institutions to make action more possible? (Moral support? Material support? Familial support? Collaboration?)

Where do your loyalties lie? Where do you end, and other creatures begin? What will be your legacy? What do you want to leave behind?

What do you need, and what do you have to give up, to make that happen? And if you don't do it, who will?

Knowing the answers to those questions, having discarded the paralyzing mythologies of those in power, choose your future, and fight for it.

THE LIVING

The hottest places in Hell are reserved for those who in a period of moral crisis maintain their neutrality. —JOHN F. KENNEDY

We must always take sides. Neutrality helps the oppressor, never the victim. Silence encourages the tormentor, never the tormented.
—ELIE WIESEL

THIS BOOK BEGAN AS AN EXPLORATION of decay—how in the real, physical world, death and decay lead to more life; how life feeds off life and life feeds off death; how decay is a recirculation of compounds crucial to life; how life is a circle, not a straight line; how our bodies are the oldest and most precious gifts we give to each other—and it ends with the looming collapse of this linear culture, this culture that has been functionally and systematically doomed from its beginning.

There is no meaningful immortality in the fabrication of materials that do not decay. You can manufacture thousands upon thousands—millions upon millions—of plastic forks or plastic billiard balls or plastic cars or computers, and when you die, you will still be dead. And what have you left behind? Your body will have become as toxic as the industrial processes in which you participated.

There is another way, that leads not to the pseudo-immortality of plastic forks and plasticized bodies, and that emerges not from this literally and physically insane stronger identification with this culture, with the artificial, with the not-alive yet not-dead status of so many of this culture's fabrications and so many of this culture's members, than with life, than with the processes of birth, life, death, decay, than with the real, physical world.

399

The other way is, simply put, to reidentify with the living, with life, with the real, physical world. And to live your life and to give your life for the benefit of the living, the benefit of those who give life to you and those you love.

/ / /

I walk across the land we in the neighborhood came together to try to save. I hear the stream flowing through remains of the forest. I hear birds, though far fewer than last year, before the trees were cut.

The place still looks like the site of a massacre, most probably because that's exactly what it is. But even in this first spring I see grasses trying to reclaim the edges of clearings, where the soil is not quite so compacted. If this land were left alone, I think, in time it could start to grow back, to become again a forest.

But only if there remains enough life for it to recover, and only if it's not further wounded.

I walk the roads they pushed in last year. The roads are far wider than necessary for log trucks, just the right size—no big surprise here—to serve a subdivision.

The "developer" has not yet come back. The court order says he can't until this fall. If he comes back then, we'll fight him again, and again, and again, and then again.

But I don't think he'll come back, at least for now. The housing market has collapsed, and is continuing to collapse, with no end in sight.

This collapse might be the big one, the one from which this culture never recovers, the one where the debts for so long pushed into the future begin not only to come due—as these debts have for so long been paid by passenger pigeons, great auks, salmon, bison, prairie dogs, indigenous humans, forests, oceans, by all of this world that (or rather, who) has by this culture been relegated to a source of raw materials and a dumping ground for wastes—but to boomerang back so hard that even those who have tried so desperately to insulate themselves from the effects of their actions begin to feel the pain they have for so long inflicted upon all others. If this collapse is the big one, then our actions to save this land might very well have saved it forever, which, if immortality is our goal, is certainly an "immortality" far more desirable than any number of plastic forks, or even well-written and well-produced books designed not to degrade for centuries. If not, and if this culture—and in this case the housing market—rallies temporarily, rallies strong enough to cause this "developer" to once again have financial incentives to destroy this forest, well, then, we in this neighborhood will fight him again, and we will hold him off until the next housing market collapse. If that one is the big one, then we will have saved this land. If not, then we will fight him again.

At some point, sooner or later—and by now rather much sooner than later—the collapse *will* be the big one, and this culture and its unspeakable destructiveness will no longer have such power to destroy. At that point, we—all of us who identify with and who have aligned ourselves with the real, physical world, and all of us who are part of the real, physical world, save those of this culture who pretend this culture is more important than life on this planet—can begin, ourselves, to rest and recover. Until then, it is not time to rest, not time to hold back, not time to not fight back with all of who and what we are, not time to not give everything we have to this struggle against this culture that destroys everyone, everything, and everywhere. Until this culture collapses—until we align with the natural world to help bring it down, to help make this collapse, or the next one, or the one after that, the final collapse of this culture—it is time to hold on tight in our hearts to what and whom we love, and to push as hard as we can against this culture's relentless omnicide, and then to push harder than we ever thought we could, and then harder, and harder still. Until this culture collapses it is time to push against and past our own fears and other self-imposed limitations, to push again, to push harder, and to push harder still, and to fight, to fight for our own lives and more importantly the lives of those we love, for the life of the land and air and water and soil and trees and fish and birds and frogs and slugs and salamanders and all others who give us life. Until this culture collapses it is time to fight, and to fight harder than we ever thought we could, and then harder, and harder still, and to not give up, but to keep holding tight to what and whom we love, to keep pushing, to keep fighting. Because someday, sooner or later—and by now, rather much sooner than later—we're going to win.

NOTES

1. We feel comfortable talking about the book speciating because books, like other pieces of art, like dreams, are alive.
2. Based on L. A. Fahm, *The Waste of Nations* (Montclair, NJ: Allanheld, Osmun & Co., 1980), p. 40, adjusted to current world population.
3. For an examination of how piss came to be a dirty word, see Derrick Jensen, *The Culture of Make Believe* (White River Junction: Chelsea Green, 2002, 2004), p. 541.
4. Krasilnikov, N. A., "Soil Microorganisms and Higher Plants," Academy of Sciences of the USSR, 1958.
5. For information on metabolism and excretion of these antibiotics, see their product monographs, available online at places like http://www.rxlist.com/cgi/generic/levoflox_cp.htm and http://www.rxlist.com/cgi/generic/cipro_cp.htm.
6. In hydroponic growing, plants do grow in a medium of sand or vermiculite, which doesn't contain a community of bacteria. The reason they can survive is because the nutrients they need for growth are produced in a chemical factory in a form simple enough for the plants to take up.
7. Elizabeth Royte, *Garbage Land: On the Secret Trail of Trash* (New York: Back Bay Books, 2006), p. 193.
8. Ibid.
9. Michal Jessen, "The Bag Beast," Environmentally Speaking, http://www.nelsonbc.ca/pages/jessen/The_Bag_Beast.htm, visited March 22, 2007.
10. There are 31,536,000 seconds in a year. Assuming a life span of eighty years, you could observe 2,552,880,000 plastic bags being produced in your life span. That number goes into 5 trillion 1,982 times.
11. Royte, *Garbage Land*, p. 193.
12. Ibid.
13. "Reusable Bags—Trends From Around The World," Reusablebags.com, http://www.reusablebags.com/facts.php?id=9, site visited March 23, 2007.
14. Baldev Chauhan, "Indian state outlaws plastic bags," BBC News, August 7, 2003, http://news.bbc.co.uk/2/hi/south_asia/3132387.stm, site visited March 17, 2007.
15. Franklin and Associates, "Paper vs. Plastic Bags," Institute for Lifecycle Environmental Assessment, 1990, http://www.ilea.org/lcas/franklin1990.html, site visited March 27, 2007.
16. Judith Walzer Leavitt, *The Healthiest City: Milwaukee and the Politics of Health Reform* (Princeton, NJ: Princeton University Press, 1982), p. 4, 125–6.
17. Martin V. Melosi, *Garbage in the Cities: Reuse, Reform, and the Environment, 1880–1980* (Chicago: Dorsey, 1981).
18. "Milestones in Garbage," US Environmental Protection Agency, January 3, 2008, http://www.epa.gov/msw/timeline_alt.htm, site visited February 2, 2008.

19. "A Garbage Timeline," Rotten Truth (About Garbage), 1998, http://www.astc.org/exhibitions/rotten/timeline.htm, site visited April 21, 2007.

20. Ibid.

21. Ibid.

22. See the federal Ocean Dumping Act of 1972, and the Ocean Dumping Ban Act of 1988, respectively.

23. Lydia Maria Child, *The American Frugal Housewife*, 16th ed., enlarged and corrected (Boston: Carter, Hendee, 1835).

24. Rotten Truth, "A Garbage Timeline."

25. Susan Strasser, *Waste and Want: A Social History of Trash* (New York: Holt Paperbacks, 2000), p. 30.

26. Rotten Truth, "A Garbage Timeline."

27. US Food Administration, Garbage Utilization: With Particular Reference to Utilization by Feeding (Washington: GPO, 1918), p. 11.

28. I never knew I could get tired of cashews, but I did. So did my mom. So did all my friends. So did all my friends' neighbors. So did my dogs. So did the chickens. So did the ducks and geese and turkeys. I became a regular Cashew Santa Claus.

29. W. Atkinson and R. New, "An Overview of the Impact of Source Separation Schemes on the Domestic Waste Stream in the UK and Their Relevance to the Government's Recycling Target," Warren Spring Laboratory, Stevenage, Herts., Strategy Unit, 2002, http://www.wasteonline.org.uk/resources/InformationSheets/HistoryofWaste.htm.

30. Rotten Truth, "A Garbage Timeline."

31. Ibid.

32. Ibid.

33. Daniel Thoreau Sicular, "Currents in the Waste Stream: A History of Refuse Management and Resource Recovery in America," MA thesis, Department of Geography, University of California, Berkeley, 1984, p. 76.

34. Melosi, *Garbage in the Cities*, p. 105–06

35. A term of British origin.

36. "History of waste and recycling," Waste Watch, October, 2004, http://www.wasteonline.org.uk/resources/InformationSheets/HistoryofWaste.htm, site visited March 17, 2007.

37. Rotten Truth, "A Garbage Timeline."

38. Helen Campbell, "As to Ashes and Rubbish," *American Kitchen Magazine*, 12 (August 1900), p. 174–176.

39. Rotten Truth, "A Garbage Timeline."

40. W. F. Morse, "The Collection and Disposal of Municipal Waste," *The Municipal Journal and Engineer*, New York, 1908.

41. Helen Spiegelman and Bill Sheehan, "Unintended Consequences: Municipal Solid Waste Management and the Throwaway Society," Product Policy Institute, March 2005.

42. "MSW Combustion," Cardiff University Waste Research Station, http://www.wasteresearch.co.uk/ade/efw/mswcombustion.htm, site visited February 18, 2008.

43. Michael Warhurst and Anna Watson, "Dirty truths: Incineration and climate change," Friends of the Earth, May 2006, http://www.foe.co.uk/resource/briefings/dirty_truths.pdf.

44. D. Hogg, et al, "Greenhouse Gas Balances of Waste Management Scenarios," Report for the Greater London Authority, Eunomia, 2008.

45. Dr. Paul Connett, quoted from a press conference at Queen's Park, Toronto, which took place March 29, 2007. Video of the press conference is available online at http://www.youtube.com/watch?v=XB5iOtxlpCs. You can also read one of Dr. Connett's

PowerPoint presentations at http://www.recycle.ab.ca/2006Proceedings/PaulConnett_Zero_waste.pdf.

46. "Evaluation of Emissions from the Open Burning of Household Waste in Barrels," Volume 2, Appendices A-G, U.S. EPA, Control Technology Center, November 1997, http://www.epa.gov/ttn/catc/dir1/barlbrn2.pdf.

47. Spiegelman and Sheehan, "Unintended Consequences," p. 14.

48. Amanda Claridge, *Rome: An Oxford Archaeological Guide* (Oxford, UK: Oxford University Press, 1998).

49. It's worth noting that women, who kept and sorted the items of interest for barter, were usually in charge of negotiating with peddlers, selecting the items they wanted to receive in exchange, and spending the cash they would sometimes receive. The eventual decline of this barter system had a corresponding impact on the economic influence of women in general.

50. Strasser, *Waste and Want*, p. 70.

51. By the year 1860, humans all over the world extracted only half a million barrels of oil a year, the amount of oil the US now uses in thirty-five minutes ("The Politics of Oil," University of Houston Digital History, April 21, 2007, http://www.digitalhistory.uh.edu/historyonline/oil.cfm, site visited April 21, 2007).

   For data on current US oil consumption, see "Petroleum Basic Data," Energy Information Administration, July 2007, http://www.eia.doe.gov/neic/quickfacts/quickoil.html, site visited September 2, 2007.

52. "Milestones in Garbage," US Environmental Protection Agency.

53. Judith A. McGaw, *Most Wonderful Machine: Mechanization and Social Change in Berkshire Paper Making, 1801–1885* (Princeton, NJ: Princeton University Press, 1987), p. 40.

54. Lyman Horace Weeks, "A History of Paper-Manufacturing in the United States, 1690–1916," *Lockwood Trade Journal*, New York, 1916, p. 66–67.

55. Ibid, p. 61–62.

56. Reuters, "Accommodating an army of garbage pickers," CNN.com, March 26, 2003, http://www.cnn.com/2003/WORLD/americas/03/26/argentina.train.reut/.

57. Ibid.

58. Elijah Zarwan, "Buenos Aires: The Ghost Train of the *Cartoneros*," Worldpress.org, August 2002, http://www.worldpress.org/photo_essays/cartoneros/, site visited April 3, 2007.

59. Ibid.

60. Reuters, "Accommodating an army of garbage pickers."

61. Luis Alberto Urrea, *By the Lake of Sleeping Children: The Secret Life of the Mexican Border* (New York: Anchor Books, 1996), p. 40.

62. Ibid.

63. Kinsee Morlan, "Home sweet dump: Tijuana landfills are home to hundreds of people who depend on garbage for their livelihood," *San Diego CityBeat*, January 15, 2008, http://www.sdcitybeat.com/cms/story/detail/home_sweet_dump/6551/.

64. "Finding hope among the ruins," Shawna Bell.

65. Urrea, p. 41.

66. "Home sweet dump," Kinsee Morlan.

67. Charles Shaw, "The Priest and the Prophet: Can Industrial Civilization Really Become Sustainable? Should it?" *Grist*, August 15, 2006, http://www.grist.org/comments/soapbox/2006/08/15/shaw/, site visited December 13, 2006.

68. William McDonough, "Executive Summary: Designs celebrating the joy and creativity of the human spirit and the abundance of nature," http://www.mcdonoughpartners.com/executive_summary.shtm, site visited December 13, 2006.

69. Never mind that these "closed-loop cycles" will never happen on anything approaching the scale that McDonough and others are talking about. Of course, a truly closed-loop economy would require industries to cease manufacturing any materials that could or would not be safely, completely, and indefinitely recycled. More than that, it would require that recycling materials (such as vulcanized tires) be more economically feasible than simply making them anew from raw materials and exporting the waste. Which is to say that closed-loop recycling practices would have to be technically possibly, economically rewarding, and legally imperative. Moreover, vast quantities of cheap energy would have to be available for the intensive material processing.

    Currently, none of those conditions are met. And we have no reason to believe they *will* be met. The myth of a cradle-to-cradle industrial future depends on the fantasy that either every material now manufactured will eventually be fully and economically recyclable, or that a suitable replacement can be made for every material, or both. This fantasy is completely divorced from any semblance of plausibility or reality, but it does serve to offer a rose-colored diversion, a technocratic rationalization for the waste caused by the current system: *Don't worry if we keep churning out industrial technology, because eventually it will be painlessly and seamlessly replaced by a harmless analogue.*

    Further, it's absolutely extraordinary how McDonough consistently attempts to conflate machines and living beings. He says that materials "become food for biological *cycles,*" (my emphasis) as opposed to becoming food for actual beings: slugs, soil, trees. Yet he calls industrial inputs "nutrients" (definition: "any substance that can be metabolized by an organism to give energy and build tissue"). This language is 100 percent reversed.

    Even his notion that there exist industrial equivalents for waste equaling food is based on the same faulty conceit that industry is in some sense natural. But there are fundamental differences between nature and industry, and there are fundamental differences between shit and industrial wastes, not the least of which is that beings evolved for billions of years eating each other and eating each other's by-products. Beings have not evolved over billions of years eating industrial sludge, which means members of the natural world cannot reasonably be expected to metabolize these wastes. And so far as industry reusing these wastes, yes, of course it's better to recycle as many industrial wastes (not "nutrients") as possible, but A) we need to be honest and acknowledge that they can't be recycled "forever," as McDonough puts it elsewhere, and it's disingenuous and/or stupid (or at the very least incredibly sloppy) to pretend they can; and B) it's far better to not make these wastes in the first place.

    I see what McDonough is trying to do with all of this—he's attempting to slur the distinction between what is natural and what is industrial, or more to the point, what is living and what is a machine—but to do so is inaccurate, inapt, unwise, and harmful. Living beings are living beings, and factories are factories.

70. Note that one of his designs was a Nike headquarters in Europe, which, in his words from a different essay, is "an exceptionally pleasant place to work." The same cannot be said for the sweatshops where poor young brown people (especially women) slave to make Nike products. The extraordinarily racist and classist priorities inherent in making "delightful places" for the owning classes to administer the companies that exploit the poor is the same old colonialism that has pauperized most of the planet so that the owning classes can enjoy "delightful places" as McDonough puts it, or "comforts or elegancies," as a nineteenth-century pro-slavery philosopher put it, with a new veneer of lofty rhetoric covering the same old rationalizations.

71. These last two statements are precisely 180 degrees from the truth. For crying out loud, McDonough designs things like truck factories, Nike headquarters, and airports. Of course McDonough's work consists of damage management strategies and retrofitting

destructive systems. And of course those are important things to do. But even more important is that we be honest with ourselves and others, and not claim to be doing things we're not and at the same time claim to not be doing things we are.

72. McDonough repeats this refrain more or less constantly. Each time, he states that he wants to eliminate the *concept* of waste—as opposed to eliminating waste (in this case the toxic products that are one of the absolutely physically necessary and inevitable hallmarks of an industrial economy)—which makes clear his choice of words is intentional. This seems significant to me, in part because his work really does end up being far more about *concepts* of sustainability than about sustainability itself.

73. It's important to keep in mind while reading McDonough's stirring rhetoric that what he's really talking about is designing things like truck factories, Nike headquarters, and airports. No matter how much anybody gets paid for it, no matter how much wishing takes place—no matter how strong the denial—and no matter how well designed are the truck factories, Nike headquarters, and airports, by definition they will not and cannot "restore and support nature." And certainly Nike, Ford, and other transnational corporations have done tremendous harm to "human societies" across the planet.

74. William McDonough, "Leading Change Toward Sustainability," http://www.mcdonough.com/writings/leading_change.htm, site visited December 13, 2006. Note that he replaced—significantly replaced—the most important word in this last sentence. He doesn't really mean "human footprint," but rather "industrial footprint." He is once again and as always conflating "human" and "industrial," a conflation made by many who share civilization's narcissistic delusion that it really is the only way of life worth living—the only thing worth saving, the only thing that fully exists. But "human" and "industrial" are *not* the same, no matter how much industrialized people may claim and believe it to be so. There have existed, and still exist, many human cultures which are not industrialized.

This conflation is one more example of McDonough's probably sometimes conscious and sometimes unconscious conflation of living beings and machines, in this case probably (and I'm guessing here based on his word choices and more generally his writing) stemming from him personally identifying far more strongly as a member of industrial society—I would say a servant of the Machine than as a human animal requiring habitat. This identification is a fatal error, fatal for those suffering under the delusion that there are not fundamental differences between living beings and machines, between humans and industry; and fatal as well for those who have the misfortune to encounter those so deluded.

To be honest, I'm not even sure McDonough really means the word "positive" or he would not be able to work as he does with companies like Nike, especially without mentioning even once the atrocious nature of the labor conditions for many people who make these shoes.

75. McDonough, "Leading Change." He is of course correct with this statement. Unfortunately, despite his rhetoric, his work *is* aimed at "making a fundamentally destructive system more efficient." It is physically impossible, for reasons I lay out in *Endgame, Volume I* and elsewhere, for all of the "products and processes" of industrial civilization to be "safe, healthful and regenerative." Further, we have to ask, "safe, healthful and regenerative" for whom? For polar bears? For blue whales? How are a truck factory, a Nike headquarters, an airport, and the products they create and economic processes they facilitate "regenerative" for Kootenai sturgeon, clouded salamanders, hammerhead sharks? And we also have to ask, "safe, healthful and regenerative" as compared to what? As compared to older factories, maybe (on the surface, certainly yes, but modern factories are also often more efficient; that is, they more quickly convert raw materials into products; in other words, more quickly convert the living to the dead; in other words,

more quickly consume the planet, which means that, to the degree that more efficient factories increase production and consumption, the more harmful they are to the living world). But compared to a native forest or meadow? Of course not. McDonough is not a stupid man. I find it hard to believe that he does not understand that industrial processes are not compatible with healthy landbases.

76. William McDonough, "Buildings Like Trees, Cities Like Forests," http://www.mcdo-nough.com/writings/buildings_like_trees.htm, site visited December 13, 2006. The article first appeared in *The Catalog of the Future* (Pearson Press, 2002). Perhaps he means "other living beings." In any case, through this lush description he doesn't mention where he gets the concrete, steel, and so on for the building, nor does he mention where the computers used in this building are made, nor what will happen to these computers next year when they become obsolete. Nor does he mention the purpose of the building. It could, for all we know, be a prison. It could be a death camp. It could be a zoo. It could be an abattoir. It could be a truck factory. It could be a Nike headquarters. It could be an airport. None of these are sustainable, and we should not pretend they are. It is misleading at best for McDonough to speak of sustainable architecture without taking into account what the buildings are *for*.

77. William McDonough, "Full," http://www.mcdonough.com/full.htm, site visited December 13, 2006. Of course changing the design of the world is precisely the problem. The world is designed perfectly well, and does not need to be redesigned. It's the culture that needs to change, not the world. And I'll tell you a secret that most civilized people do not allow themselves to acknowledge: this culture is not the world. The world and this culture are vastly different things. This is essentially the same problem of misidentification I pointed out earlier, where McDonough conflated "human" and "industrial." Here the author conflates "world" and "western culture," or more precisely "world" and "factory." In either case it's a fatal and unforgivable misidentification.

78. William McDonough, "Ford Rouge Dearborn Truck Plant," http://www.mcdonough-partners.com/projects/ford-dtp/default.asp?projID=ford-dtp, site visited December 16, 2006. Note his use of the word *optimizes*, as opposed to the word *minimizes*. *Optimizes* implies that the "impact" of this industrial activity on the external environment is positive: one does not, for example, "optimize" the impacts of cancer on one's body. It's precisely this sort of dishonesty that bothers me most about McDonough's work. McDonough is writing here, we should never forget, about a truck factory. One can minimize the harm done by this factory, but one cannot optimize ("to make as perfect, effective, or functional as possible"; "to make the most of") its impact, since the "impact" will be negative. Of *course* if one is going to have truck factories one should attempt to minimize their damage, make them slightly less unsustainable. But one should not pretend that one is doing more than one is. One should not pretend—nor cause (nor even allow) others to believe the lie—that a truck factory, a Nike headquarters, or an airport can be sustainable. That's wrong. It's harmful. It wastes time we do not have. It facilitates and rationalizes denial and the maintenance of an exploitative and destructive system. It is a lie.

79. McDonough, "Ford Rouge."

80. Actually, Nike's business revolves around desperately poor, dispossessed people slaving at sweatshops, getting paid barely enough to survive, sometimes being beaten when they don't work fast enough, certainly being fired (or jailed or tortured or killed) if they try to unionize.

81. McDonough's use of the word *habitat* is more of that same inaccurate conflation of natural and industrial. My dictionary defines *habitat* as "the natural home of a plant or animal." An office building is not our natural home. It is not our habitat.

82. As George Draffan and I made clear in our book *Strangely Like War: The Global Assault on Forests*, "sustainably harvested wood" is a scam, in precisely the same vein as the "green capitalism" and "sustainable development" scams. Good grief, both Weyerhaeuser (the largest clearcutter in the world) and Plum Creek (which even a Republican legislator called the "Darth Vader of the timber industry") have been certified as green. The phrase is just new words for the same old exploitation and ecocide. Welcome to the rhetoric used to rationalize and facilitate quite possibly the final dismemberment of the planet.

83. William McDonough, "Nike European Headquarters," http://www.mcdonoughpartners.com/projects/nike/default.asp?projID=nike, site visited December 17, 2006.

84. William McDonough and Michael Braungart, "From Inspiration to Innovation: Nike's Giant Steps Toward Sustainability," originally in the July–August 2002 issue of green@work. I saw it by going to www.williammcdonough.com, then clicking on "Writings," then clicking on "Cradle to Cradle Case Studies," and then clicking on the article. Site visited December 17, 2006.

    I find it outrageous and obscene that McDonough has the gall to use the phrase "social equity" when speaking of Nike: in Vietnam, Nike publicly denies violating the legal minimum wage of forty-five dollars per month even though its own secret studies (and also pay stubs) makes clear it does; some of Nike's contract factories in China illegally require workers to deposit their first month's salary with the factory owner, who then keeps their "deposit" if they leave within a year; workers at Nike's contract factories in Malaysia have been forced to run until they collapse because they wore "nonregulation" shoes; workers in Nike's contract factories have been beaten (with a Nike shoe) to the point of hospitalization for "poor sewing" ("Facts and FAQs About Nike's Labor Abuses," http://www.personal.umich.edu/~lormand/poli/nike/nikelabor.htm, site visited December 19, 2006). Social equity? Nike is paying Tiger Woods 100 million dollars over five years. Average Nike workers in Thailand make between three and four dollars per day. The average Thai worker would have to labor for 25 or 30 million days (70,000 or 80,000 years) to earn what Woods will receive. A Thai worker would have to labor for thirty-five or forty years to make what Woods receives in one day (Thai Labor Campaign, "Woods Meets Nike Protesters; Tiger Woods Was Escorted Through An Angry Crowd," 1world media, http://www.1worldcommunication.org/labornews.htm#A%20Letter%20to%20Tiger%2oW, site visited December 20, 2006). Social equity? Nike's European headquarters cost somewhere between 65 and 72 million dollars. Architects usually receive from 5 to 15 percent of total construction costs in fees, which would put fees for McDonough's company somewhere between 3.5 and 10.5 million dollars, or, taking a figure in the middle, as much as one of Nike's Thai workers would make in about 5,500 years. Social equity? Nike cofounder Phil Knight's net worth is 7.9 billion dollars (Jack Forbes, *Columbus and Other Cannibals*, [Brooklyn: Autonomedia, 1992], p. 400), up 500 million dollars from just two years ago. You can do the math on how long it would take those who actually produce the shoes to make this much money.

85. Note the phrase "concept of waste" once again.

86. This is a nice sleight of phrase, but it's meaningless. How precisely has his work with Nike resolved this conflict between nature and commerce? The way this culture is structured, commerce requires the consumption of nature. For proofs of this, please see all my other books, but most notably *Endgame, Volume I; The Culture of Make Believe;* and *Strangely Like War.* Or more to the point, step into real wild nature. Ask a passenger pigeon. Ask a sea turtle. Ask a swordfish. McDonough has not resolved the conflict between nature and commerce. It cannot be resolved within an industrial system. Indus-

trial systems (and, more broadly, civilizations) destroy their landbases. It's not helpful, but rather extremely harmful, to pretend otherwise.

87. Name one "highly productive [industrial] facility" that has positive effects on the natural world, both locally and more distant. You can't, because industrial facilities are destructive locally, and they require materials to be brought in and consumed. I suspect McDonough understands this, because you'll note that he writes, "on their surroundings," as opposed to "on the natural world." Even if an industrial facility had a positive effect on the local surroundings (which it won't: if you ask the land at the Ford Rouge plant if it would rather be a truck factory with a "living roof" or a native forest or meadow, I'm certain which it would rather), it would still require raw materials to be brought in from elsewhere. This routine importation of materials destroys the landbases where those materials originate. For a thorough exploration of this, see Jensen, *Endgame*.

88. No matter how many times he says this, it's still not going to happen.

89. McDonough and Braungart. Of course those who are more or less enslaved in sweatshop factories making Nike products may not agree that anything about Nike is particularly life-affirming or positive.

Further, industrial systems are rapidly destroying what is left of this "abundance of nature" he mentions.

90. McDonough and Braungart, "From Inspiration to Innovation."

91. Once again, I don't see how nature benefits by the construction of this Nike headquarters, no matter how beautiful, no matter how state-of-the-art, no matter how groovy. If you asked that piece of land if it would rather be a Nike headquarters or a forest, we all know how it would answer. To make the point perfectly clear: it is entirely possible for humans to have "healthy, [mutually] beneficial relationships" with their landbases, or more precisely with the larger communities of their nonhuman neighbors. Indeed, that is the only way to live sustainably, and many cultures lived in place ten thousand years or longer. But it is entirely impossible for an industrial economy (or for that matter a civilized economy, as I define it in *Endgame, Volume I*) to have a "healthy, [mutually] beneficial relationship" with a landbase. Utterly impossible.

92. "Facts and FAQs."

93. McDonough and Braungart, "From Inspiration to Innovation."

94. Vietnam Labor Watch, "Nike Labor Practices in Vietnam," http://www.saigon.com/~nike/reports/report1.html, site visited December 24, 2006.

95. Clean Clothes Campaign, "Labour Rights in Indonesia: What is Menstruation Leave," Newsletter 13, November 2000, http://www.cleanclothes.org/news/newsletter13-indon.htm, site visited December 24, 2006.

96. "Facts and FAQs."

97. Ibid.

98. "Financial Compensation for Nazi Slave Laborers," http://www.religioustolerance.org/fin_nazi.htm, site visited December 24, 2006.

99. I'm guessing the answer is yes.

100. "Facts and FAQs."

101. It really bothers me that he talks about "placeness" in a description of an airport. Air travel disrespects "placeness" in part by disrespecting distance, and certainly also disrespects "placeness" by facilitating the homogenization of culture through routine travel (for the globally wealthy).

102. McDonough, "Corporate Flight."

103. Functionally necessary for civilization, *not* functionally necessary for humanity's survival. In fact, civilization is incompatible with human survival, as the poor already see, and the rest of us shall see very soon.

104. For my most thorough articulation of this, please see *Endgame*.

105. I am grateful to Stephanie McMillan for the preceding analysis.
106. Derrick Jensen, *A Language Older Than Words* (White River Junction, Vermont: Chelsea Green, 2000).
107. I guess McDonough might say that the doctors *optimized* the effects of the camps on the inmates.
108. Habitat, as in the *real world*, not as in McDonough's habitat™ that is nothing more than a well-designed office building.
109. My thanks to Tiiu Ruben.
110. Oh, I'm sorry. I forgot. Bush was attempting to bring freedom™ and democracy™ to the region. By the way, I've figured out how to listen to speeches by US politicians without going insane. Just do simple word substitutions. Every time they say "freedom" they really mean "fascism." Every time they say "democracy" they really mean "corporate control." So when they say, "We're bringing freedom and democracy to the Middle East," what they really mean is, well, you can do the substitutions.
111. Unless they have already attacked me or someone I care about.
112. William R. Catton, Jr., *Overshoot: The Ecological Basis for Revolutionary Change* (Urbana: University of Illinois Press, 1982), p. 273.
113. Rationalization can be defined as the deliberate elimination of information unnecessary to achieving an immediate task. Oscar H. Gandy, Jr., *The Panoptic Sort: A Political Economy of Personal Information* (Boulder: Westview Press, 1993), p. 42.
114. Frances Densmore, *Teton Sioux Music*, Bulletin 61, Bureau of American Ethnology, Smithsonian Institution, 1918, p 172. My apologies for the sexual specificity of this language.
115. Derrick Jensen, "Where the Buffalo Go: How Science Ignores the Living World: An Interview with Vine Deloria," *Sun*, July 2000.
116. Or rather, one that doesn't rot. And that's the point.
117. Susan Casey (photographs by Gregg Segal), "Our Oceans Are Turning Into Plastic . . . Are We?" *Best Life Magazine*, May 11, 2007, http://www.bestlifeonline.com/cms/publish/health-fitness/Our_oceans_are_turning_into_plastic_are_we_2.shtml, site visited June 11, 2007.
118. Richard Shears, "Is this the world's most polluted river," *Daily Mail*, http://www.dailymail.co.uk/pages/live/articles/news/worldnews.html?in_article_id=460077&in_page_id=1811, site visited June 11, 2007.
119. Casey, "Our Oceans."
120. Alan Weisman, "Polymers Are Forever," *Orion*, May/June 2007, http://www.orionmagazine.org/index.php/articles/article/270/, site visited June 11, 2007.
121. Casey, "Our Oceans."
122. Ibid.
123. Ibid.
124. Okay, I'm lying. I took a few detours, like one to www.mybodypart.com, which even has a mybodypart.com store (where, sadly, I can't buy body parts, much as I may want an extra appendage here or there, though instead I can buy hot and cold breast compresses for use after "augmentation," "skin rejuvenation lotion," "camouflage cream," and so on). They also have "free makeover software" you can download so that you can "See what you can be! It's never been this easy to instantly see what plastic surgery, permanent makeup, or different hairstyles could look like using pictures of you." I was sorely tempted to see how I would look with Mick Jagger lips, a Tom Cruise nose, and Scarlett Johansson breasts (declared by TMZ.com the most beautiful breasts in Hollywood), but in order to get the software I had to supply too much personal information, and the last thing I want is sales calls and spam from plastic surgeons (Lord knows I already get enough spam for penis enlargement (or rather pen*s enlarg*m*ent), not, I hasten to

add, that I need *that*: harrumph). So, regretfully, I moseyed away from websites about plastic ("marked by artificiality or superficiality"), and toward websites about plastic ("generic name for certain synthetic or semisynthetic materials that can be molded or extruded into objects or films or filaments or used for making e.g. coatings and adhesives"). Free dictionary.

125. "What is plastic made of," *Wikianswers*, http://wiki.answers.com/Q/What_is_plastic_made_out_of, site visited June 18, 2007.

126. "Ivory Billiard Balls," Smithsonian Museum of Natural History, http://americanhistory.si.edu/collections/object.cfm?key=35&objkey=185, site visited June 18, 2007.

127. Unfortunately for Hyatt, the billiard industry seems to have stiffed him. "Billiard Ball," Wikipedia.com, http://en.wikipedia.org/wiki/Billiard_ball, site visited June 18, 2007.

128. "What is plastic," *Wikianswers*.

129. Between the time I wrote this and the time I'm editing this—about a year—the forest here has become nearly silent because songbird populations have crashed.

130. For an exploration of toxic mimics, see Derrick Jensen, *Endgame, Volume I*, New York: Seven Stories Press, 2006), p. 164–165.

131. Note I said continuation, not immortality. There's a difference.

132. Casey, "Our Oceans."

133. Lisa Stiffler, "PDBEs: They are everywhere, they accumulate and they spread: Chemical flame retardants pose threat to humans, environment," *Seattle PI*, March 29, 2007, http://seattlepi.nwsource.com/local/309169_pbde28.html, site visited June 23, 2007.

134. Casey, "Our Oceans."

135. "Bisphenol A," *Wikipedia*, http://en.wikipedia.org/wiki/Bisphenol_A, site visited June 20, 2007.

136. Cobbled together from the following: L. Hardell, M. J. Walker, B. Walhjalt, L. S. Friedman, and E. D. Richter, "Secret ties to industry and conflicting interests in cancer research," *American Journal of Industrial Medicine*, 2006, found at *Our Stolen Future*, http://www.ourstolenfuture.org/Industry/2006/2006-1103hardelletal.html, site visited June 21, 2007; Bill Hogan, "Paging Dr. Ross," *Mother Jones*, November/December 2005, http://www.motherjones.com/news/outfront/2005/11/paging_dr_ross.html, site visited June 21, 2007; Frederick vom Saal and W. Welshons, "Large effects from small exposures. II. The importance of positive controls in low-dose research on bisphenol A," *Environmental Research*, 100: 50:76, found at *Our Stolen Future*, http://www.ourstolenfuture.org/NewScience/oncompounds/bisphenola/2006/2006-0101vomsaalandwelshons.html, site visited June 21, 2007; and Jody Woodruff, "Plastic Bottles and Can Liners Under Scrutiny Again," *Sentient Times*, April/May 2007, http://www.sentienttimes.com/07/07_apr_may/plastic.html, site visited June 21, 2007.

137. "PVC Facts," Healthy Building Network, http://www.healthybuilding.net/pvc/facts.html, site visited June 20, 2007.

138. "Dioxin," from http://en.wikipedia.org/wiki/Dioxin, http://www.ejnet.org/dioxin/, and http://www.cqs.com/edioxin.htm, all sites visited June 20, 2007.

139. Isamu Igura, "Half-life of each dioxin and PCB congener in the human body," *Organohalogen Compounds*, 66 (2004), 3376–3384, http://dioxin2004.abstract-management.de/pdf/p311.pdf, site visited June 20, 2007.

140. Joe Thornton, *Dioxin From Cradle to Grave*, April 1997, http://www.greenpeace.org/raw/content/usa/press/reports/the-pvc-lifecycle-dioxin-from.pdf, p. 41–42, site visited June 20, 2007.

141. Casey, "Our Oceans."

142. Ibid.

143. Ibid.

144. List slightly modified from Paul Goettlich, "The Sixth Basic Food Group," *Mindfully.org*, November 29, 2004, http://www.mindfully.org/Plastic/6th-Basic-Food-Group.htm, site visited June 21, 2007.

145. Ibid.

146. Paul Goettlich, "Plastic in the Sea," *Mindfully.org*, October 5, 2005, http://www.mindfully.org/Plastic/Ocean/Sea-Plastic-LN-PG5oct05.htm, site visited June 22, 2007.

147. Stiffler, "PDBEs."

148. Casey, "Our Oceans."

149. Paula Bock, "Oceans of Waste: Waves of junk are flowing into the food chain," *Seattle PI*, April 23, 2006, http://seattletimes.nwsource.com/html/pacificnw04232006/coverstory.html, site visited June 22, 2007.

150. "Tailings," Wikipedia, http://en.wikipedia.org/wiki/Tailings, site visited July 6, 2007.

151. "APP of Mine Overburden," Colorado School of Mines, "CH 126 Experiment #6, Acid Producing Potential of Mine Overburden," http://www.mines.edu/fs_home/jhoran/ch126/app.htm, site visited July 6, 2007.

152. "Microbial Influences: Thiobacillus ferrooxidans," Acid Mine Drain Drainage Experiments at CSM, http://www.mines.edu/fs_home/jhoran/ch126/microbia.htm, site visited July 6, 2007.

153. Ibid.

154. Ibid.

155. Although the main leak was methyl isocyanate (note the last three syllables), the reaction that led to the release of MIC produced cyanide and other toxic chemicals as well.

156. In this case cyanide is the complexant.

157. All of this is from the film *The Treasure of the Sierra Madre*.

158. "Cyanide Leach Mining Packet," *Mineral Policy Center*, Washington, DC, 2006.

159. Ibid.

160. Jennifer Bogo, "Crying Rivers (Romania gold mine spills cyanide into rivers) (Brief Article)," *High Beam Encyclopedia*, May 1, 2000, http://www.encyclopedia.com/doc/1G1-62298547.html, site visited July 13, 2007.

161. Erin Klauk, "Environmental Impacts at Fort Belknap from Gold Mining," *Impacts of Resource Development on Native American Lands*, http://serc.carleton.edu/research_education/nativelands/fibelknap/environmental.html, site visited July 13, 2007.

162. "Summitville, the Exxon Valdez of the Mining Industry," *Sprol*, October 17, 2005, http://www.sprol.com/?p=268m site visited July 13, 2007.

163. "Medical Waste," US Environmental Protection Agency, March 4, 2008, http://www.epa.gov/epaoswer/other/medical/basic.htm, site visited March 10, 2008.

164. "Guidance Document For Regulated Medical Waste," New Jersey Department of Environmental Protection, Solid Waste Management Program, December 2007, http://www.nj.gov/dep/dshw/rrtp/rmw.htm, site visited March 10, 2008.

165. "Issues in medical waste management," Office of Technology Assessment, US Congress, 1988.

166. "Medical Waste," US Environmental Protection Agency.

167. Converted from 50,000 metric tonnes. "PharmEng Announces Exclusive Agreement With Chinese Manufacturer for Distribution of Acetaminophen to North America Resulting in Significant Raw Material Cost Savings," Yahoo Business News, March 26, 2007, http://biz.yahoo.com/ccn/070326/200703260380180001.html?.v=1, site viewed March 2, 2008.

168. Ibid.

169. "Pharmaceutical Market Trends, 2007–2011: Key market forecasts and growth opportunities (2nd Edition), URCH Publishing, November 2007.

170. "Pharmaceuticals, Hormones, and Other Organic Wastewater Contaminants in U.S. Streams," USGS Fact Sheet FS-027-02, June 2002, http://toxics.usgs.gov/pubs/FS-027-02/index.html.

171. For 100 million, see "Hormonal Contraception Does Not Appear To Increase HIV Risk," *ScienceDaily*, December 7, 2006, http://www.sciencedaily.com/releases/2006/12/061207161016.htm, site visited March 14, 2007. For the 10 million figure, see J. C. Abma, A. Chandra, W. D. Mosher, L. S. Peterson, and L. J. Piccinino, "Fertility, family planning, and women's health: new data from the 1995 National Survey of Family Growth," *Vital Health Stat* 23 (19), 1997: 1–114.

172. "Researchers Compare Chicken, Human Genomes," National Institutes of Health, February 26, 2008, http://www.genome.gov/12514316, site visited March 15, 2008.

173. Susan Jobling, Richard Williams, Andrew Johnson, Ayesha Taylor, Melanie Gross-Sorokin, Monique Nolan, Charles R. Tyler, Ronny van Aerle, Eduarda Santos, and Geoff Brighty, "Predicted Exposures to Steroid Estrogens in UK Rivers Correlate with Widespread Sexual Disruption in Wild Fish Populations," *Environmental Health Perspectives*, April 2006.

174. Nicole Silk and Kristine Ciruna, "A Practitioner's Guide to Freshwater Biodiversity Conservation," The Nature Conservancy, July 2004.

175. Elizabeth Cheriathundam and Alvito P. Alvares, "Species differences in the renal toxicity of the antiarthritic drug, gold sodium thiomalate," *Journal of Biochemical Toxicology*, 2 (4): 175–181.

176. Rachel C. Veldhoen, Skirrow, Heather Osachoff, Heidi Wigmore, David J. Clapson, Mark P. Gunderson, Graham Van Aggelen and Caren C. Helbing, "The bactericidal agent triclosan modulates thyroid hormone-associated gene expression and disrupts postembryonic anuran development," *Aquatic Toxicology* 80 (3), December 2006: 217–227.

177. Antonia M. Calafat, Larry L. Needham, Manori J. Silva, and George Lambert, "Exposure to Di-(2-Ethylhexyl) Phthalate Among Premature Neonates in a Neonatal Intensive Care Unit," *Pediatrics* 113 (5), May 2004: e429–e434.

178. This was what was occurring in the specific study cited above.

179. John L. Konefes and Michael K. McGee, "Old Cemeteries, Arsenic, and Health Safety," *Water Industry News*, http://waterindustry.org/arsenic-3.htm, site visited February 10, 2008.

180. Ibid.

181. Material Safety Data Sheets are conveniently available on some embalming fluid manufacturer websites, which list their products and the product's hazardous constituents. See, for example, http://www.dodgeco.com/msds/msds/msds_frame_public.cfm.

182. "History of Embalming," Wyoming Funeral Directors Association, 2001, http://www.wyfda.org/basics_3.html, site visited February 11, 2008.

183. Joe Sehee, "Green Burial: It's Only Natural," PERC Reports, 25 (4), Winter 2007, http://www.perc.org/perc.php?subsection=5&id=1015, site viewed December 30, 2007.

184. Methanol is the substance in improperly made moonshine that causes blindness by destroying the optic nerve.

185. See http://www.champion-newera.com/CHAMP.PDFS/encyclo648.pdf and other articles at http://www.champion-newera.com/encyclo.html.

186. See http://www.champion-newera.com/CHAMP.PDFS/encyclo655.pdf.

187. Compiled from statistics by Casket and Funeral Association of America, Cremation Association of North America, Doric Inc., The Rainforest Action Network, and Mary Woodsen, Pre-Posthumous Society.

188. Seth Faison, "Lirong Journal; Tibetans, and Vultures, Keep Ancient Burial Rite," *New York Times*, July 3, 1999.

189. See, for example, http://www.greenburials.org/.
190. "Outside the Box," *CBC Newsworld*, 2005, http://video.google.ca/videoplay?docid=95631496866916379d&q.
191. See Marla Cone, *Silent Snow: The Slow Poisoning of the Arctic* (New York: Grove Press, 2006). This story is also told in her article "Dozens of Words of Snow, None for Pollution," *Mother Jones Magazine*, January/February 2005, http://www.motherjones.com/news/feature/2005/01/12_402.html.
192. Ibid.
193. The executive summary is available at http://archive.ewg.org/reports/bodyburden2/execsumm.php.
194. Ibid.
195. J. Onstot, R. Ayling, and J. Stanley, *Characterization of HRGC/MS Unidentified Peaks from the Analysis of Human Adipose Tissue. Volume 1: Technical Approach*, US Environmental Protection Agency Office of Toxic Substances, Washington, DC, 1987.
196. For further discussion of this point, see http://www.chemicalbodyburden.org/whatisbb.htm.
197. I got the phrase "nature loves a community" from Paul Stamets.
198. Here are some numbers to back up what we already know, from an article entitled "Wilderness almost non-existent on planet Earth: study," *Breibart.com*, http://www.breitbart.com/article.php?id=070628185002.qck4e5qz&show_article=1&catnum=0, site visited July 31, 2007:

    "As of 1995, only 17 percent of the world's land area remained truly wild—with no human populations, crops, road access or night-time light detectable by satellite, the authors reported.

    "Half of the world's surface area is used for crops or grazing; more than half of all forests have been lost to land conversion; the largest land mammals on several continents have been eliminated; shipping lanes crisscross the oceans, according to the paper.

    "In Europe, 22,000 kilometers of coastline are paved.

    "Due to extensive damming, nearly six times as much water is held in artificial storage worldwide as is free-flowing, according to the article.

    "Beyond the obvious signs of human influence, other, more subtle changes are evident everywhere, Kareiva said.

    "Natural selection has been supplanted by human selection, meaning that certain species—such as companion pets—thrive, while others—such as river trout [sic]—have been altered specifically for human consumption, often to their detriment."
199. E. Cobham Brewer, "Dictionary of Phrase and Fable," http://www.bartleby.com/81/4834.html, site visited July 26, 2007.
200. Bob Blaisdell, ed., *Great Speeches by Native Americans* (Mineola: Dover Publications, 2000), p. 84–85.
201. Forbes, *Columbus*, p. 26–27.
202. Dee Brown, *Bury My Heart at Wounded Knee: An Indian History of the American West* (New York: Holt, Rinehart, and Winston, 1970), p. 273, 449.
203. Just last night a neighbor said, "I hope all of his hair falls out. That's what I wish for those I really don't like." She is far kinder in her fantasies than I.
204. Or maybe not. Long after I wrote this section, and even having walked that part of the stream too many times to remember, I may still be wrong. The stream might flow through into Elk Creek, at least seasonally. Here's how I found out. Late in the process Sawyer and Schultz sprang another lie on us: that the stream went underground between two points that they named; this stretch of stream was probably a quarter mile from Elk Creek, and seemed to be chosen specifically to help their case. As mentioned, I thought it went underground elsewhere, but nowhere near where they stated. So I

walked the stream there. Walked isn't really the right word, since the forest is so thick. Crawled works, as does climbed (as in over, under, between branches), as does scootched. It was also clear even to my extremely incompetent eye for tracking that I was the only human who had walked (crawled, climbed, scootched) that stretch for quite a while: I could not help but cause a fair amount of damage, and I had seen the havoc Sawyer caused when he walked a stretch, hacking down everything in sight, yet there was no damage at all. In no place did the stream go underground. I walked it at the driest time of year. Now here's the point: as I ducked under a branch, I happened to see a slight movement out of the corner of my eye. Looking closer, I saw a dark four-inch fish facing me, right in the middle of the stream. It remained a moment, motionless, then darted into the dark at the edge of the stream. It was either a juvenile cutthroat or coho, or an adult sculpin. The fact that the stream has fish in it implies to me that it does go through to Elk Creek. And please note, by the way, that even if the stream didn't go into Elk Creek, because the barrier isn't natural, it would in any case have to be treated as coho habitat.

205. Which, like other governmental agencies associated with resource extraction, has been captured by the industry it is supposed to oversee, and acts in servitude to that industry, acting also as a barrier between that industry and the public that industry harms: "Gosh, if the California Department of Forestry approves this Timber Harvest Plan, and if Fish and Game signs off on it, then I guess it must be fine: these organizations wouldn't lie to serve industry, would they?"

206. The owls had already nearly been exterminated by industrial logging; in fact they were down to a couple of dozen pairs. Then the decision maker at the BC Ministry of Forests determined that more logging would be good for the owls because it would "thin" the forests (never mind that spotted owls require old growth, and that every study shows a direct correlation between logging and spotted owl decline), and further, the decision maker said, by logging in the habitat of the few remaining pairs, the province would gain valuable data on the impacts of such logging. No, I'm not making this up. A scientific panel had been convened to determine what should be done to allow the birds to recover. Not surprisingly, the panel determined that no logging should be allowed in owl habitat. Equally unsurprisingly, the province asked the BC Supreme Court to ignore the panel's report, and instead funded a panel of timber company biopimpologists who came up with their own plan. For yet a third time we should not be surprised, as this august panel determined that up to one third of the remaining owl habitat could be logged (for now, with more to follow), and that half of the entire region could be deforested. The logging has included the felling of actual nest trees. Any concern about the fate of the owls (and the forest) is dismissed, because, as with Smith and Blackhawk, the corporations must make a profit. Or, as Myke Chutter, the province's bird specialist, said, discussions over owl recovery have "centred on what [timber] licensees may or may not be willing to do rather than what the owl needed." Gosh, where have we heard that before? And finally, British Columbia has determined that the best way to save the province's last spotted owls is to capture them and place them in a tourist attraction (perhaps after seeing the owls, you can go on a helicopter tour ($169 for twenty minutes), then go tandem paragliding, and to top it all off you can watch the resort's "World Famous Lumberjack Show": "Recreating the historic world of a logging camp, two lumberjacks will dazzle and amaze you with their skills, tricks and showmanship. Among the challenges are a 60-foot tree climb, an axe-throwing competition, and the always exhilarating log-rolling."). At that point, with the owls safely exterminated in the wild (which of course was what the province and timber corporations wanted all along), British Columbia and the corporations it serves will be free to log the rest of the forests. But truly, we don't need to worry; we just need to trust Canadian Environment Minister

Rona Ambrose when she says there is "no imminent threat" of extirpation for spotted owls, even though the number of spotted owls in Canada had dropped again, from twenty-two to seventeen. I'm not sure precisely how low the number would have to go before she would consider it an imminent threat. And we all know she wouldn't lie to us, would she?

207. Hank Sims, "Campbell's Account: Are the Fortuna mayor's undisclosed checks from Pacific Lumber a problem?" *North Coast Journal*, September 20, 2007, http://www.northcoastjournal.com/092007/cover0920.html, site visited September 23, 2007.

208. Brown, *Bury My Heart*, p. 273, 449.

209. Sharon Duggan and Tara Mueller, Guide to the Forest Practice Act and Related Laws: Regulation of Timber Harvesting on Private Lands in California, (Point Arena, CA: Solano Press, 2005).

210. The other side said that we didn't care about the land at all, but that we merely wanted to protect this playground where we'd ridden our ORVs so extensively that we'd created trails (ummm, none of us even *have* ORVs, and double ummm, I thought those weren't ORV trails, but rather "pre-existing roads" that Smith was "maintaining"?).

211. Lundy Bancroft, *Why Does He Do That? Inside the Minds of Angry and Controlling Men* (New York: Berkley Books, 2002), p. 34–35. Bold and italics in original.

212. Ibid.

213. I went back and forth whether to call it manufacturing or growing: the former would be more true to the intents and processes of the "farmers" who oversee these factory farms, while the latter would honor the plants as living beings, which is why I chose the latter; fuck the corporate owners of factory farms and their deathly perspective.

214. Brown, *Bury My Heart*, p. 273, 449.

215. Dr. Greg Laughlin: "The technology is not at all far-fetched. . . . [W]e just require delicacy of planning and manoeuvring" (Robin McKie, "Nasa aims to move Earth," *Guardian Unlimited*, June 10, 2001, http://observer.guardian.co.uk/international/story/0,,504486,00.html, site visited November 9, 2007). And what's wrong with these NASA scientists? Did they never see that classic television program *Space 1999* and learn that strange things happen when a celestial body is blown out of orbit? People could end up duplicated and come face to face with their future selves; or they could answer a distress signal and be forced to crash land on a distant moon which will turn out to be a penal colony, where our commander will find himself a prisoner of the beautiful Elizia and her equally beautiful prison guards; or a mysterious power from an alien planet could take control, luring the inhabitants to a paradise of eternal peace but living death. (Oh, wait, we already have that one: it's called industrial civilization, except we have neither the paradise nor eternal peace.)

216. Please note that he's doing the same damn thing that McDonough does, which is to try to naturalize capitalism, make us think capitalism is somehow natural or compatible with the natural world.

217. Peter Montague, "Rachels' Democracy and Health News #932," November 8, 2007, http://www.precaution.org/lib/07/ht071108.htm, site visited November 10, 2007.

218. Associated Press, "UN chief calls for action on climate change," *San Francisco Chronicle*, November 8, 2007, http://www.chron.com/disp/story.mpl/world/5287278.html, site visited November 13, 2007.

219. The journalistic equivalents of biopimpologists: since neither of the words *journopimps* or *pimpalists* really cuts it, how about *pimpwriters*.

220. Note that he doesn't bother to get the population right, which is closer to 7 billion.

221. The lemming thing is a lie created by Walt Disney's filmmakers for the 1958 Disney movie *True-Life* [sic] *Adventures: White Wilderness*. As David Perlman reported in the *San Francisco Chronicle*, "Disney's camera crew was filming in Alberta where lemmings don't

live. So the crew imported their own lemming extras, herded them together to simulate a mass migration and then drove them into a Hollywood-style panic to the edge of [a] . . . cliff, from which they leaped to their deaths en masse" (David Perlman, "Lemmings' Death Wish Nothing but a Tale—Told by Disney: Predators, Not Mass Suicide, Account for Arctic Rodents' Population Cycles, Study Finds," *San Francisco Chronicle*, October 31, 2003). Walt Disney was making a snuff film. Further, even if the metaphor worked, which it doesn't, not only because lemmings aren't metaphors, but rather actual beings, and not only because for a metaphor to work, it has to be true not just symbolically but literally—if it's not true literally it isn't grounded—the metaphor works much better to describe civilization's march to the end of the world. Many people have described the current culture as driving headlong over a cliff. I interviewed Jan Lundberg years ago, and he said: "We're going to live through an 'economic and political discontinuity of historic proportions,' as one analyst puts it, or the crash, as we more often refer to it. I like the language of oil industry geologist Dr Walter Youngquist: 'My observations in some seventy countries over about fifty years of travel and work tell me that we are clearly already over the cliff. The momentum of population growth and resource consumption is so great that a collision course with disaster is inevitable.'"

222. I need to add that in the next paragraph he gave his methodology for arriving at his own solutions to global warming: "Most of my ideas are based on (or originated from) a wide variety of prophetic writings concerning the design of a new city in the Bible." His primary "realistic and practical" idea is the following, and I swear I've not made up a single word (I couldn't; I don't have this much, er, creativity): "The solution would have to rethink, redesign the entire city concept. For example, if our transportation were powered by electricity, instead of fossil fuel, then at the very least that opens the potential for a much cleaner city. Electric vehicles (I am not talking about personal automobiles, but something more akin to a bus or subway) are currently economic while traveling along flat roads or downhill. They are not economic going uphill. So, in my design the road is downhill for 99% of the way. . . ." Yes, you read that correctly: his "realistic and practical" idea is for roads to nearly always go downhill.

223. Dennis Gabor, *Inventing the Future* (London: Secker & Warburg, 1963), p. 162.

224. Carl Jung, *Memories, Dreams, Reflections* (New York: Random House, 1961), p. 235–237.

225. Greg Palast, "War Paint and Lawyers: Rainforest Indians versus Big Oil," *BBC Newsnight*, http://www.gregpalast.com/war-paint-and-lawyers-rainforest-indians-versus-big-oil, site visited December 10, 2007.

226. Cary Tennis, "I'm a brilliant scientist and I fear for the world's fate: Since You Asked. . . ." *Salon*, http://www.salon.com/mwt/col/tenn/2008/01/24/planet_death/, site visited January 25, 2008.

227. Then today I got another note from yet another person saying this same fucking thing. He wrote, "I do not agree with the notion that the earth will be killed by civilization. The earth will in the scheme of the cosmos survive our insanity." He went on to talk about how this means that "the Creator" will take care of everything. I get this shit *all the time*. It's like they teach these clichés in school, or something. Well, now that I think about it. . . . But we really don't need to worry. If "the Creator" doesn't take care of the problems, surely the Easter Bunny will.

228. Frederick Winslow Taylor, *Principles of Scientific Management, Comprising Shop Management, The Principles of Scientific Management and Testimony Before the Special House Committee* (New York: Harper & Row, 1911).

229. "The Laws of Magic," The Deoxyribonucleic Hyperdimension, http://deoxy.org/lawsofmagic.htm, site visited November 24, 2007.

230. Later studies suggested that the motions weren't totally made up, but were related to the birds' natural foraging motions.

231. Elisabet Sahtouris, "ELISABET SAHTOURIS BIOS and PRESS PHOTOS," http://www.sahtouris.com/INFO/, site visited March 13, 2008.

232. "After Darwin," speech by Elisabet Sahtouris, http://www.ratical.org/LifeWeb/Articles/AfterDarwin.html, site visited March 13, 2008. Of course the natural world has built countless viable butterflies, and this culture systematically exterminates them, but we shouldn't talk about that, since those are only real butterflies and not metaphorical ones that rationalize inexcusable behavior.

233. "Baring Witness," http://www.baringwitness.org/, site visited December 28, 2007.

234. "Global Orgasm," http://www.globalorgasm.org/, site visited December 28, 2007.

235. "Green Circle," http://www.witchvox.com/vn/vn_detail/dt_ev.html?a=usma&id=47508, site visited December 28, 2007.

236. Belinda Gore, *Ecstatic Body Postures* (Santa Fe: Bear and Company, 1995).

237. The Nazis did the same thing to the Jews, presenting them with false choices (for example, "Would I be safer with a blue or red identity card?" when both colors led to ghettos then gas chambers) to keep them meaninglessly occupied instead of rejecting, then dismantling, the whole murderous system.

238. I am grateful to Stephanie McMillan for the preceding analysis.

239. I am grateful to Lierre Keith for bringing this to my attention, and for her sterling analysis.

240. Lewis Mumford, *The Myth of the Machine: The Pentagon of Power* (New York: Harcourt Brace Jovanovich, 1970), p. 435.

241. See Paul Goettlich, "The Sixth Basic Food Group," and also Hillary Mayell, "Ocean Litter Gives Alien Species an Easy Ride," *National Geographic News*, April 29, 2002.

242. "A billion cell phones old in 2006," *Inquirer*, March 5, 2007, http://www.theinquirer.net/en/inquirer/news/2007/03/05/a-billion-cell-phones-sold-in-2006.

243. Tim Lehnert, "Your Desktop could be a Time Bomb: Making computers—and disposing of them—exacts a harsh environmental cost," *Phoenix*, November 29, 2006, http://thephoenix.com/article_ektid28678.aspx, site visited September 12, 2007.

244. This is a commonly used figure from the EPA (see, for example http://www.epa .gov/osw/nonhaz/industrial/guide/index.htm). This number was generated during the 1980s, so it's very possibly much larger now.

245. Based on garbage density figures from "How long would it take to fill the Grand Canyon with trash?" *Earth & Sky Radio*, http://www.earthsky.org/radioshows/48503/how-long-would-it-take-to-fill-the-grand-canyon-with-trash, site visited June 20, 2007.

246. United Nations Development Programme, *Human Development Report 1992*, http://hdr.undp.org/en/reports/global/hdr1992/.

247. Organisation for Economic Co-operation and Development Environmental Data Compendium, 2002. Waste generation statistics available online at http://www.nationmaster .com/graph/env_was_gen-environment-waste-generation.

248. R. B. Gordon, M. Bertram, and T. E. Graedel, "Metal stocks and sustainability," *Proceedings of the National Academy of Sciences*, 13, (5), January 31, 2006: 1209–1214.

249. Richard Heinberg, in his book *The Party's Over*, conclusively debunks the possibility of maintaining the current economic system through the use of so-called alternative energies.

250. Fossil fuels are being created continuously in very small amounts, so it could be argued that they are renewable if your entire society requires only infinitesimally small amounts. In nature processes, one gallon of gasoline requires about 98 tons of gasoline to produce. This means that to produce the amount of oil used by society each year would require all of the plant matter that the Earth could grow over a period of more than four hundred years. See http://www.eurekalert.org/pub_releases/2003-10/uou-bm9102603.php.

251. We are not including indigenous warfare, which has very little in common with civilized and especially industrial warfare. Indigenous warfare, which often simply resembles fairly violent sports, can and often is sustainable, and in fact arguments have been put forward that it can facilitate sustainability by making sure that small groups remain small, and do not join into larger, less sustainable groups.

252. Doug Rokke, "Depleted Uranium: Uses and Hazards," January 2001, http://www.ratical.org/radiation/DU/DUuse+hazard.pdf .

253. Ibid.

254. Mike Barber, "First Gulf War still claims lives," *Seattle Post-Intelligencer*, January 16, 2006, http://seattlepi.nwsource.com/local/255812_gulfvets16.html, site visited February 20, 2008.

255. Doug Rokke, "The War Against Ourselves," November 2002, http://www.yesmagazine.org/article.asp?ID=594, site viewed February 20, 2008.

256. "The Final Report to the Prosecutor by the Committee Established to Review the NATO Bombing Campaign Against the Federal Republic of Yugoslavia: Use of Depleted Uranium Projectiles," United Nations, 2001. See part IV, A, ii.

257. See Karen Parker, JD, "The Illegality of DU Weaponry," A paper prepared for the International Uranium Weapons Conference, Hamburg, Germany, October 16–19, 2003. See http://www.webcom.com/hrin/parker/du2003.doc. You can read other briefs and documents on Parker's website, http://www.webcom.com/hrin/parker.html.

258. Ibid.

259. "Depleted uranium," World Health Organization Fact sheet #257, January 2003, http://www.who.int/mediacentre/factsheets/fs257/en/, site viewed February 21, 2008.

260. Regarding commercial fluorine production, see "Fluorine," *The Columbia Encyclopedia*, Sixth Edition (New York: Columbia University Press, 2007).

261. *Dr. Strangelove or: How I Learned to Stop Worrying and Love the Bomb*. Stanley Kubrick. 1964. If you aren't familiar with that scene, go watch the movie—it's still one of the greatest satires in film history.

262. Dan Fagin, "Second Thoughts on Fluoride," *Scientific American*, January 2008, http://www.sciam.com/article.cfm?id=second-thoughts-on-fluoride, site visited February 22, 2008.

263. Correspondence between Rebecca Hanmer, EPA Deputy Assistant Administrator for Water, and Dr. Leslie Russell, a dentist in Newtonville Massachusetts, March 30, 1983. Available online at http://www.dartmouth.edu/%7Ermasters/AHABS/docs/Hanmer-ToRussell.pdf.

264. Myron J. Coplan and Roger D. Masters, "Silicofluorides and Fluoridation," *Fluoride*, 34 (3), 2001: 161–164.

265. Canadian Broadcasting Company, *Air of Death*, 1967. A transcript with screen captures is available at http://www.fluoridealert.org/cbc-transcript.htm. This was a controversial documentary in some quarters, which triggered a lawsuit against the CBC. The CBC won the suit.

266. US Department of Agriculture, "Air Pollutants Affecting the Performance of Domestic Animals," Agricultural Handbook No. 380, Revised, 1972.

267. George Glasser, "Fluoride and the Phosphate Connection," *Earth Island Journal*, Summer 1998, http://earthisland.org/eijournal/fluoride/fluoride_phosphates.html.

268. If you are interested in reading critiques of municipal water fluoridation, please see the following sources: "Fluoride: Worse Than We Thought" by Andreas Schuld (http://www.westonaprice.org/envtoxins/fluoride.html), the Fluoride Action Network (http://www.fluorideaction.net/), and Scientific American's "Second Thoughts on Fluoride" by Dan Fagin (http://www.sciam.com/article.cfm?id=second-thoughts-on-fluoride).

269. Here's another example of a bad idea: Julian Simon was an economist and former White House economic advisor. He was an ardent advocate of cornucopianism—the belief that Earth has effectively unlimited resources, and that there are no limits to the growth and expansion of human civilization. The cornucopian and the technotopian have a lot of philosophical overlap in terms of a shared belief in the ability of technology (and usually capitalism) to "meet all human needs." Evidently an intact biosphere is not considered a "human need." What people like Julian Simon are really talking about is an ever-growing industrial infrastructure into the future, and understanding why this idea is or is not feasible will help us to understand why a technotopia is or is not feasible.

Julian Simon is often quoted for assertions like the following: "We have in our hands now—actually in our libraries—the technology to feed, clothe, and supply energy to an ever-growing population for the next 7 billion years."

Statements like this betray a fundamental lack of understand of exponential growth, as well as technology and cosmology. In the last forty years, the human population has doubled. Imagine if tomorrow morning NASA began a massive space emigration program, and began to send out spaceships in all directions at near the speed of light. Imagine a sphere of humans expanding in all directions at as close to the speed of light as they can get.

If the population continued to double every forty years, in four centuries there would be a thousand people for every person alive today. In eight centuries there would be a million people for every person alive today. In sixteen centuries there would be more than a trillion people for everyone alive today. It's safe to say that your neighborhood is starting to get a little crowded at this point.

Around the year 8850 C.E., the sphere of humanity would be nearly fourteen thousand light years across, and every single cubic inch would be filled with humans. The population, extreme dieting aside, simply won't be able to double anymore because there won't be enough room. In fact, the collective mass of all humans would now far exceed the mass of the visible universe. (Moreover, getting anywhere near that point is impossible not just because of resource limitations, but because any object that massive would have long ago collapsed into a black hole!) The fact that the population reaches such a point within less than one millionth of the time Julian Simon allocated for indefinite growth only underscores the absurdity of his position.

270. Andrea Tone, "Contraceptive consumers: gender and the political economy of birth control in the 1930s," *Journal of Social History*, Spring, 1996, http://findarticles.com/p/articles/mi_m2005/is_n3_v29/ai_18498205.

271. "Lysol douche ad, 1948," The Museum of Menstruation and Women's Health, http://www.mum.org/Lysol48.htm, site viewed August 12, 2007.

272. Rachel Lynn Palmer and Sarah K. Greenberg, *Facts and Frauds in Woman's Hygiene* (New York; The Sun Dial Press, 1936). See "Lysol and Zonite," p. 150–151.

273. Ibid.

274. For more on this point, see Andrea Tone, *Devices and Desires: A History of Contraceptives in America* (New York: Hill and Wang, 2001).

275. From *Ladies' Home Journal*, November 1927, p. 79.

276. The executive quoted was Albert Lasker, head of the Lord & Thomas agency. Quoted by John Gunther, *Taken at the Flood: The Story of Albert D. Lasker* (New York: Harper, 1960), p. 154.

277. Lloyd Stouffer, quoted in "Plastics in disposables and expendables," *Modern Plastics*, 34 (8), April 1957: 93.

278. Ibid.

279. Stuart Ewen, *Captains of Consciousness: Advertising and the Social Roots of the Consumer Culture* (New York: Metropolitan Books, Henry Holt and Company, 1999).

280. Clive Ponting, *A Green History of the World* (London: Penguin, 1991), p. 337.

281. Richardson Wright, "The Decay of Tinker Recalls Olden Days of Repairing," *House & Garden*, August 1930, p. 48.

282. Katharine Mieszkowski, "Plastic Bags are killing us," *Salon*, August 10, 2007, http://sacdcweb07.salon.com/news/feature/2007/08/10/plastic_bags/index1.html, site viewed March 2, 2008.

283. Daniel Imhoff, "Thinking Outside of the Box," *Whole Earth*, Winter 2002, p. 13.

284. This, of course, only considers the person actually *operating* the machine, and ignores the people required to build the machine, to mine and refine the metals the machine is built from, to extract the oil to fuel it, and so on. It also ignores the people who are displaced or killed by the culture that builds the machine.

285. This is from a manuscript of Lierre Keith's upcoming book *The Vegetarian Myth*, one of the best books ever written about agriculture, ecology, and diet.

286. See "Contraceptive on your hip" in Chellis Glendinning, *My Name is Chellis and I'm in Recovery from Western Civilization* (Boston: Shambhala, 1994).

287. Lewis Mumford, "Authoritarian and Democratic Technics," 1964.

288. "Jessica Taylor talks to veterans of Oak Ridge," *Guardian*, July 5, 2006, http://www.guardian.co.uk/secondworldwar/story/0,,1812911,00.html, site visited December 13, 2007.

289. Ibid.

290. "Steam Locomotive Builders," steamlocomotive.com, January 30, 2008, http://www.steamlocomotive.com/builders/, site viewed February 12, 2008.

291. Jensen, *Culture*, 441. I've changed it from "our culture" to "this culture" because I think it's important to separate ourselves from this culture. My thinking has of course evolved since writing that earlier book.

292. Mumford, "Authoritarian and Democratic Technics."

293. Ibid.

294. Rudy Rucker, "Fundamental Limits to Virtual Reality," March 3, 2008, http://www.rudyrucker.com/blog/2008/03/03/fundamental-limits-to-virtual-reality/, site viewed March 4, 2008. For a longer essay on some of the same topics, see his essay "The Great Awakening," in the August 2008 edition of *Asimov's SF* or the *Year Million* anthology published by Atlas Books.

295. Joseph Tainter, *The Collapse of Complex Societies* (Cambridge, UK: Cambridge University Press, 2003), p. 4.

296. See, for example, the recent rapid climb in grain prices worldwide.

297. Of course, there will always be some exceptions to this. During the Great Depression, many in the entertainment industry still profited, because the difficult social and economic conditions encouraged escapism.

298. "Rails missing; it's hard to keep track . . ." Reuters, February 6, 2006, http://today.reuters.com/news/newsArticle.aspx?type=oddlyEnoughNews&storyID=2006-02-03T185855Z_01_L03713130_RTRUKOC_0_US-GERMANY-RAIL.xml, site visited December 3, 2007.

299. Tom Zytaurk, "Metal theft ring wreaks havoc across Lower Mainland," *Now Newspaper*, March 11, 2004, http://www.thenownewspaper.com/issues03/094103/features .html, site visited November 2, 2007.

300. Stephen Wickens, "The case of the disappearing pop cans," *Globe and Mail*, January 28, 2006, http://www.theglobeandmail.com/servlet/ArticleNews/TPStory/LAC/20060128/ALUMINUM28/TPNational/Toronto, site visited November 2, 2007.

301. Mark Skaer, "Have Copper, Will Steal Units," *Air Conditioning, Heating & Refrigeration News*, October 30, 2006, http://www.achrnews.com/Articles/Cover_Story/fae418d13458e010Vgn VCM100000f932a8c0, site visited November 2, 2007.

302. "Eskom Safety Z Card," Eskom, February 18, 2005, http://www.eskomdsm.co.za/images/EskomZCardEng.pdf.

303. Rahul Mahakani, "Wired world: Tapping power or funeral pyre?" *Times of India*, June 29, 2003, http://timesofindia.indiatimes.com/articleshow/49256.cms, site visited November 4, 2007.

304. Mark Gregory, "India struggles with power theft," *BBC News*, March 15, 2006, http://news.bbc.co.uk/2/hi/business/4802248.stm, site visited November 3, 2007.

305. Isuri Kaviratne, "Electricity piracy adds to CEB's woes," *Sunday Times*, February 25, 2007, www.sundaytimes.lk/070225/News/107news.html, site visited March 2, 2008.

306. "Gas Flaring in Nigeria: A Human Rights, Environmental and Economic Monstrosity," Climate Justice Programme and Environmental Rights Action / Friends of the Earth Nigeria, June 2005, www.foe.co.uk/resource/reports/gas_flaring_nigeria.pdf.

307. "Probe ordered after Nigeria blast," *BBC News*, May 13, 2006, http://news.bbc.co.uk/2/hi/africa/4768159.stm, site visited March 4, 2008.

308. Daniel Howden, "Shell may pull out of Niger Delta after 17 die in boat raid," *The Independent* (UK), January 17, 2006, http://www.corpwatch.org/article.php?id=13121, site visited March 4, 2008.

309. Hester Le Roux, "Evidence and Analysis: The Role of Natural Resources in Fuelling and Funding Conflict in Africa," London, September 2004, http://www.commission-forafrica.org/english/report/background/leroux_background.pdf.

310. Tainter, *Collapse*, p. 188.

311. See various discussions and sources from *The Overworked American: The Unexpected Decline of Leisure*, by Juliet B. Schor, online here: http://www.swiss.ai.mit.edu/~rauch/worktime/hours_workweek.html.

312. Neal Stephenson, *Snow Crash* (New York: Bantam Books, 1992).

313. Kate Clifford Larson, *Bound For the Promised Land: Harriet Tubman, Portrait of an American Hero* (New York: Ballantine Books, 2004).

314. A modern version might add that they want automobiles without oil drilling, or computers without mining, or "green" future that they can buy at Wal-Mart.

315. Frederick Douglass, "The Significance of Emancipation in the West Indies," Speech, Canandaigua, New York, August 3, 1857; in *The Frederick Douglass Papers, Series One: Speeches, Debates, and Interviews*, 3: 1855–63, edited by John W. Blassingame (New Haven: Yale University Press). Emphasis added.

316. See http://www.inthewake/keith1.html.

317. Brian Glick, *War at Home: Covert Action Against us Activists and What We Can Do About It* (Boston: South End Press, 1989).

318. For detailed discussion of COINTELPRO and its sequelae, see Ward Churchill and Jim Vander Wall, *The COINTELPRO Papers: Documents from the FBI's Secret Wars Against Domestic Dissent* (Boston: South End Press, 1990), and Ward Churchill and James Vander Wall, *Agents of Repression: The FBI's Secret Wars against the Black Panther Party and the American Indian Movement*, (Boston: South End Press, 1988).

319. Who tells us this? The state and its supporters, of course, but would you agree that the Nazi government had a legitimate monopoly on violence? What about the Soviet Union? What about British colonial powers? Were the Founding Fathers wrong to take up arms against taxation without representation?

320. For example, see J. K. van Ginneken and M. Wiegers, "Various causes of the 1994 genocide in Rwanda with emphasis on the role of population pressure," presented at the 2005 Annual Meeting of the Population Association of America, Philadelphia, Pennsylvania, March 31–April 2, 2005; and *Collapse* by Jared Diamond.

321. Erich Fromm, *The Anatomy of Human Destructiveness* (New York: Basic Books, 1986), p. 195.

322. Bruno Bettelheim, *The Informed Heart: Autonomy in a Mass Age* (New York: Macmillan Free Press, 1960). Emphasis in original.
323. This, of course, doesn't mean that would shouldn't try to take care of each other—of course we should. And it doesn't mean that we should avoid or impugn healing when it can happen. Rather, the healing we get now may be more like the healing of a combat medic—to be patched up and get back into the fight—rather than a permanent recuperation.
324. Thank you to Lierre Keith for the first three questions, which come from Allan G. Johnson's book *The Gender Knot*.

BIBLIOGRAPHY

Abma, J. C., A. Chandra, W. D. Mosher, L. S. Peterson, and L. J. Piccinino. "Fertility, family planning, and women's health: new data from the 1995 National Survey of Family Growth." *Vital Health Stat* 23 (1997): 1–114.

Agence France-Presse. "Wilderness almost non-existent on planet Earth: study." *Breitbart.com*, June 28, 2007. http://www.breitbart.com/article.php?id=070628185002.qek4e5qz&show _article=1&catnum=0.

Associated Press. "UN chief calls for action on climate change." *San Francisco Chronicle*, November 8, 2007.

Association of Science-Technology Centers Incorporated. "A Garbage Timeline." *The Rotten Truth (About Garbage)*. http://www.astc.org/exhibitions/rotten/timeline.htm (accessed April 21, 2007).

Atkinson, W., and R. New. "An Overview of the Impact of Source Separation Schemes on the Domestic Waste Stream in the UK and Their Relevance to the Government's Recycling Target." Waste Online. http://www.wasteonline.org.uk/resources/InformationSheets/History-ofWaste.htm.

Bancroft, Lundy. *Why Does He Do That? Inside the Minds of Angry and Controlling Men*. New York: Berkeley Books, 2002.

Barber, Mike. "First Gulf War still claims lives." *Seattle Post-Intelligencer*, January 16, 2006. http://seattlepi.nwsource.com/local/255812_gulfvets16.html.

BBC News. "Probe ordered after Nigeria blast." May 13, 2006. http://news.bbc.co.uk/2/hi/africa/4768159.stm.

Bell, Shawna. "Finding hope among the ruins." *The Durango Telegraph*, March 13, 2008. http://www.durangotelegraph.com/telegraph.php?inc=/07-03-29/coverstory.htm.

Bettelheim, Bruno. *The Informed Heart: Autonomy in a Mass Age*. New York: Macmillan Free Press, 1960.

Blaisdell, Bob, ed. *Great Speeches by Native Americans*. Mineola: Dover Publications, 2000.

Blunt, Zoe. "'No Threat' to Spotted Owls in Canada: Govt." *Guerilla News Network*, August 27, 2006, http://zoeblunt.gnn.tv/articles/2498/_No_Threat_to_Spotted_Owls_in_Canada_Govt (accessed September 16, 2007).

————. "Killing Spotted Owls with Chainsaws: Has government-approved logging doomed Canada's spotted owl?" *Guerilla News Network*, July 26, 2006, http://zoeblunt.gnn.tv/articles/2438/Killing_Spotted_Owls_with_Chainsaws (accessed September 16, 2007).

Board of Forestry and Fire Protection. "Disciplinary Actions: Case Number 217." *Licensing News* 19, no. 1, (2000):16. http://www.fire.ca.gov/CDFBOFDB/pdfs/LicensingNewsMaster5-001.pdf.

Bock, Paula. "Oceans of Waste: Waves of junk are flowing into the food chain." *Seattle PI*, April 23, 2006, http://seattletimes.nwsource.com/html/pacificnwo4232006/coverstory.html (accessed June 22, 2007).

Bogo, Jennifer. "Crying Rivers (Romania gold mine spills cyanide into rivers) (Brief Article)." *High Beam Encyclopedia*, May 1, 2000, http://www.encyclopedia.com/doc/1G1-62298547.html (accessed July 13, 2007).

Boje, David. "Indonesia." *Academics Studying Nike, Reebok, and Adidas-Indonesia Subcontract Factories.* http://business.nmsu.edu/~dboje/nike/indonesia.html (accessed December 19, 2006).

Bonewits, P. E. I. "The Laws of Magic." *The Deoxyribonucleic Hyperdimension.* http://deoxy.org/lawsofmagic.htm (accessed November 24, 2007).

Borenstein, Sean. "Populations of 20 Common Birds Declining." *San Francisco Chronicle*, June 14, 2007, http://www.sfgate.com/cgi-bin/article.cgi?f=/n/a/2007/06/14/national/a182901D95 .DTL&hw=chickadee&sn=001&sc=1000 (accessed June 18, 2007).

Brewer, E. Cobham. "Deluge." *Dictionary of Phrase and Fable.* Bartleby.com. http://www.bartleby.com/81/4834.html (accessed July 26, 2007).

Brown, Dee. *Bury My Heart at Wounded Knee: An Indian History of the American West.* Holt, Rinehart, and Winston, New York, 1970.

Bureau of American Ethnology. "Bulletin 61: Teton Sioux Music." Washington, DC: Smithsonian Institution, 1918.

Buxton, Herbert T., and Dana W. Kolpin. "Pharmaceuticals, Hormones, and Other Organic Wastewater Contaminants in U.S. Streams." US Geological Survey, June 2002. http://toxics .usgs.gov/pubs/FS-027-02/index.html.

Byrd, Deborah, and Joel Block. "How long would it take to fill the Grand Canyon with trash?" *Earth & Sky Radio.* EarthSky Communications. http://www.earthsky.org/radioshows/48503/ how-long-would-it-take-to-fill-the-grand-canyon-with-trash.

Calafat, Antonia M., Larry L. Needham, Manori J. Silva, and George Lambert. "Exposure to Di- (2-Ethylhexyl) Phthalate Among Premature Neonates in a Neonatal Intensive Care Unit." *PEDIATRICS* 113, no. 5, May 2004: e429–e434.

Campbell, Helen. "As to Ashes and Rubbish," *American Kitchen Magazine* 12, August 1900.

Campbell, Jonathan A. "What Is Dioxin?" Natural Therapy Virtual Clinic. http://www.cqs.com/ edioxin.htm (accessed June 20, 2007).

Cardiff University Waste Research Station. "MSW Combustion." http://www.wasteresearch.co .uk/ade/efw/mswcombustion.htm.

Casey, Susan, and Gregg Segal. "Our oceans are turning into plastic. . . . Are we?" *Best Life Magazine,* May 11, 2007. http://www.bestlifeonline.com/cms/publish/health-fitness/Our _oceans_are_turning_into_plastic_are_we_2.shtml. (accessed June 11, 2007).

Catton, William R., Jr. *Overshoot: The Ecological Basis for Revolutionary Change.* University of Illinois Press: Urbana, 1982.

Chauhan, Baldev. "Indian state outlaws plastic bags." *BBC News,* August 7, 2003. http://news .bbc.co.uk/2/hi/south_asia/3132387.stm.

Cheriathundam, Elizabeth, and Alvito P. Alvares. "Species differences in the renal toxicity of the antiarthritic drug, gold sodium thiomalate." *Journal of Biochemical Toxicology* 11, no. 4: 175–181.

Child, Lydia Maria. *The American Frugal Housewife,* 16th ed., enlarged and corrected. Boston: Carter, Hendee, 1835.

Claridge, Amanda. *Rome: An Oxford Archaeological Guide.* Oxford: Oxford University Press, 1998.

Clean Clothes Campaign. "Labour Rights in Indonesia: What is Menstruation Leave." Newsletter 13, November 2000. http://www.cleanclothes.org/news/newsletter13-indon.htm.

Climate Justice Programme and Environmental Rights Action / Friends of the Earth Nigeria. "Gas Flaring in Nigeria: A Human Rights, Environmental and Economic Monstrosity." June 2005. http://www.foe.co.uk/resource/reports/gas_flaring_nigeria.pdf

Colorado School of Mines. "Acid Producing Potential of Mine Overburden." http://www.mines .edu/fs_home/jhoran/ch126/app htm (accessed July 6, 2007).

———. "Microbial Influences: Thiobacillus ferrooxidans," *Acid Mine Drain Drainage Experiments at CSM.* http://www.mines.edu/fs_home/jhoran/ch126/microbia.htm (accessed July 6, 2007).

Columbia Encyclopedia, 6th ed., s.v. "Flourine."

Conant, Compton, and Urey. "Memorandum to Brigadier General L.R. Groves." *Mindfully.org.* http://www.mindfully.org/Nucs/Groves-Memo-Manhattan30oct43.htm.

Cone, Marla. "Dozens of Words for Snow, None for Pollution." Mother Jones Magazine, January/February 2005. http://www.motherjones.com/news/feature/2005/01/12_402.html.

———. *Silent Snow: The Slow Poisoning of the Arctic.* New York: Grove Press, 2006.

Coplan, Myron J., and Roger D. Masters. "Silicofluorides and Fluoridation." *Fluoride* 34, no. 3 (2001): 161–164.

Douglass, Frederick. "The Significance of Emancipation in the West Indies." In *The Frederick Douglass Papers, Series One: Speeches, Debates, and Interviews.* vol. 3, edited by John W. Blassingame. New Haven: Yale University Press, 1985.

Duggan, Sharon, and Tara Mueller. *Guide to the Forest Practice Act and Related Laws: Regulation of Timber Harvesting on Private Lands in California.* Point Arena: Solano Press, 2005.

Ejnet.org. "Dioxin Facts." Web Resources For Environmental Justice Activists. http://www.ejnet .org/dioxin/ (accessed June 20, 2007).

Energy Information Administration. "Petroleum Basic Data." Department of Energy, July 2007. http://www.eia.doe.gov/neic/quickfacts/quickoil.html (accessed September 2, 2007).

Environmental Protection Agency. *Evaluation of Emissions from the Open Burning of Household Waste in Barrels.* Vol. 2, *Appendices A-G.* Control Technology Center, November 1997. http://www.epa.gov/ttn/catc/dir1/barlbrn2.pdf

Eskom. "Eskom Safety Z Card." February 18, 2005. http://www.eskomdsm.co.za/images/ EskomZCardEng.pdf

Fagin, Dan. "Second Thoughts on Fluoride." *Scientific American,* January, 2008. http://www.sciam.com/article.cfm?id=second-thoughts-on-fluoride.

Fahm, L. A. *The Waste of Nations.* Montclair: Allanheld, Osmun & Co., 1980.

Faison, Seth. "Lirong Journal: Tibetans, and Vultures, Keep Ancient Burial Rite." *New York Times,* July 3, 1999.

Finley, Harry. "Lysol ad from March 1948." *Museum of Menstruation and Women's Health.* http://www.mum.org/Lysol48.htm (accessed August 12, 2007).

Forbes, Jack D. *Columbus and Other Cannibals.* Brooklyn: Autonomedia, 1992.

Forbes. "The 400 Richest Americans #30: Philip H Knight." *Forbes,* September 21, 2006. http://www.forbes.com/lists/2006/54/biz_06rich400_Philip-H-Knight_2KZ5.html.

Franklin and Associates. "Paper vs. Plastic Bags." Institute for Lifecycle Environmental Assess-ment, 1990. http://www.ilea.org/lcas/franklin1990.html.

Free Dictionary, s.v. "Plastic." http://www.thefreedictionary.com/plastic (accessed June 15, 2007).

Friends of the Earth. "Gas Flaring in Nigeria." *Friends of the Earth Media Briefing,* October 2004. http://www.remembersarowiwa.com/pdfs/gasflaringinnigeria.pdf

Fromm, Erich. *The Anatomy of Human Destructiveness.* New York: Basic Books, 1986.

Gabor, Dennis. *Inventing the Future.* London: Secker & Warburg, 1963.

Gandy, Oscar H., Jr. *The Panoptic Sort: A Political Economy of Personal Information.* Boulder: West-view Press, 1993.

Glasser, George. "Fluoride and the Phosphate Connection." *Earth Island Journal,* Summer, 1998. http://earthisland.org/eijournal/fluoride/fluoride_phosphates.html.

Glendinning, Chellis. *My Name is Chellis and I'm in Recovery from Western Civilization.* Boston: Shambhala, 1994.

Glick, Brian. *War at Home: Covert Action Against Us Activists and What We Can Do About It.* Boston: South End Press, 1989.

Global Orgasm. http://www.globalorgasm.org (accessed December 28, 2007).

Goettlich, Paul. "Plastic in the Sea," *Mindfully.org*, Oct 5, 2005. http://www.mindfully.org/Plastic/Ocean/Sea-Plastic-LN-PG5oct05.htm (accessed June 22, 2007).

———. "The Sixth Basic Food Group." *Mindfully.org*, November 29, 2004. http://www.mindfully.org/Plastic/6th-Basic-Food-Group.htm (accessed June 21, 2007).

Gordon, R. B., M. Bertram, and T. E. Graedel. "Metal stocks and sustainability." *Proceedings of the National Academy of Sciences* 13, no. 5 (2006): 1209–1214.

Gore, Belinda. *Ecstatic Body Postures*. Santa Fe: Bear and Company, 1995.

Gregory, Mark. "India struggles with power theft." *BBC News*, March 15, 2006. http://news.bbc.co.uk/2/hi/business/4802248.stm.

Grouse Mountain Vancouver Tourism. http://www.grousemountain.com/welcome.cfm (accessed September 16, 2007).

Gunther, John. *Taken at the Flood: The Story of Albert D. Lasker*. New York: Harper, 1960.

Hadfield, Doug. "Resourcex Introduction to Mining and Mining Investment." *Mining Industry News Channel*, http://www.huliq.com/21238/resourcex-reports-an-introduction-to-mining-and-mining-investment (accessed September 1, 2007).

Hardell, L, M. J. Walker, B. Walhjalt, L. S. Friedman, and E. D. Richter. "Secret ties to industry and conflicting interests in cancer research." *Our Stolen Future*, November 3, 2006. http://www.ourstolenfuture.org/Industry/2006/2006-1103hardelletal.html.

Hartmark-Dounas, Laura. "Summitville, the Exxon Valdez of the Mining Industry," *Sprol*, October 17, 2005. http://www.sprol.com/?p=268m.

Healthy Building Network. "PVC Facts," http://www.healthybuilding.net/pvc/facts.html.

Hogan, Bill. "Paging Dr Ross." *Mother Jones*, November/December 2005. http://www.motherjones.com/news/outfront/2005/11/paging_dr_ross.html.

Hogg, D., et al. "Greenhouse Gas Balances of Waste Management Scenarios." *Eunomia*, 2008. London: Report for the Greater London Authority, 2008.

Holdsworth, Andy, Cindy Hale, and Lee Frelich. "Invasive Earthworms in our Forests: Contain Those Crawlers." *Minnesota Department of Natural Resources*. http://www.dnr.state.mn.us/invasives/terrestrialanimals/earthworms/index.html (accessed May 8, 2007).

Howden, Daniel. "Shell may pull out of Niger Delta after 17 die in boat raid." *The Independent*, January 17, 2006. http://www.corpwatch.org/article.php?id=13121.

Igura, Isamu. "Half-life of each dioxin and PCB congener in the human body." *Organohalogen Compounds* 66 (2004): 3376–3384. http://dioxin2004.abstract-management.de/pdf/p311.pdf.

Inquirer Staff. "A billion cell phones sold in 2006." *The Inquirer*, March 5, 2007. http://www.theinquirer.net/en/inquirer/news/2007/03/05/a-billion-cell-phones-sold-in-2006.

Jensen, Derrick, and George Draffan. *Strangely Like War: The Global Assault on Forests*. White River Junction: Chelsea Green, 2003.

———. *Welcome to the Machine: Science, Surveillance, and the Culture of Control*. White River Junction: Chelsea Green, 2004.

Jensen, Derrick. "Where the Buffalo Go: How Science Ignores the Living World: An Interview with Vine Deloria." *The Sun*, July 2000.

———. *A Language Older Than Words*. White River Junction: Chelsea Green, 2000.

———. *The Culture of Make Believe*. White River Junction: Chelsea Green, 2004.

———. *Endgame*. New York: Seven Stories Press, 2006.

Jessen, Michael. "The Bag Beast." *Environmentally Speaking: Michael Jessen Speaks Up For The Environment*. http://www.nelsonbc.ca/pages/jessen/The_Bag_Beast.htm.

Jobling, Susan, et al. "Predicted Exposures to Steroid Estrogens in U.K. Rivers Correlate with Widespread Sexual Disruption in Wild Fish Populations." *Environmental Health Perspectives*, April 2006.

Josephson, Matthew. *The Robber Barons: The Great American Capitalists, 1861–1901*. New York: Harcourt Brace, 1934.

Jung, Carl. *Memories, Dreams, Reflections*. New York: Random House, 1961.

Kayiratne, Isuri. "Electricity piracy adds to CEB's woes." *The Sunday Times*. Sunday, February 25, 2007. www.sundaytimes.lk/070225/News/107news.html.

Klauk, Erin. "Environmental Impacts at Fort Belknap from Gold Mining." *Impacts of Resource Development on Native American Lands*, Carelton College. http://serc.carleton.edu/research _education/nativelands/ftbelknap/environmental.html (accessed July 13, 2007).

Konefes, John L., and Michael K. McGee. "Old Cemeteries, Arsenic, and Health Safety." *Water Industry News*, Environmental Market Analysis. http://waterindustry.org/arsenic-3.htm.

Laing, R. D. *The Politics of Experience*. New York: Ballantine, 1967.

Larson, Kate C. *Bound For the Promised Land: Harriet Tubman, Portrait of an American Hero*. New York: Ballantine Books, 2004.

Le Roux, Hester. "Evidence and Analysis: The Role of Natural Resources in Fuelling and Funding Conflict in Africa." *Commission for Africa*. http://www.commission-forafrica.org/english/report/background/leroux_background.pdf

Leavitt, Judith W. *The Healthiest City: Milwaukee and the Politics of Health Reform*. Princeton: Princeton University Press, 1982.

Lehnert, Tim. "Your Desktop could be a Time Bomb: Making computers—and disposing of them—exacts a harsh environmental cost." *The Phoenix*, November 29, 2006. http://thephoenix.com/article_ektid28678.aspx.

Lormand, Eric. "Facts and FAQs About Nike's Labor Abuses." http://www-personal.umich.edu/~lormand/poli/nike/nikelabor.htm (accessed December 19, 2006).

Mahakani, Rahul. "Wired world: Tapping power or funeral pyre?" *The Times of India*. June 29, 2003. http://timesofindia.indiatimes.com/articleshow/49256.cms.

Marshall, Jessica. "War of the Worms." *New Scientist*, March 1, 2007. http://environment.newscientist.com/channel/earth/mg19325931.600-war-of-the-worms.html.

Mayell, Hillary. "Ocean Litter Gives Alien Species an Easy Ride." *National Geographic News*, April 29, 2002.

McDonough, William, and Michael Braungart. "From Inspiration to Innovation: Nike's Giant Steps Toward Sustainability." http://www.mcdonough.com/writings/inspiration_innovation.htm.

McDonough, William. "Buildings Like Trees, Cities Like Forests." http://www.mcdonough.com/writings/buildings_like_trees.htm.

———. "Corporate Flight Center." http://www.mcdonoughpartners.com/projects/corpflight/default.asp?projID=corpflight (accessed December 28, 2006).

———. "Executive Summary: Designs celebrating the joy and creativity of the human spirit and the abundance of nature." http://www.mcdonoughpartners.com/executive_summary.shtm.

———. "Ford Rouge Dearborn Truck Plant." http://www.mcdonoughpartners.com/projects/ford-dtp/default.asp?projID=ford-dtp (accessed December 16, 2006).

———. "Full." http://www.mcdonough.com/full.htm.

———. "Leading Change Toward Sustainability." http://www.mcdonough.com/writings/leading_change.htm.

———. "Nike European Headquarters." http://www.mcdonoughpartners.com/projects/nike/default.asp?projID=nike (accessed December 17, 2006).

McGaw, Judith A. *Most Wonderful Machine: Mechanization and Social Change in Berkshire Paper Making, 1801–1885.* Princeton. Princeton University Press, 1987.

McKie, Robin. "NASA aims to move Earth." *Guardian Unlimited*, June 10, 2001. http://observer.guardian.co.uk/international/story/0,,504486,00.html.

Melosi, Martin V. *Garbage in the Cities: Reuse, Reform, and the Environment, 1880–1980.* Chicago: Dorsey, 1981.

Mieszkowski, Katharine. "Plastic Bags are killing us. " *Salon.com.* http://www.salon.com/news/feature/2007/08/10/plastic_bags.

Mineral Policy Center. "Cyanide Leach Mining Packet." Washington, DC: GPO, 2006.

Montague, Peter. "Rachels' Democracy and Health News #932." Rachel's News, *Environmental Research Foundation*, November 8, 2007. http://www.precaution.org/lib/07/ht071108.htm.

Morlan, Kinsee. "Home sweet dump: Tijuana landfills are home to hundreds of people who depend on garbage for their livelihood." *San Diego CityBeat*, January 15, 2008. http://www.sdcitybeat.com/cms/story/detail/home_sweet_dump/6551/.

Morse, W. F. *The Collection and Disposal of Municipal Waste.* New York: Municipal Journal and Engineer, 1908.

Mumford, Lewis. "Authoritarian and Democratic Technics." 1964.

———. *The Myth of the Machine: The Pentagon of Power.* New York: Harcourt Brace Jovanovich, 1970.

National Human Genome Research Institute. "Researchers Compare Chicken, Human Genomes." http://www.genome.gov/12514316.

National Museum of American History. "Ivory Billiard Balls." *Smithsonian Institute.* http://americanhistory.si.edu/collections/object.cfm?key=35&objkey=185 (accessed June 18, 2007).

NationMaster.com. "Environment Statistics: Waste generation (most recent) by county." http://www.nationmaster.com/graph/env_was_gen-environment-waste-generation.

New Jersey Department of Environmental Protection. "Guidance Document For Regulated Medical Waste." December, 2007. http://www.nj.gov/dep/dshw/rrtp/rmw.htm (accessed March 10, 2008).

Office of Technology Assessment. "Issues in medical waste management." Washington DC: United States Congress, 1988.

Onstot, J, R. Ayling, and J. Stanley. *Characterization of HRGC/MS Unidentified Peaks from the Analysis of Human Adipose Tissue. Volume 1: Technical Approach.* Washington, DC: US Environmental Protection Agency Office of Toxic Substances, 1987.

"Outside the Box." Online video. *CBC Newsworld,* 2005. http://video.google.ca/videoplay?docid=9563149686691637968&q

Palast, Greg. "War Paint and Lawyers: Rainforest Indians versus Big Oil." *BBC Newsnight.* http://www.gregpalast.com/war-paint-and-lawyers-rainforest-indians-versus-big-oil/

Palmer, Rachel Lynn, and Sarah K. Greenberg. *Facts and Frauds in Woman's Hygiene.* New York: The Sun Dial Press, 1936.

Perlman, David. "Lemmings' Death Wish Nothing but a Tale Told by Disney: Predators, Not Mass Suicide, Account for Arctic Rodents' Population Cycles, Study Finds." *San Francisco Chronicle,* October 31, 2003.

Pharmaceutical Market Trends, 2007–2011: Key market forecasts and growth opportunities. 2nd ed. London: URCH Publishing, 2007.

Ponting, Clive. *A Green History of the World.* London: Penguin, 1991.

Reusablebags.com. "Reusable Bags—Trends From Around The World." http://www.reusablebags.com/facts.php?id=9 (accessed March 23, 2007).

Reuters. "Accommodating an army of garbage pickers." *CNN.com,* March 26, 2003. http://www.cnn.com/2003/WORLD/americas/03/26/argentina.train.reut/ (accessed April 4, 2007).

————. "Rails missing; it's hard to keep track . . ." February 6, 2006. http://today.reuters.com/news/newsArticle.aspx?type=oddlyEnoughNews&storyID=2006-02-03T185855Z_01_L03713130_RTRUKOC_0_US-GERMANY-RAIL.xml.

Robinson, B. A. "Financial Compensation for Nazi Slave Laborers." Ontario Consultants on Religious Tolerance, December 2006. http://www.religioustolerance.org/fin_nazi.htm.

Rokke, Doug. "Depleted Uranium: Uses and Hazards." Rat Haus Reality: Ratical Branch. January 2001. http://www.ratical.org/radiation/DU/DUuse+hazard.pdf

———. Interview by Sunny Miller. "The War Against Ourselves." *Yes Magazine,* Spring 2003. http://www.yesmagazine.org/article.asp?ID=594.

Royte, Elizabeth. *Garbage Land: On the Secret Trail of Trash.* London: Back Bay, 2006.

Rucker, Rudy. "Fundamental Limits to Virtual Reality." *Rudy's Blog,* March 3, 2008. http://www.rudyrucker.com/blog/2008/03/03/fundamental-limits-to-virtual-reality.

Sahtouris, Elisabet. "After Darwin." *LifeWeb.* Rat Haus Reality: Ratical Branch. http://www.rat-ical.org/LifeWeb/Articles/AfterDarwin.html (site visited March 13, 2008).

———. "Elisabet Sahtouris Bio and Press Photos." http://www.sahtouris.com/INFO/ (accessed March 13, 2008).

ScienceDaily. "Hormonal Contraception Does Not Appear To Increase HIV Risk." http://www.sciencedaily.com/releases/2006/12/061207161016.htm.

Sehee, Joe. "Green Burial: It's Only Natural." *PERC Reports* 25, no. 4 (2007). http://www.perc .org/perc.php?subsection=5&id=1015 (accessed December 30, 2007).

Shaw, Charles. "The Priest and the Prophet: Can Industrial Civilization Really Become Sustainable? Should it?" *Grist,* 15 August 2006. http://www.grist.org/comments/soapbox/2006/08/15/shaw/.

Shears, Richard. "Is this the world's most polluted river?" *Daily Mail,* June 5, 2007. http://www.dailymail.co.uk/news/article-460077/Is-worlds-polluted-river.html.

Sheehan, D., and P. Reffell. *Baring Witness.* http://www.baringwitness.org/ (accessed December 28, 2007).

Sicular, Daniel Thoreau. "Currents in the Waste Stream: A History of Refuse Management and Resource Recovery in America." M.A. thesis, University of California Berkeley, 1984.

Silk, Nicole, and Kristine Ciruna. *A Practitioner's Guide to Freshwater Biodiversity Conservation.* Arlington: The Nature Conservancy, July 2004.

Sims, Hank. "Campbell's Account: Are the Fortuna mayor's undisclosed checks from Pacific Lumber a problem?" *North Coast Journal,* September 20, 2007. http://www.north-coastjournal.com/092007/cover0920.html.

Skaer, Mark. "Have Copper, Will Steal Units." *Air Conditioning, Heating & Refrigeration News,* October 30, 2006. http://www.achrnews.com/Articles/Cover_Story/fae418d13458e010Vgn VCM100000f932a8c0 (accessed November 2, 2007).

Spiegelman, Helen, and Bill Sheehan. "Unintended Consequences: Municipal Solid Waste Management and the Throwaway Society." *Product Policy Institute,* March 2005.

Stamets, Paul. *Mycelium Running: How Mushrooms Can Help Save the World.* Berkeley: Ten Speed Press, 2005.

Steamlocomotive.com. "Steam Locomotive Builders." http://www.steamlocomotive.com/builders/ (accessed February 12, 2008).

Stephenson, Neal. *Snow Crash.* New York: Bantam Books, 1992.

Stiffler, Lisa. "PDBEs: They are everywhere, they accumulate and they spread: Chemical flame retardants pose threat to humans, environment," *Seattle PI*, March 29, 2007. http://seattlepi.nwsource.com/local/309169_pbde28.html.

————. "The message in the (plastic) bottle is dire: Material absorbs pollutants and becomes part of the ecosystem." *Seattle PI*, October 10, 2006. http://seattlepi.nwsource.com/specials/brokenpromises/288097_plastic10.asp.

Strasser, Susan. *Waste and Want: A Social History of Trash*. New York: Holt Paperbacks, 2000.

Tainter, Joseph A. *The Collapse of Complex Societies*. Cambridge: Cambridge University Press, 2003.

Taylor, Frederick Winslow. *Principles of Scientific Management, Comprising Shop Management, The Principles of Scientific Management and Testimony Before the Special House Committee*. New York: Harper & Row, 1911.

Taylor, Jessica. "Let's build a bomb." *The Guardian*, July 5, 2006. http://www.guardian.co.uk/secondworldwar/story/0,,1812911,00.html.

Tennis, Cary. "I'm a brilliant scientist and I fear for the world's fate: Since You Asked . . ." *Salon.com*, January 24, 2008. http://www.salon.com/mwt/col/tenn/2008/01/24/planet_death.

Thai Labor Campaign. "Woods Meets Nike Protesters; Tiger Woods Was Escorted Through An Angry Crowd." *1World Media*. http://www.1worldcommunication.org/labornews.htm#A%20Letter%20to%20Tiger%20W (accessed December 20, 2006).

Thornton, Joe. "Dioxin From Cradle to Grave." *Greenpeace*. http://www.greenpeace.org/ raw/content/usa/press/reports/the-pvc-lifecycle-dioxin-from.pdf.

Tone, Andrea. "Contraceptive consumers: gender and the political economy of birth control in the 1930s." *Journal of Social History*, Spring 1996. http://findarticles.com/p/ articles/mi_m2005/is_n3_v29/ai_18498205.

US Department of Agriculture. *Air Pollutants Affecting the Performance of Domestic Animals*. Revised ed. Washington, DC: GPO, 1972.

US Environmental Protection Agency. "Medical Waste." http://www.epa.gov/epaoswer/ other/medical/basic.htm (accessed March 10, 2008).

————. "Milestones in Garbage." http://www.epa.gov/msw/timeline_alt.htm (accessed February 2, 2008).

US Food Administration. *Garbage Utilization: With Particular Reference to Utilization by Feeding*, Washington, DC: GPO, 1918.

United Nations Development Programme. *Human Development Report 1992*. http://hdr.undp .org/en/reports/global/hdr1992/.

United Nations. "The Final Report to the Prosecutor by the Committee Established to Review the NATO Bombing Campaign Against the Federal Republic of Yugoslavia: Use of Depleted Uranium Projectiles," United Nations, 2001.

University of Houston. "The Politics of Oil." *Digital History*. http://www.digitalhistory.uh.edu/ historyonline/oil.cfm (accessed April 21, 2007).

Urrea, Luis Alberto. *By the Lake of Sleeping Children: The Secret Life of the Mexican Border.* New York: Anchor Books, 1996.

Van Ginneken, J. K., and M. Wiegers. "Various causes of the 1994 genocide in Rwanda with emphasis on the role of population pressure." Presented at the 2005 Annual Meeting of the Population Association of America, Philadelphia, Pennsylvania, March 31–April 2, 2005.

Veldhoen, Rachel C. Skirrow, et al. "The bactericidal agent triclosan modulates thyroid hormone-associated gene expression and disrupts postembryonic anuran development." *Aquatic Toxicology* 80, no. 3 (2006): 217–227.

Vietnam Labor Watch. "Nike Labor Practices in Vietnam." http://www.saigon.com/~nike/reports/report1.html.

Vom Saal, Frank, and W. Welshons. "Large effects from small exposures. II. The importance of positive controls in low-dose research on bisphenol A," *Environmental Research* 100, no. 50: 76. http://www.ourstolenfuture.org/NewScience/oncompounds/bisphenola/2006/2006-0101vom-saalandwelshons.html.

Warhurst, Michael, and Watson, Anna. "Dirty truths: Incineration and climate change." Friends of the Earth, May 2006. http://www.foe.co.uk/resource/briefings/dirty_truths.pdf.

Waste Watch. "History of waste and recycling." http://www.wasteonline.org.uk/resources/InformationSheets/HistoryofWaste.htm, (accessed March 17, 2007).

Weeks, Lyman Horace. *A History of Paper-Manufacturing in the United States, 1690–1916,* New York: Lockwood Trade Journal, 1916.

Weisman, Alan. "Polymers Are Forever." *Orion,* May/June 2007. http://www.orion-magazine.org/index.php/articles/article/270/.

Wickens, Stephen. "The case of the disappearing pop cans." *Globe and Mail,* January 28, 2006. http://www.theglobeandmail.com/servlet/ArticleNews/TPStory/LAC/20060128/ALU-MINUM28/TPNational/Toronto.

Wikianswers. "What is plastic made of?" http://wiki.answers.com/Q/What_is_plastic_made_out_of (accessed June 18, 2007).

Wikipedia. "Billiard Ball." Wikimedia Foundation. http://en.wikipedia.org/wiki/Billiard_ball (accessed June 18, 2007).

———. "Bisphenol A." Wikimedia Foundation. http://en.wikipedia.org/wiki/Bisphenol_A (accessed June 20, 2007).

———. "Dioxin." Wikimedia Foundation. http://en.wikipedia.org/wiki/Dioxin (accessed June 20, 2007).

———. "Tailings." Wikimedia Foundation. http://en.wikipedia.org/wiki/Tailings (accessed July 6, 2007).

Witch's Voice. "Green Circle." http://www.witchvox.com/vn/vn_detail/dt_ev.html?a=usma&id=47508 (accessed December 28, 2007).

Woodruff, Jody. "Plastic Bottles and Can Liners Under Scrutiny Again." *Sentient Times,* April/May 2007. http://www.sentienttimes.com/07/07_apr_may/plastic.html.

World Health Organization. "Depleted uranium." January 2003. http://www.who.int/media-centre/factsheets/fs257/en/ (accessed February 21, 2008).

Wright, Richardson. "The Decay of Tinker Recalls Olden Days of Repairing." *House & Garden*, August 1930.

Wyoming Funeral Directors Association. "History of Embalming." http://www.wyfda.org/basics_3.html.

Yahoo.com. "PharmEng Announces Exclusive Agreement With Chinese Manufacturer for Distribution of Acetaminophen to North America Resulting in Significant Raw Material Cost Savings." *Yahoo Business News*, March 26, 2007. http://biz.yahoo.com/ccn/070326/2007032603801800o1.html?.v=1 (accessed March 2, 2008).

Zarwan, Elijah, and Andrea Di Martino. "Buenos Aires: The Ghost Train of the *Cartoneros*." Worldpress.org. http://www.worldpress.org/photo_essays/cartoneros/ (accessed April 3, 2007).

Zytaruk, Tom. "Metal theft ring wreaks havoc across Lower Mainland." *The Now Newspaper*. March 11, 2004. http://www.thenownewspaper.com/issues03/094103/features.html

INDEX

abused children, 274–75
acetaminophen, 132
Acronym of Death (AoD), 253
acrylonitrile butadiene rubber, 136–37
action v. inaction, 384–85
activism, 323–24
    action v. inaction in, 384–85
    California redwood forest, 264
    child-like passivity in, 275
    effective strategy needed in, 385
    forests saved through, 195
    as infantile, 263
    as lonely, 262
    ocean needing, 272
acts of commission, 384
acts of omission, 384
addictions, 200
advanced technologies, 245–47
advice columnist, 213–14
Agent Orange, 114
agents of change, 231–32
agricultural raw materials, 304–5
agriculture, monotechnic, 341–42
Akosombo Dam, 92
Alamosa River, 125–26
aluminum, 366
Amaru (dog), 83–85
Amazon River, 57
Ambrose, Rona, 416n205
The American Frugal Housewife, 31
anaerobic organisms, 321
The Anatomy of Human Destructiveness
    (Fromm), 394
Ancient Egyptian mummification, 139–40
ancient procedure, recycling, 45
anger, 180–81
animals, organic matter fed to, 31
Antebellum slaves, 396

antibiotics, 11
AoD. *See* Acronym of Death
applied technologies, 207
Après nous le deluge, 150
aquatic caddisfly larva, 310
aquatic life, 125–26
Arab-Israel war, 315
architect, sustainable development, 61–62,
    408n75
Arctic
    POPs in, 144
    women living in, 144
Argentina, local recycling in, 50
Armillaria ostoyae, 96
Armstrong, Jeannette, 243
Aronowitz, Stanley, 239
arterial embalming, 140–41
asphalt, roads covered with, 160–61
asymmetric warfare, 373–273
Audubon, 225
authoritarian system, 351, 377
    social/economic disruptions and, 391
authoritarian technics, 344–45, 351
autolysis, 12
autonomy, 344–45
awards, 62
Axelrod, Herbert R., 57

bacteria
    digestion specialty of, 313
    hydroponic plants not needing, 403n5
    plant's roots with, 14
    in soil, 13
Baekeland, Leo Hendrik, 103
Baia Mare gold mine, 124
Bakelite, 103
Bancroft, Lundy, 185–86
Ban Ki-moon, 203

Baring Witness, 232–33
bartering, 46–47
   materials and, 47
   women and, 405n48
bathtub, overflowing, 287
berm, stream blocked by, 160
Bertel, Rosalie, 99
Bettelheim, Bruno, 394, 395
bilirubin, 7
billiard balls, 103–4
bin Laden, Osama, 371
biodegradable plastics, 304, 305–6
biological *cycles*, 406n68
biologically-derived materials, 300
biologically synthesized fuels, 306–7
biopimpologist, 163–64, 170
bioplastic, 301–2
   as biodegradable, 305–6
   energy required producing, 302–3
   environmental impact producing, 303
   manufacturing, 302
   scale matters in, 308–9
biosphere, 307
birds, magical thinking of, 223
birth control pills, 133
bisphenol A (BPA), 109, 112
Black Hawk, 153–54
board of supervisors, 169–70
body burden
   of chemicals, 145
   persistent, 145
Bonhoeffer, Dietrich, 378, 382
bottom ash, 43
BPA. *See* bisphenol A
brain cells, digesting selves, 12
Brave Buffalo, 86–87
breast milk, 108
bullets, 315
burn barrels, 43
Bush, George W., 73
business as usual scenario, 288–97, 296–97
butterflies, 419n231
Butterfly: A Tiny Tale of Great Transformation
   (Huddle), 231
by-product waste, 289

California Department of Forestry, 173–74
California Forest Practice Act, 175
cancer, 114–15
capitalism, 229–30, 258–59
car
   culture, 249
   new smell of, 108
carbohydrates, 302

carbon
   appropriation, 307
   in cellulose, 313–14
   cycle, 45
   offset schemes, 228
   sequestration mechanism, 303
carbon dioxide, 119
carcinogenic, 111
carrying capacity
   of human beings, 392–93
   of rabbits, 272–73
cartoneros, 50
cascara leaves, 22
caste attitudes, 50
caterpillars, 230
   transformation of, 229
cattle, 320
Catton, William R., 78
causality, 238
causation, correlation equal, 238
C. difficile infection, 14
cell phone batteries, 289
cells
   brain, 12
   new molecules in, 311
cellucotton, 329
cellulose
   carbon in, 313–14
   indigestability of, 312–13
   plants building block of, 312
cellulose nitrate, 104
centralized decision-making, 345
centralized dumping, 30
centralizing control/externalizing
   consequences, 358
ceremonies, of indigenous people, 243–44
chanterelles (mushrooms), 96
Chase, Richard Trenton, 185
cheap energy, 296
chemicals, 114–15
   body burden of, 145
   human body with, 143–44
   industrial, 145–46
Cheney, Dick, 171
Chevron, 212
chickens
   waste eaten by, 32
   worm density from, 37–38
childish behaviors, 274–75
child-like passivity, 275
children
   abused, 274–75
   ensalved, 260–61
   ill-behaved, 273–74

laws protecting, 269
   self-destructive behaviors of, 268
   sheltering, 269
   swill, 25–26
children's health advocates conference, 74–76
chlorine, PVC largest use of, 110–11
Churchill, Ward, 76
Chutter, Myke, 416n205
cipro, 11
citizens, 175
   capitalism redefining, 258–59
   collapse faced by, 377–78
   earth's defense by, 380–81
   fluoride ingested by, 320
   land and, 72
civilizations. *See also* industrial civilization;
   real physical world
   collapse common outcome of, 360
   complex society difference with, 365
   constructed history of, 17
   early metal recycling of, 45–46
   as expansionist, 357
   externalizing consequences, 359
   globalization of, 229–30
   human survival in, 410n102
   living outside of, 382
   modern, 350
   as monopolies, 364–65
   sociopolitical complexity loss in, 360–61
   sustainable economics of, 374
   unsustainable, 363
civilized selves, 260–61
Clarke, Arthur C., 241
clean coal technologies, 228
cleanliness, 330
closed-loop cycles, 405n68
coal burning, 40
coal/wood ash, 39–40
Cofan people, 212
COINTELPRO (FBI counter intelligence
   program), 388–89
collapse, 358, 361
   asymmetric warfare and, 372–73
   citizens faces with, 377–78
   of civilization's common outcome, 360
   complexity costs causing, 364
   of industrial civilization, 277–78
   industrial civilization's ecological, vii
   industrial/factors for, 361
   as one big system breaking down, 359–60
   real world, 366
   of Roman Empire, 374–76
   sociopolitical complexity loss causing,
      360–61

of this culture, 376
   war accelerating, 370
   waste production decrease/industrial, 362
The Collapse of Complex Societies (Tainter),
   360, 374
collective mass, of human beings, 421n268
colonialism, 406n69
Colorado, 251
columnist, advice, 213–14
commerce/nature conflict, 409n85
commercial fisherman, 252–53
communities
   large-scale, 239
   waste decisions of, 26–27
compartmentalization, 81
   in concentration camps, 82
   interviewer with, 91–92
complex societies, 363
   civilizations difference with, 365
   complexity costs collapsing, 364
computers
   electronic waste from, 49
   sub-atomic level optimizing, 355
computronium, 355
concentration camps, 70, 395
   compartmentalization in, 82
   of Nazis, 393
Cone, Marla, 144
Confederacy of the Haudenosaunee, 381
Connett, Paul, 43
conquerors, resource, 370
conquest, 347
consumer goods, more expensive, 361–62
consumers
   capitalism defining, 258–59
   consumption reduced by, 379, 386
   environmental issues and, 379
   garbage byproducts of, 285
consumption
   consumers reducing, 379, 386
   decreasing, 361–62
   glorifying, 332
   population and, 300
contraceptives, 134
copper, 366
corn, 304
cornucopia, 421n268
corporate airport, 67–68
corporate flight center, 67
corporations, 331
   eliminating, 349–50
   exploitation by, 337
   profits sought by, 336
   railways modern, 348–49

transnational, 407n72
correlation, equal causation, 238
Cosmeticism, 78
cows, 56–57
crap, 8
creatures
   decay mechanism of, 337–38
   relationships of, 239
cremators, 42
Crisis Coalition, 78–79
Crohn's disease, 94–96
this culture. *See also* human cultures;
      industrial civilization
   abandoning, 381–82
   business as usual scenario for, 288–97
   business as usual unsustainable of,
      296–97
   collapse of, 376
   compartmentalization by, 81
   conquest basis of, 247
   culture valued over life in, 254–56
   damage control for, 357
   diversions in, 200–201
   does not give back, 148–50
   earth being destroyed by, 78, 285
   earth more important than, 278
   earth will only be changed by, 213–15
   exploit/destroy right of, 71
   fighting back/against, 382–83, 401
   future scenarios for, 284–85, 287–88
   good citizen of, 175
   human-first value structure of, 215–16
   immortality quest of, 107
   information/theory concentration of, 87
   internal consistency of, 185–86
   is killing the planet, 99
   less destructive, 68–69
   linear thinking of, 86
   living simply in, 195
   lying to each other, 156–57
   magical thinking of, 219–20
   morality of, 197–98
   pain from, 253–54
   permanence valued by, 58–59
   plastic fabrication of, 399–400
   profits/morality in, 188
   real physical world and, 204
   resource extraction basis of, 164–65
   romantic relationship with, 261
   scalped the forests, 158–59
   sheltering children from, 269
   short attention span of, 273
   short term decisions of, 333
   this world is not, 408n76

tidal wave of destructiveness in, 251
   waste production of, 327
The Culture of Make Believe (Jensen), 348
culture of resistance, 386–87, 388, 389–90
   building, 395–96
   in concentration camps, 395
   fighting back with, 396–97
   long term thinking needed in, 397
   risks involved in, 397–98
   threat represented by, 393–94
cyanide
   mines using, 122–23
   river poisoned with, 124
   uses of, 121–22

damage
   control, 357
   of earth, 217–18
dams, 92
   on Elk Creek, 159
   free-flowing water and, 415n197
   for logging, 159–60
   rivers killed by, 89–90
The Dark Ages, 375
death
   -camp culture, 82
   fear of, 201
   of horses, 39
   scavengers eating from, 12–13
decay
   creatures mechanism for, 337–38
   experiment on, 3
   fascination with, 4–5
   materials resisting, 311
   recirculation of compounds in, 399
   stopping body from, 141
decision-making, centralized, 345
decolonization, 250
decomposing garbage, 51–52
decomposing organic matter, 28
deforestation, 162–63
Deloria, Vine, 87
the Delta, explosions in, 368
delusional thinking, 220–21
democracy, 351
   participatory, 381
democratic technics, 343
demonstrations, 264–65
Denver Nuggets, 221
depleted uranium (DU), 315–16
   Gulf War veterans symptoms from, 316
   tests applied to, 317
   Yugoslavia's use of, 316
despair, 250

destruction, 172
    activities of, 182–83
    developer's, 172
    of diversity, vii–viii
    earth's, 217
    factories and, 74
    government and, 174
    government protecting, 182–83
    by industrial civilization, 70, 195
    by industrial economy, 283–84
    industrial facilities', 409n86
    of life, 174
    message of, 70
    pain caused by, 253–54
    real physical world, 253
    this culture's, 68–69, 251
developer(s)
    environmental/planning regulations
        avoided by, 162–63
    as greedy/destructive, 172
    intimidation by, 174–75
    judicial system and, 176–77
    keep fighting, 401
    landowner desires and, 165–66
    lies of, 156–57, 172–73
    neighborhood fighting, 155–56
    planet dismembered by, 156
    settlement violation of, 180
    title company and, 161
development, 168
Dewailly, Eric, 144
Diamond, Stanley, 374
dictionary, translation, 204
digestion, 313
dioxin, 44, 111
disposal, garbage, 28
diversions, 200–201
diversity, destruction of, vii–viii
Dixie Cup, 330
double-bind, 259–60
Douglas fir, 95
Douglas, Frederick, 383
downcycle, 334–35
Draffan, George, 408n81
drilling rights, 212
DU. *See* depleted uranium
Dudok, Willem, 64
dumping
    centralized, 30
    ocean, 30
    strategies of, 28–29
dumpster diving
    food recovered by, 33
    for fun, 35–36

    as political act, 36–37
dumpsters, grocery stores locking, 34
Dunnville, Ontario, 319

earth. *See also* real physical world
    better because I was born, 278–79
    citizens defending, 380–81
    creature relationships on, 239
    damage caused of, 217–18
    developers dismembering, 156
    enlightened stance toward, 226–27
    God giving dominion over, 93
    greenhouse gases on, 368
    harm reduction of, 258
    healing the, 235
    healthier stronger, 191–92
    human beings killing, 117–18
    industrial civilization destroying, 18
    industrial facilities destructive to, 409n86
    listening to, 216
    natural evolution of, 320–21
    personalizing destruction of, 217
    photosynthesis budget of, 307–8
    this culture and, 278
    this culture destroying, 78, 199–200, 285
    this culture is not, 408n76
    this culture will only change, 213–15
    in trouble, 324
    truly wild percentage of, 415n197
    unlimited resources of, 421n268
earthworms, 14
Eastern Garbage Patch, 101
eaves trough, 20
Ebbesmeyer, Curtis, 116
ecological apocalypse, 192
The Ecological Basis for Revolutionary Change
    (Catton), 78
ecological principles, 272–73
economic disruptions, 391
economic power, 384
economic system
    conquest basis of, 247
    salmon killed by, 270
    trees as resources in, 186–87
effective writing, 155–56
Eichmann, Adolf, 166
electricity tapping (abuse), 367
electronic waste, 49
elephants, killing of, 103–4
Elk Creek
    berm blocking stream to, 160
    small dam on, 159
    stream running into, 415n203
email exchange, 217

embalming, 139–40
    arterial, 140–41
    formaldehyde used in, 141
emissions, 42–43
endangered species, 165–66
Endgame (Jensen), 37
endocrine disruptors, 112
enemy infrastructure, 372
energy
    bioplastic requiring, 302–3
    cheap, 296
    declining/industrial complexity and,
        369–70
    gasoline needed to produce, 419n249
    green, 43
    industrial agriculture needing, 339–40
    paper industry requiring, 305
    waste to, schemes, 43
enlightened stance, earth, 226–27
enlightenment, 276
enslaved child, 260–61
entertainment industry, 422n296
entitlement, 186–87
environment
    bioplastic production impacting, 303
    consumers addressing, 379
    modern movement of, 130
    PDBEs dumped in, 108
    plastic bags impact on, 19
    problems not real, 89
    socially responsible management of,
        306
    styrofoam fast-food containers impact
        on, 305
Environmental Defense Fund, 225
environmental regulations, 162–63
environmental test, 317
ephemeralization, 352–53
essence of life, 148–49
estrogen, fish exposed to, 133–34
ethereal, 90
ethnocentrism, 392
excrement, 8
expansionism, 357
exploitation, 66
    by corporations/government, 337
    culture's right of, 71
    lifestyle of, 288–89
    rationalized, 85
    by Roman Empire, 375
exporting, garbage, 286
externalizing consequences, 358, 359
external morality, 188–89
extraction chemicals, for gold, 121

extralegal force (violence), 389
exudates, 13

factories
    farm, 417n212
    McDonough's work concerning, 74
Facts and Frauds in Woman's Hygiene
        (Palmer/Greenberg), 329
fairness, guiding principle of, 198
fair-trade products, 225
fantasies
    realities v., 206–9
    technologies fulfilling, 209
fascism, 391
fat, 112
fear
    of death, 201
    people oppressed by, 254
feces, 8
feminine hygiene, 328–29
feminine power, 233
feminist theorist, 264
fertilizer, 302, 319
fighting back, 389–90
    culture of resistance, 396–97
    against developers, 401
    resistance groups, 263
    against this culture, 382–83, 401
Fish and Wildlife, 182
fish, estrogen exposure of, 133–34
floods, 92
fluoride
    cattle influenced by, 320
    in drinking water, 318–19
food
    dumpster diving for, 33
    throwing away, 33–34
foraging, 32
Forbes, Jack, 154
Ford Rouge Dearborn Truck Plant, 62–64,
        408n77
forests
    activism for, 195, 264
    in eaves trough, 20
    fungus in, 96–97
    organization of, 88
    shitting in, 6
    this culture scalping, 158–59
formaldehyde, 141
free-flowing water, 415n197
free garbage collection systems, 44
free-market environmentalism, 228
Fresh Kills Landfill, 130
Friendly Fascism, 394

frogs
    red-legged, 171
    slugs eaten by, 5–6
Fromm, Erich, 394
fuels, biologically synthesized, 306–7
Fuller, Buckminster, 352
fungi
    Douglas fir with, 95
    in forest, 96–97
    soil's relationship with, 14
furans, 44
future scenarios, this culture, 284–85, 287–
    88

Gabor, Dennis, 206
Gaia theory, 357
garbage, 9
    burning of, 42
    byproducts of, 285
    coal/wood ash as, 39–40
    collection, 44
    decomposing, 51–52
    disposal strategies of, 28
    exporting, 286
    history of, 26–27
    large quantities of, 27–28
    pickers, rules of, 52–53
    piling up, 333
    real physical world's growing, 285
    strategies for, 27
    trash workers and, 49–50
garbage collection system
    free, 44
    Milwaukee's, 25
    municipal, 39–40, 292
    urban, 44
gas flaring, 368
gasoline, 419n249
Gearhart, Sally Miller, 264
genetic engineering, 240
genocide, 393
Germans, 392
    middle class, 394–95
Gerngross, Tillman, 301, 303
ghost net, 100–101
gift of life, 147–48
Gingrich, Newt, 202
Glick, Brian, 388
global garbage production, 295
globalization
    of humanity, 229
    industrial civilization basis of, 229–
        30
global production, 18–19

global warming
    Crisis Coalition details of, 78–79
    industrial civilization causing, 203
    solutions suggested for, 201–2, 418n221
goal of life, Native Americans, 87
God, 93
gold, extraction process of, 121
Goldman, Emma, 225
Goldstein, Marc, 112
Gore, Al, 69, 201–2
Gottleich, Paul, 114
government
    cultures of resistance dangerous to, 388
    destructive activities protected by, 182–83
    destructive to life, 174
    exploitation by, 337
    resource extraction by, 416n204
    sponsored terrorism, 385
    wastefulness encouraged by, 332
Grand Canyon, 295
grease/oils, 31–32
Great Blue Heron Rookery, 163
Great Depression, 422n296
greedy, 172
Greenberg, Sarah K., 329
green energy, 43
A Green History of the World (Ponting), 332
greenhouse gases, 368
Greenpeace UK, 254
green plastics, 304–5
green technology, 342
grocery stores, 34
growing up, 270
guerillas, 372–73, 388
Gulf War, 316

habitat (reduction), 322, 408n80
Hair, Jay, 380
Haldane, J.B.S., 309
harassment, legal, 389
harm reduction, 258
Harris, Mike, 371
harvested wood, 408n81
Hawken, Paul, 69, 202
health, 52
    care facilities, 129–30
    of land, 5
    plastics and, 114–15
heavy metals, 119
hegemony, 86
helminths. *See* whipworms
hermit crab, 310
Hero for the Planet, 68
Hippocratic oath, 70

history, of ancient civilizations, 17
Hitler, Adolf, 392
Holmes, Thomas, 140
Holocene mass extinction, 322
horizontal hostility, 77
household waste, 38–39, 291
    of US, 41
Huddle, Norie, 231
human beings
    carry capacity of, 392–93
    cellulose undigestable by, 312–13
    civilizations survival of, 410n102
    collective mass of, 421n268
    conditions improvement for, 352
    don't need, 353
    ecological principles exemption of, 272–73
    fear oppressing, 254
    gift of life, 147–48
    impermeable boundaries of, 85
    indigenous people meeting needs of, 243
    industrial waste and, 406n68
    killing earth, 117–18
    landbase's beneficial relationship with,
        410n90
    more special notion of, 271–72
    most intelligent notion of, 272
    plastics and, 113
    real physical world better off without, 259
    social system benefiting, 227
    system valued over, 216
human body
    chemical groups in, 143–44
    decay and, 141
    decomposition of, 12
    gift of life, 147–48
human cultures
    industrial footprint in, 407n73
    landbase not damaged by, 196
    linear progressive, 337
    morality of, 153–54
    sustainability primary value in, 300
humaneness test, 317
human-first value structure, 215–16
human waste, 9
hundredth monkey story, 228–29
hunter-gatherer society, 29, 30–31
Hunza Valley, 248
Hurricane Katrina, 376–77
Hyatt, John Wesley, 104
hydrogen sulfide, 7
hydroponic plants, 403n5

Idaho, 251
IDPs. *See* Internally Displace Persons

ill-behaved children, 273–74
imaginal cells, 229, 231–32
imaginal discs, 231
immobilization, 14
immortality, 107
impermeable boundaries, 85
inaction v. action, 384–85
incentives, 334–35
incinerators, 42
    emissions of, 42–43
    higher temperatures of, 43–44
    as least efficient, 43
    medical waste in, 131–32
indefinite performance, 56
Indians. *See* Native Americans
indigenous people
    advanced technologies of, 245–47
    ceremonies of, 243–44
    cleaning themselves, 22
    human culture's morality comments of,
        153–54
    human needs met by, 243
    industrial infrastructure replaced by, 354
    oil production influencing, 368
    technologies and, 242
    technologies of, 243–45
    warfare of, 420n250
industrial agriculture, 339–40
industrial capitalism
    not sustainable, 204–6
    real physical world and, 202–3
    real world of, 201
    as social construct, 205
    solutions promoting, 286–87
industrial chemicals, 145–46
industrial civilization
    BPA pumped out by, 109
    business as usual scenario for, 288–97
    centralizing control in, 358
    delusion of sustainability of, 74
    destructive message of, 70
    earth being killed by, 69
    earth better off without, 259
    earth laid to waste by, 18
    ecological collapse in, vii
    everyone loses in, 259–60
    globalization and, 229–30
    global warming caused by, 203
    habitat reduction of, 322
    impending crash of, 277–78
    inherently destructive, 195
    laws of ecology denied in, 356
    machine dependent, 339
    mass extinctions and, 322

modern day, 369
nature/commerce conflict in, 409n85
our civilized selves must die getting rid of, 260–61
oxygen reduction from, 321–22
as planet's natural evolution, 320–21
problem solving drawbacks of, 324–25
safe healthy regenerative, 407n74
unsustainable technologies basis of, 342
waste determined by, 335
waste production of, 289–94
industrial economy
getting rid of, 260
as inherently destructive, 283–84
omnicide basis of, 259
Industrialization, 78
industrial manufactures, 47
industry, 290–92
declining energy and, 369–70
facilities of, 409n86
footprint of, 407n73
human beings/waste of, 406n68
infrastructure of, 353–54, 354
scale/waste of, 309, 331
stopping waste production of, 44
sustainability of, 63
inferior races, 392
infiltration, 388
information, 87
In Search of the Primitive (Diamond), 374
insects, 14
intelligence, 272
internal consistency, 185–86
Internally Displace Persons (IDPs), 367
internal morality, 188–89
slavery and, 186–87
international humanitarian law, 317
internet
modern civilizations use of, 350
obsessions/addictions and, 200
interviewer
compartmentalized thinking of, 91–92
"real world" perspective of, 89–90
intimidation, 174–75
involuntary recycling, 366
Iraq, invasion of, 73

jellyfish, 100
Jensen, Derrick, 35, 37, 211, 348, 376
jhator, 142–43
judicial system, 176–77
Jung, Carl, 207
junk, 9
juvenile hormone, 232

Keith, Lierre, 77, 324, 341, 383, 385
Kimberly-Clark, 329
King, Martin Luther, Jr., 397
Klamath River, 181
Kotex sanitary napkins, 329–30

laboratory experiments, 239–40
labor efficiency, 339–41
Laing, R. D., 185
The Lake of Sleeping Children (Urrea), 51
land
access to, 36
health of, 5
people accommodating to, 72
landbase
human cultures not damaging, 196
human's healthy beneficial relationship with, 410n90
no good reason destroying, 126
sustainability of, 56–57
waste products and, 99
landowner, 165–66
A Language Older Than Words (Jensen), 35
large-scale communities, 239
latex gloves, 136
latrine, 8
lavatory, 8
laws
children protected by, 269
non-existent, 176
laws of ecology, 356
leakage, 84
Lebensraum (living space), 392
legacy
relationships better off as, 193–94
what will you leave behind as, 191–92
legal harassment, 389
legal system, 180
lemmings, 417n220
levaquin, 11
lies, developer and, 156–57, 172–73
life
-affirming, 260
-birth-death-decay, 399–400
destruction of, 174
lifestyle, of exploitation, 288–89
lifetime of experience, 87
Lifton, Robert Jay, 70
lignin (wood component), 313–14
linear thinking, 86, 88
listening, 216
litter, 9
living morality, 187
living simply, 195, 258

logging
    dams used for, 159–60
    spotted owl correlation with, 416n205
logistical support, 387–88
long term thinking, 397
Louis XV, king, 150
Lovelock, James, 202, 357
Lovins, Amory, 69
Lynch, Kevin, 285
Lysol, 328

machines
    ephemeralization of, 352–53
    industrial civilization dependent on, 339
    metals required for, 422n283
Madame de Pompadour, 150
magic, 242
    power relationships in, 241–42
    science v., 248
magical technologies, 244
magical thinking, 233
    birds manifesting, 223
    forms of, 224–28
    megalomania underlying, 237–38
    not delusional, 220–21
    rendered powerless by, 224
    in technotopia, 323–24
    this culture's practice of, 219–20
    wishful thinking and, 221–22
magnificent bribe, 370
mainstream environmentalist, 106
Manhattan Project, 345
Manifest Destiny, 186
manipulation, 391–92
manufacturing
    biologically-derived materials in, 300
    bioplastic, 302
    by-product waste from, 289
    closed-loop cycles and, 405n68
maquiladoras, 51
mass extinctions, 322
mass movement, 396
materials
    agricultural raw, 304–5
    bartering of, 47
    decay resistance of, 311
    as food for biological *cycles*, 406n68
    non-toxic biodegradable, 301
    reusing, 46
    war needing, 314–15
matter, 355
maturity, 87
Maze, 13
McDonald's, 305

McDonough, William, 61–78, 314
    awards won by, 62
    closed-loop cycles from, 405n68
    factories less destructive by, 74
    as Hero for the Planet, 68
    industrialization and, 78
    Nike European headquarters by, 64–66
    optimize used by, 408n72
    social equity phrase used by, 409n83
    solutions presented by, 202
    superficial cleaning up of, 73
    sustainability concepts of, 407n71
    as sustainable development architect,
      61–62, 408n75
    truck plant designed by, 62–64
McMillan, Stephanie, 410n104
medical waste, 129–30
medications, 132
meditation, 235
megalomania, 237–38
megamachines, 344, 348
MEND. *See* Movement for the
    Emancipation of the Niger Delta
metabolites, 132
metal recycling, 300
    in early civilizations, 45–46
    in technotopia, 314–15
metals
    machines requiring, 422n283
    mining of, 346
methanol, 141, 414n183
Mexico, subsistence farmers of, 34–35
middle class status, 394–95
military
    technologies, 314
    US aggression of, 233–34
mill ore, 123
Milwaukee, garbage collection systems, 25
mineralization, 12
mines
    Baia Mare gold, 124
    cyanide used in, 122–23
    metal, 346
    phasing out of, 301
    railways developed for, 347
    river/cyanide from, 124
    steam engine used in, 346–47
    streams killed by, 120–21
    Summitville, 125
    Zortman-Landusky, 124–25
mine tailings
    heaps of, 123–25
    toxic minerals in, 119–20
Misseldine, Carol, 334

molecules
  cells with, 311
  mineralization breaking down, 12
Monbiot, George, 307
monopolies, 364–65
monotechnic, 341
Moore, Charles, 116, 286
Moore, Kathleen Dean, 274
morality, 153–54
  foundation of, 192–93
  internal/external, 188–89
  living, 187
  of this culture, 197–98
  this culture's profit, 188
Morlan, Kinsee, 53
Movement for the Emancipation of the
    Niger Delta (MEND), 368
movements
  mass, 396
  modern environmental, 130
  natural burial, 143–44
  radical, 383
  resistance, 77, 387–88, 390
MSWM. *See* Municipal Solid Waste
    Management
Mullen, Herb, 185
Mumford, Lewis, 276, 341, 343, 370
  democratic technics comments of, 343
  technological progress thoughts of, 351
municipal dump, 29, 285
  health issues from, 52
  in Tijuana, 51, 53
municipal garbage collection systems, 39–
    40, 292
municipal solid waste, 289–90
Municipal Solid Waste Management
    (MSWM), 41
municipal waste production, 293–94
mushrooms, 96–97
  chanterelles, 96
  turkey tails, 96–97
mybodypart.com, 411n123
*The Myth of the Machine: The Pentagon of
    Power* (Mumford), 276

nakedness, 233
narcissistic notion, 271
Narcissus, 105
National City Lines, 332
national highway system, 44
National Socialism, 394
National Wildlife Federation, 225, 380
Native Americans
  goal of life of, 87

  religion of, 154
  sacredness/values of, 154
Natron, 139
natural burial, 143–44
natural gas infrastructures, 39–40
natural polymers, 314
natural selection, 415n197
nature benefits, 410n90
nature/commerce conflict, 409n85
The Nazi Doctors (Lifton), 70
Nazis
  concentration camps of, 393
  manipulation used by, 391–92
  middle class status and, 394–95
  slave factories of, 66–67, 393
  superior Germans and, 392
neighborhood
  developer fought by, 155–56
  public hearings complaints of, 173–74
net primary production (NPP), 307
New Jersey, 130
New York city, dead horses in, 39
Nike European headquarters, 64–66
  nature benefits of, 410n90
  negative aspects of, 65–66
  sweatshops and, 406n69, 408n79,
    409n83
nitrile rubber gloves, 135
nitrogen, 14
nonindustrialized nations, 294–96
nonviolent action, 384
NPP. *See* net primary production
nuclear technologies, 345
nurdles (resin pellets), 115–16
nutrients, 14

Oak Ridge, 345
obsessions, 200
oceans
  activism needed for, 272
  dumping in, 30
  plastics in, 101–2, 286
off-gassing, 108
oilfields, control of, 373
oil production
  gas flaring during, 368
  peaked, 361, 376–77
oil supply, 368–69
omnicide, 259
Operation Whitecoat, 318
Oppenheimer, J. Robert, 166
optimize/minimize, 408n72
organic matter
  animals fed, 31

decomposing, 28
extracting grease/oils from, 31–32
hunter-gathers discarding, 30–31
increased waste of, 40
tossing away, 29
organization, of forest, 88
overhead expenses, 371–72
oxygen
industrial civilization reducing, 321–22
mask/breaking down of, 135–36
Oxygen Revolution, 321

Pacific Electric system, 332
pacifism, 384
packaging, disposable, 330
pain
growing up requiring, 270
from this culture, 268
this cultures destructiveness causing,
253–54
Palmer, Rachel Lynn, 329
paper industry
energy intensive, 305
large-scale propaganda campaigns of, 48
wood pulp used in, 48–49
Parker, Karen, 317
Parkes, Alexander, 103
Parks, Rosa, 385
participatory democracy, 381
Paul, Apostle, 270
PAX. *See* Potassium Amyl Xanthate
PBDEs. *See* poly-brominated diphenyl ethers
Peace Pilgrim, 228
peak oil, 361, 376–77
*Peak Oil Survival: Preparing for Life After
Gridcrash* (Jensen), 376
peddlers, 46–47
People for the Ethical Treatment of
Animals (PETA), 75
perception
of powerlessness, 224
slavery and, 186–87
Perlman, David, 417n220
permanence, 58–59
persistent body burden, 145
persistent organic pollutants (POPs), 144
personal change, 257
personal consumption, 294
perspective
skewed, 185–86
waste matter of, 10
PETA. *See* People for the Ethical Treatment
of Animals
pharmaceuticals, 134

PHAs. *See* polyhydroxyalkanoates
phosphate plant, 319–20
photographic evidence, 169–70
photosynthesis
biologically synthesized fuels needing,
306–7
earth's budget of, 307–8
finite capacity of, 307
warm-blooded animals possible from, 321
phthalates, 135
physical reality, 224
phytoplankton, 101
piggeries, 31
pigs
street-cleaning activities of, 31
whipworms supplied by, 94–96
piss, 8
placeness, 410n100
planet. *See* earth
plants
bacteria/roots of, 14
cellulose building block of, 312
exudates released by, 13
hydroponic, 403n5
roots of, 13–14
soil bacteria's relationship with, 13
waxy coating of, 311–12
PLAs. *See* polylactic acids
plastic bags
downcycling of, 334–35
environmental impact of, 19
mistaken as jellyfish, 100
menace of, 18
production of, 403n9
plastics, 107–8. *See also* bioplastic
biodegradable, 304, 305–6
defining, 102–3
earliest form of, 103
fork made of, 17
green, 304–5
health issues from, 114–15
humans living without, 113
in oceans, 101–2, 286
phthalates softening, 135
in phytoplankton, 101
polymers, 102–3, 312
production rate of, 331–32
in sea gulls, 100
in sperm whale, 136
this cultures fabrication of, 399–400
turtle deformed by, 99–100
US production of, 116
poisonous, 111
poker tournament, 179–80

political act
    of dumpster diving, 36–37
    living simply as, 258
political speeches, 411n109
The Politics of Experience (Laing), 185
The Politics of Nonviolent Action (Sharp), 384
poly-brominated diphenyl ethers (PBDEs),
    108
polyhydroxyalkanoates (PHAs), 302
polylactic acids (PLAs), 302
polymers, 102–3
    cellulose/plastics are, 312
    natural, 314
polytechnic, 341
polyvinylchloride (PVC), 110–11
Ponting, Clive, 332
POPs. *See* persistent organic pollutants
populations
    consumption and, 300
    songbirds collapsing, 106
    waste increase along with, 290
Posada, Alfonso López, 53
positive thinking, 220
Potassium Amyl Xanthate (PAX), 127
Potemkin, Grigori Aleksandrovich, 72
Potemkin villages, 72
pottery business, 363–65
power
    demand overcoming, 383–84
    diversity of tactics against, 383
    economic, 384
    feminine, 233
    magic and, 241–42
    passivity toward, 275
    privilege opposing, 397
    relationships, 241–42
    of symbols, 220
    white man's goal of, 154–55
powerlessness, 224, 236
printer ink, 328
prisoners, 142
privileged, 397
problem solving drawbacks, 324–25
problems, roots of, 386
production rate, 331–32
product waste, 41
profits, 188, 336
propaganda campaigns, 48
prostate infection, 238
pseudodespair, 275–76
pseudo-solutions, 69
psychic distress, 84
psychological warfare, 388–89
pthalates, 108–9

public hearings, 173–74
putrefaction, 12
PVC. *See* polyvinylchloride
pyrite, 120

rabbits, 272–73
radical movements, 383
radioactive wastes, 119
Raffensperger, Carolyn, 110
rage, 252
rags, 47–48
    collection of, 47
    recycling, 49
    routes, 48
railways
    mines developing, 347
    as modern corporation, 348–49
    resource extraction made possible by, 348
    standardization from, 348
Ramos-Horta, José, 67
rapture story, 230
rationalization, 411n112
rationalized exploitation, 85
realities, 84, 206–9, 356–57
real physical world
    better off without humans, 259
    destroy what's destroying, 205–6
    stopping destruction of, 253
    fighting for, 263
    garbage growing in, 285
    identify with living in, 401
    industrial capitalism and, 201, 202–3
    life-birth-death-decay in, 399–400
    power of symbols in, 220
    rigged processes in, 252
    sustainability of, 203
    this culture more important than, 204
real world
    collapse, 366
    interviewer's perspective of, 89–90
reciprocity, 148–49
recirculation of compounds, 399
recycling, 225
    ancient procedure of, 45–46
    in Argentina, 50
    involuntary, 366
    large-scale, 46
    metal, 300, 314–15
    rags, 49
red-backed voles, 196
Red Cloud, 154–55
red-legged frog, 171
reduction plants, 31–32
reforms, 208, 397

regulations, environmental, 162–63
relationships
  better off from, 193–94
  long term mutually beneficial, 247–48
  with this culture, 261
religion, of Native Americans, 154
remicade, 246
Republican Party, 163
resin pellets (nurdles), 115–16
resistance, 76. *See also* culture of resistance
  anger/sorrow from, 180–81
  consequences of, 254
  culture of, 386–87, 389–90
  effective, 251–52
  forms of, 259
  goals/group norms shared in, 387
  groups fighting back in, 263
  logistical support roles in, 387–88
  movement, 77, 390
  organizing movement of, 387–88
  self-deception in, 265
resource(s)
  conquerors fighting for, 370
  earth's unlimited, 421n268
  guerillas fighting for, 372–73
  trees as, 186–87
  US invading for, 370–71
  world consisting of, 272
resource extraction, 188
  by government, 416n204
  railways assisting in, 348
  rewarding of, 164–65
reuse, of materials, 45–46
risks, culture of resistance, 397–98
Rittenhouse Mill, 47
river lampreys, 158
rivers. *See also* streams
  Alamosa, 125–26
  cyanide in, 124
  dams killing, 89–90
  domestic waste in, 100
Rivers and Harbors Act, 30
roads, asphalt, 160–61
Roaring Twenties, 390–91
robber economy, 374
Rokke, Doug, 316
Roman Empire
  collapse of, 374–76
  exploitation by, 375
rubbish, 9
rubble, 9
Rucker, Rudy, 355

Saal, Frederick vom, 112

sacredness, 154
Sahtouris, Elisabet, 229–31
salamanders
  destroying of, 157
  slender, 172
  torrent, 197
salmon
  coho, 158
  economic system killing, 270
sanitary landfill, 28
Saro-Wiwa, Ken, 367
Sawyer, Jim, 156
scale
  industrial, 309
  matter of, 308–9
scavengers, 12–13
Schultz, Ed, 156
Schultz, George, 171
science
  cause/effect in, 240
  magic v., 248
science assignment, 4
sea gulls, 100
self-deception, 265
self-destructive behaviors, 268
settlement violation, 180
SEX. *See* sodium ethyl xanthate
Sharp, Gene, 384
Shatner, William, 371
Sheehan, Bill, 44
Shistar, Terry, 58
shit
  breaking down, 10–11
  composition of, 6–7
  in forest, 6
  soil nutrients from, 11
  toilet paper used for, 21–22
  word origin of, 8
shopping bags, 19
short attention span, 273
short term decisions, 333
Sierra Club, 225
Silent Snow (Cone), 144
Simon, Julian, 421n268
skewed perspective, 185–86
Skinner, BF, 223
skunk cabbages, 160
skyscrapers, 292–94
slavery, 188
  internal morality/perception and, 186–87
  by Nazis, 66–67, 393
slugs, frogs eating, 5–6
Smith, Dale, 156
  board of supervisors appeal of, 169–70

as deforester, 162–63
    planned development unimpeded of, 168
snail's shell, 310
Snow Crash (Stephenson), 377
social change, 257
social construct, 205
social disruptions, 391
social equity, 409n83
socially-constructed self, 201
socially responsible, 306
social order, 176–77
social system, 227
societal changes, 372
societies. *See also* civilizations; complex
        societies
    hunter-gatherer, 29, 30–31
    real alternative options of, 365–66
    trash workers in, 49–50
sociopolitical complexity, 360–61
sodium ethyl xanthate (SEX), 126–27
soil
    bacteria in, 13
    fungi's relationship with, 14
    human shit nutrients for, 11
    living/breathing, 38
solutions
    for global warming, 418n221
    global warming suggested, 201–2
    industrial capitalism promoted by,
        286–87
    McDonough presenting, 202
songbirds, 106
sorrow, 180–81, 270
Speer, Albert, 166
sperm whale, 136
spider's web, 310–11
Spiegelman, Helen, 44
spiritual existence, 268
spotted owl, 197, 416n205
spotted owl surveys, 166–67
Sri Lanka, 367
Stamets, Paul, 415n196
standardization
    diversity destruction in, vii–viii
    railway timetables requiring, 348
steam engines, 346–47
Stephenson, Neal, 377
steroids, 132
stormwater management system, 63
strategy, 385
streams, 120–21, 415n203
Street Cleaning Commissioner, 39
street-cleaning, of pigs, 31
stressors, 59

struggle, 383–84
A Study of History (Toynbee), 374
styrofoam fast-food containers, 305
Suarez, Francisco, 50
sub-atomic components, 355
subsistence farmers, 34–35
sulfide minerals, 120
sulfuric acid, 127
Summitville Mine, 125
survival, 148
    of earth, 278
sustainability
    business as usual and, 296–97
    civilizations without, 363
    of corporate airport, 67–68
    economics of, 374
    first principle of, 71–72
    of harvested wood, 408n81
    human cultures primary value of, 300
    indefinite performance in, 56
    industrial capitalism and, 204–6
    industrial civilization, 74
    industrial civilization technologies and,
        342
    of industry, 63
    of landbase, 56–57
    McDonough concepts of, 407n71
    real physical world, 203
    right to take whatever we want and,
        271–72
    stressors influencing, 59
    transitional steps to, 68–69
sustainable development architect, 61–62,
        408n75
sweatshops, 66, 406n69, 408n79, 409n83
swill children, 25–26
systems. *See also* authoritarian system;
        economic system; garbage collection
        system
    breaking down of, 359–60
    free garbage collection, 44
    human beings and, 216
    judicial, 176–77
    legal, 180
    Milwaukee garbage collection, 25
    municipal garbage collection, 39–40, 292
    national highway, 44
    Pacific Electric, 332
    social, 227
    stormwater management, 63
    urban garbage collection, 44
    working within, 225

tactics, diversity of, 383, 384

Tainter, Joseph A., 360, 363, 374
Taylor, Frederick Winslow, 216
technics, 343–44
technocratic prison, 277
technological progress, 351
technologies, 241
   advanced, 245–47
   applied, 207
   autonomy and, 344–45
   clean coal, 228
   defining, 310
   fantasies fulfilled through, 209
   green, 342
   of indigenous people, 243–45
   indigenous people and, 242
   industrial infrastructure required for,
      353–54
   magical, 244
   military, 314
   neutral myths of, 343
   nuclear, 345
   purpose of, 247
   technotopia and, 299–300
   tools creatively used as, 309–10
technotopia
   conditions required for, 342
   cornucopian overlap with, 421n268
   magical thinking in, 323–24
   metal recycling in, 314–15
   non-toxic biodegradable materials used
      in, 301
   out of touch, 354
   problems with, 358
   technology/utopia in, 299–300
Tecumseh, 396
temporal test, 317
Tennis, Cary, 213
territorial test, 317
terrorism, government sponsored, 385
theory, 87
thimbleberry leaves, 22
Thompson, David, 221
*Thought to Exist in the Wild: Awakening From
   the Nightmare of Zoos* (Jensen), 211
THP. *See* Timber Harvest Plan
threats, *biopimpologist*, 170
thrush, 14
Tijuana, 51, 285
Timber Harvest Plan (THP), 173, 182
timetables, 348
title company, 161
toilet paper, 330
   breaking down, 23
   for human shit, 21–22

toilets, 8
Tolowa Indians, 197
Tolowa land, 57
tools, 309–10
toxic pollution
   in mine tailings, 119–20
   from phosphate plant, 319–20
Toynbee, Arnold J., 374
transformation, 229
transistors, 355
transnational corporations, 407n72
trash, 9
trash workers, 49–50
Treaty of Versailles, 391–92
trees, as resources, 186–87
trespassing, 182
truck factory, 410n86
Tubman, Harriet, 397
turkey tails (mushrooms), 96–97
turtles, 99–100
*Tuskegee Study of Untreated Syphilis in the
   Negro Male*, 318

"Unintended Consequences: Municipal
   Solid Waste Management and the
   Throwaway Society," 44
United States (US)
   acetaminophen consumption in, 132
   aluminum/copper being stolen in, 366
   DU stockpiled by, 315–16
   household waste in, 41
   industrial waste of, 290–92
   military aggression of, 233–34
   municipal solid waste production in,
      289–90
   natural burial movement in, 143–44
   plastics production in, 116
   resource invasion of, 370–71
   waste production, 295
uranium test rods, 318
urban garbage collection systems, 44
Urrea, Luis Alberto, 51–53
US. *See* United States
utilitarianism, 71
utopia, 299–300

vacation advice, 213–14
values, 154
vanguardism, 396
violence (extralegal force), 389
visualization, 220
visualizing, 264–65

Waka 'ta ka (the Great Spirit), 87

walk away, 381–82
Walt Disney, 418n220
war, 188
    Arab-Israel, 315
    collapse accelerated by, 370
    Gulf, 316
    indigenous people and, 420n250
    materials needed for, 314–15
    overhead expense of, 371–72
    societal changes required in, 372
War at Home: Covert Action Against US
    Activists and What We Can Do About It
    (Glick), 388
warfare, asymmetric, 373–273
warm-blooded animals, 321
Washington, George, 372
waste
    by-product, 289
    chickens eating, 32
    community decisions on, 26–27
    defining, 9
    domestic, 100
    electronic, 49
    elimination plan, 327
    good disposal sites of, 29
    governments encouraging, 332
    household, 38–39, 41, 291
    human, 9
    industrial, 290–92, 406n68
    industrial civilization determining, 335
    industrial-scale, 331
    landbases and, 99
    management systems, 26–27
    medical, 129–30
    minimizing, 379–80
    municipal solid, 289–90
    organic matter, 40
    matter of perspective on, 10
    population increase along with, 290
    product, 41
    radioactive, 119
    sharing of, 31
    stopping industrial, 44
    value increasing of, 362
    what is, 7–8
wastelands, 9
waste production
    of industrial civilization, 289–94
    industrial collapse decreasing, 362
    municipal, 293–94
    nonindustrialized nations, 294–96
    of this culture, 327
    US, 295
    World Trade Center measure of, 292–93

"waste to energy" schemes, 43
Wasting Away: An Explosion of Waste
    (Lynch), 285
water
    fluoride in, 318–19
    free-flowing, 415n197
    soluble compounds, 143–44
    stormwater management system of, 63
waxy coating, 311–12
wetlands, 165–66
Weyerhaeuser, 225
whipworms, 94–96
white man
    poisoning the heart, 154
    power increase goal of, 154–55
wisdom, 87
wishful thinking, 221–22
women, 233, 264
    in Arctic, 144
    in barter system, 405n48
    feminine hygiene of, 328–29
wood pulp, 48–49
World Trade Center, 292–93
World Wildlife Fund, 225
worm density, 37–38
Wright, Richardson, 332
writing
    effective, 155–56
    as propaganda, 284–85

yellow boy, 120–21
Youngquist, Walter, 418n220
Yugoslavia, 316–17

Zortman-Landusky mine, 124–25

ABOUT THE AUTHORS

Activist, philosopher, teacher, and leading voice of uncompromising dissent, DERRICK JENSEN holds degrees in creative writing and mineral engineering physics. His books include *Endgame*, volumes 1 and 2; *As the World Burns*, with Stephanie McMillan; *A Language Older Than Words*; and *The Culture of Make Believe*.

Writer, activist, and small-scale organic farmer ARIC MCBAY works to share information about community sufficiency and off-the-grid skills. He is the author of *Peak Oil Survival: Preparation for Life after Gridcrash* and creator of "In the Wake: A Collective Manual-in-progress for Outliving Civilization" (www.inthewake.org).